TROPIC ICE

BARBARA DOMBROWSKI TROPIC ICE

FOR MORITZ

GREENLAND AMAZONIA MONGOLIA KIRIBATI TANSANIA

PROLOGUE

My journey has been a long one and its connections are complex. After many encounters and experiences, I finally feel ready to summarize my long-term project in book form. It has become a very personal work. Everything I have learned from the people I have been able to visit, the deeper connections that have opened up to me, are the result of my own experiences. This has changed my perception. Our current capitalist world order, with consumerism and individualism at its heart, feels outdated to me today and absurd in its excesses. The people I visited, their spiritual ideas about life, the landscapes and earth itself reminded me of who I am, of who we are. I realized that there is no such thing as *the people* and *nature,* but that we are all interwoven and interdependent, just as the mycelium of the mushrooms in the forest is interwoven with the trees, who themselves depend on the birds that carry their seeds.

At the beginning of my project, I only saw the surface. I saw people who live in climate-relevant places and are severely affected by the climate crisis. I wanted to report on that. To accompany my pictures, I planned to include scientific texts and artistic analyses. However, the deeper I immersed myself in the work, the more I realized that it was not about an analytical view. I had to learn, and I learned slowly. At first, I didn't see in depth and didn't realize how much we are all connected and feed off each other. I didn't see the loss of connection between us humans, between humans, nature and all the creatures we share our planet with. Onward, I began to understand and remember how mankind has lived in harmony with nature for many millennia, and that we Europeans, too, were once indigenous peoples connected to our natural surroundings. It was necessary to visit five communities on five continents to help me understand what being connected with planet earth really means. It means being part of something bigger than ourselves, being part of a system in which every element, no matter how small, has its place and meaning. In this system, nothing and no one is above the other. No taking without giving. All life exists in balance.

PROLOG

Mein Weg hierher war weit, seine Zusammenhänge sind komplex. Nach vielen Begegnungen und Erlebnissen fühle ich mich endlich bereit, mein langjähriges Projekt in Buchform zusammenzufassen. Es ist zu einer sehr persönlichen Arbeit geworden. Alles was ich von den Menschen gelernt habe, die ich besuchen durfte, die tieferen Zusammenhänge, die sich mir erschlossen haben, resultiert aus eigenen Erfahrungen. Dadurch hat sich meine Wahrnehmung verändert. Unsere aktuelle kapitalistische Weltordnung, in deren Zentrum Konsum und Individualismus stehen, fühlt sich heute für mich überholt und in ihren Auswüchsen absurd an. Die von mir besuchten Menschen, ihre spirituellen Vorstellungen vom Leben, die Landschaften und die Erde selbst haben mich erinnert, wer ich bin – wer wir sind. Mir wurde begreiflich, dass es nicht *die Menschen* und *die Natur* gibt, sondern dass wir alle verwoben und voneinander abhängig sind, so wie das Myzel der Pilze im Wald mit den Bäumen verwoben ist und wie diese wiederum abhängig von den Vögeln sind, die ihre Samen weitertragen.

Zu Beginn meines Projekts sah ich nur die Oberfläche. Ich sah Menschen, die an Orten leben, an denen sie stark von der Klimakrise betroffen sind. Darüber wollte ich berichten. Begleitend zu meinen Bildern hatte ich vor, wissenschaftliche Texte und künstlerische Analysen vorzustellen. Umso tiefer ich jedoch in die Arbeit eintauchte, umso klarer wurde mir, daß es nicht um eine analytische Betrachtung geht. Ich musste lernen, und ich lernte langsam. Anfangs sah ich nicht in die Tiefe und begriff nicht, wie sehr wir alle verbunden sind und uns gegenseitig nähren. Ich sah nicht den Verlust an Bindung zwischen uns Menschen, zwischen Menschen, Natur und allen Kreaturen, mit denen wir gemeinsam die Erde bevölkern. Nach und nach jedoch begann ich zu verstehen, wie Menschen über viele Jahrtausende in Einklang mit der Natur gelebt haben, und dass auch wir Europäer einst Naturvölker waren, die mit ihrer Umwelt verbunden lebten. Letztlich habe ich fünf Gemeinschaften auf fünf Kontinenten besucht, damit in mein Bewusstsein dringen konnte, was Verbindung mit der Erde bedeutet. Es bedeutet, Teil von etwas zu sein, das größer ist als wir, Bestandteil eines Systems zu sein, in dem jedes auch noch so kleine Element seinen Platz und seine Bedeutung hat. In diesem System steht nichts und niemand über dem anderen. Kein Nehmen ohne Geben. Alles Leben existiert in Ausgewogenheit.

“Knowing that you love the earth changes you, activates you to defend and protect and celebrate. But when you feel that the earth loves you in return, that feeling transforms the relationship from a one-way street into a sacred bond.”

Robin Wall Kimmerer, Braiding Sweetgrass

EUROPE

GREENLAND

65° 53′ 23.334″ N 37° 46′ 38.986″ W

GREENLAND

I visited the Tunumiit in the villages of Tiniteqilaaq and Sermiligaaq in East Greenland.

The Tunumiit, as the East Greenlanders call themselves, are the second largest indigenous population group in Greenland, with around 3,500 members. Until the polar explorer Gustav Frederik Holm came to East Greenland on an expedition in 1884, they lived mainly in isolation from other parts of the world. Like most Inuit societies, the social life of the Tunumiit was originally relatively egalitarian in structure. There were no prisons, instead disputes were settled in public singing duels, in which the losers were punished with the ridicule of the community and in which the drum dancer played a central role. Angakkut, the shamans with their drums, were also a link between the world of humans and the world of spirits. After extensive missionary work by Europeans, almost all Greenlanders have been officially christianized since the beginning of the 20th century. In more remote areas in the north and east of Greenland, however, the traditional animistic world view has survived to this day.

Like all Greenlandic Inuit, the Tunumiit used to be hunters, fishermen or gatherers. Even today, many families on the east coast still make a living from hunting; alongside tourism, it is the main source of income. Hunting game includes seals, walruses, narwhals, cod and other fish. Agriculture is not possible due to the soil and climate.

East Greenland is dominated by two weather patterns in winter: the piteraq ice storm, which can reach gusts of more than 300 km/h, coming from the North Pole over the Greenland ice sheet, and the neqqajaaq, which brings humid air from the sea as rain and snow. When the piteraq rages, it paralyzes everything and cuts off villages and settlements. Because of the extreme weather, the cold and the darkness, most of the Greenlanders' lives take place indoors in winter. In the days when the nomadic Tunumiit retreated to earth houses in winter to withstand the weather, they had to survive for months in confined spaces, with whale oil

GRÖNLAND

Mein Besuch galt den Tunumiit in den Dörfern Tiniteqilaaq und Sermiligaaq in Ostgrönland.

Die Tunumiit, wie sich die Ostgrönländer selbst nennen, sind mit ca. 3.500 Angehörigen die zweitgrößte indigene Einwohnergruppe Grönlands. Bis 1884 der Polarforscher Gustav Frederik Holm im Zuge einer Expedition nach Ostgrönland kam, lebten sie überwiegend isoliert von anderen Weltteilen. Wie das der meisten Inuit-Gesellschaften war das Sozialleben der Tunumiit ursprünglich relativ egalitär strukturiert. Es gab keine Gefängnisse, sondern Streitfälle wurden über öffentlich Gesangsduelle ausgetragen, bei denen die Verlierer mit dem Spott der Gemeinschaft bestraft wurden und in denen den Trommeltänzern eine zentrale Bedeutung zukam. Die Schamanen, Angakkut, mit ihren Trommeln waren des Weiteren ein Bindeglied zwischen der Welt der Menschen und der Welt der Geister. Nach umfangreicher Missionierung durch europäische Akteure sind seit dem Beginn des 20. Jahrhunderts offiziell fast alle Grönländer:innen christianisiert. In abgelegeneren Gegenden im Norden und Osten Grönlands hat sich die traditionelle animistische Weltsicht allerdings bis heute erhalten.

Die Tunumiit waren, wie alle Grönland-Inuit, früher Jäger, Fischer oder Sammler. Auch heute noch leben viele Familien an der Ostküste maßgeblich von der Jagd; neben dem Tourismus ist sie dort die Haupteinnahmequelle. Jagdwild sind Robben, Walrosse, Narwale, Dorsche und sonstige Fische. Landwirtschaft ist aufgrund der Böden und des Klimas nicht möglich.

Ostgrönland wird im Winter von zwei Wetterlagen beherrscht, den Eisstürmen des Piteraq, der vom Nordpol über das grönländische Eisschild kommend in Böen mehr als 300 km/h erreichen kann, und dem Naqajaq, der feuchte Luft vom Meer als Regen und Schnee bringt. Wenn der Piteraq wütet, legt er alles lahm und schneidet Dörfer und Siedlungen ab. Wegen des extremen Wetters, der Kälte und der Dunkelheit spielt sich im Winter das Leben der Grönländer:innen größtenteils in den Häusern ab. In der Zeit, als sich die nomadischen Tunumiit im Winter in Erd-

lamps as their only source of light, and hope that the hunters would return from the cold. Anyone who has experienced the darkness, the northern lights, the cold and storms in winter understands that a fundamental belief in nature and the forces at work within it has been preserved, supported by myths and old stories. To this day, Greenlanders fear the spirits that drag people out of their homes and to their doom during winter.

Greenland's ecosystem is facing serious upheaval due to global warming. In 2015, the Arctic showed the first signs of irreversible changes. It is expected that a temperature rise of between 1° and 4° Celsius will result in the almost complete melting of Greenland's ice. As a result, sea levels are expected to rise by between one and seven meters by the end of 2100. A warmer Arctic is accompanied by numerous feedback effects, which in turn have an impact on the extent of global warming. For example, due to the loss of Arctic sea ice during the summer months in recent years, there are progressively bigger areas of seawater that absorb the sun's rays and thus further increase warming in the Arctic. The rising temperatures are causing the Arctic permafrost to thaw and release carbon dioxide and methane, which in turn contribute to global warming in the atmosphere. And finally, due to the melting of the ice sheet, more and more freshwater is flowing into the sea, leading to a decreasing density of water and thus to a slowdown in global water and temperature circulation, which also includes the Gulf Stream.

häuser zurückzogen, um der Witterung standzuhalten, mussten sie monatelang auf engstem Raum mit Waltranlampen als einziger Lichtquelle ausharren und hoffen, dass die Jäger aus der Kälte zurückkamen. Wer im Winter die Dunkelheit, die Polarlichter, die Kälte und Stürme erlebt hat versteht, dass sich, getragen von Mythen und alten Geschichten, ein fundamentaler Glaube an die Natur und in ihr wirkende Kräfte erhalten hat. Die Grönländer:innen fürchten sich bis heute vor Geistern, die im Winter die Menschen aus den Häusern nach draußen ins Verderben ziehen.

Durch die Erderwärmung steht das Ökosystem Grönlands vor schweren Umwälzungen. 2015 zeigte die Arktis erste Zeichen von irreversiblen Veränderungen. Es wird erwartet, dass ein Temperaturanstieg zwischen 1° und 4° Celsius das fast vollständige Abschmelzen des grönländischen Eises zur Folge haben wird. Dadurch wird der Meeresspiegel bis Ende 2100 voraussichtlich um einen bis sieben Meter steigen. Mit einer wärmeren Arktis gehen zahlreiche Rückkopplungen einher, die sich wiederum auf das Maß der Erderwärmung auswirken. So gibt es durch den Verlust des arktischen Meereises während der Sommermonate der letzten Jahre zunehmende Meerwasserflächen, die die Sonnenstrahlen absorbieren und somit die Erwärmung in der Arktis weiter verstärken. Durch die steigenden Temperaturen tauen die arktischen Permafrostböden und setzen Kohlendioxid und Methan frei, die in der Atmosphäre wiederum zur Erderwärmung beitragen. Und schließlich fließt bedingt durch das Abtauen des Inlandeises immer mehr Süßwasser ins Meer, was zu einer schwindenden Dichte des Wassers und damit zu einer Verlangsamung der globalen Wasser- und Temperaturzirkulation führt, zu der auch der Golfstrom gehört.

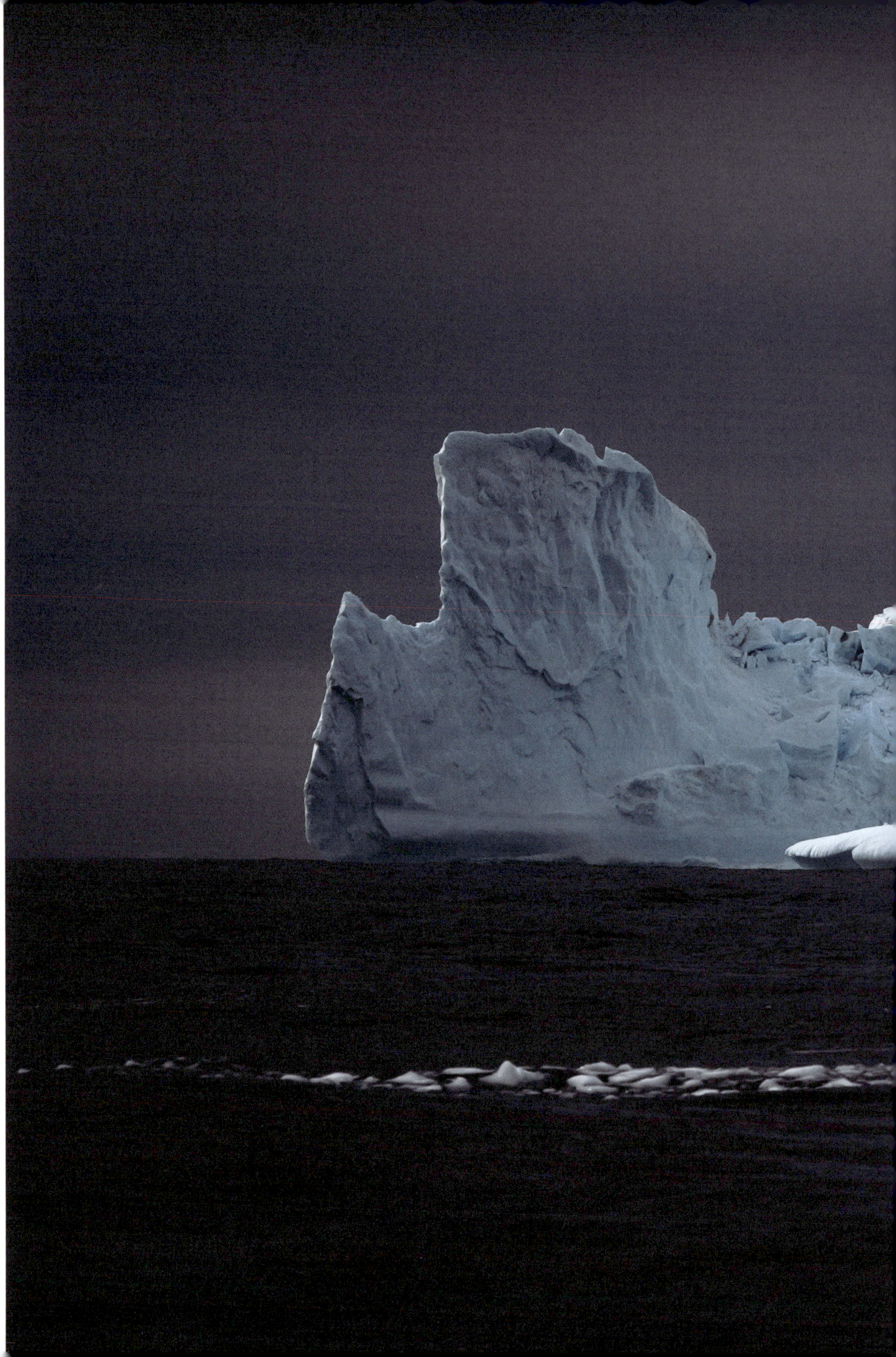

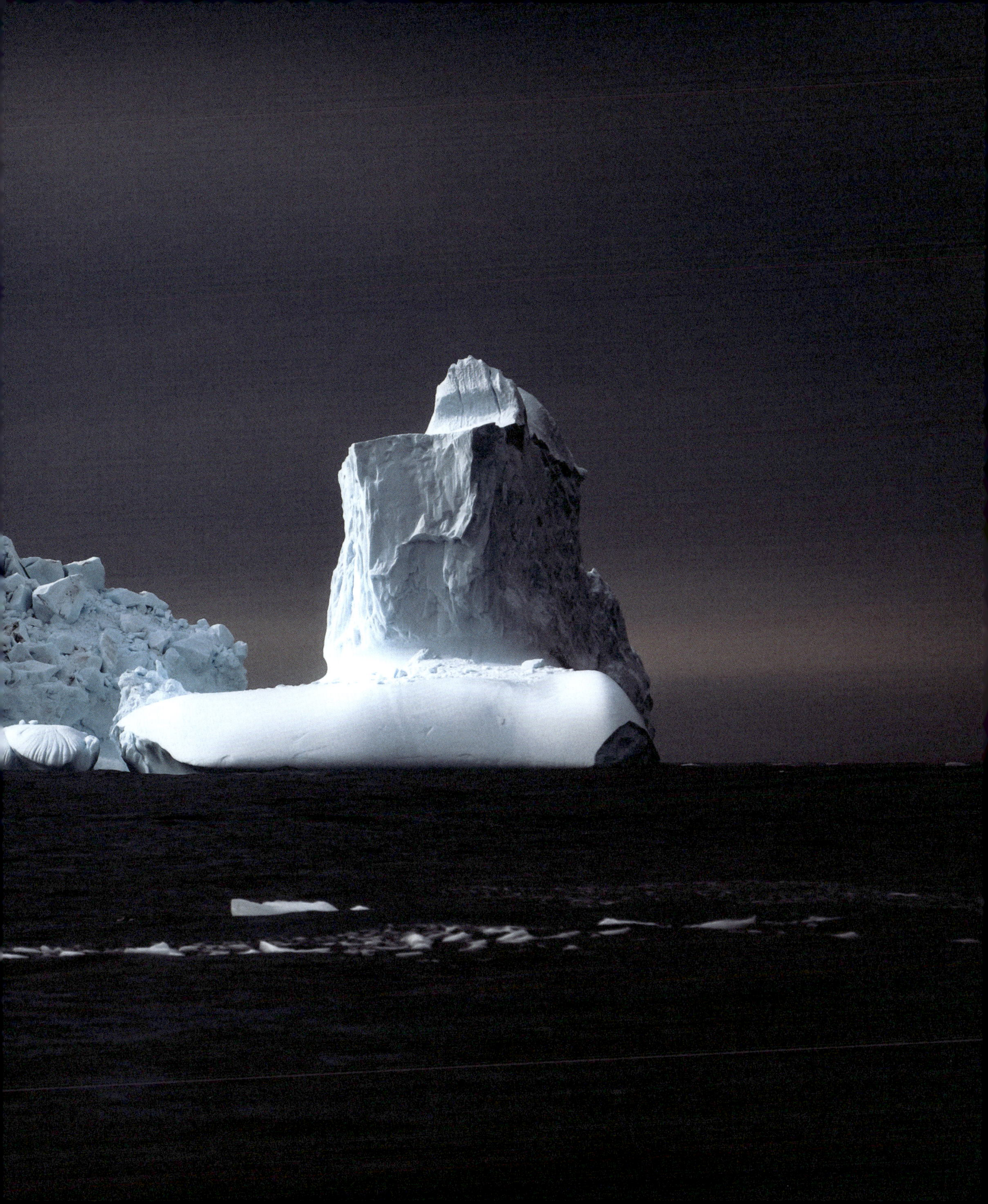

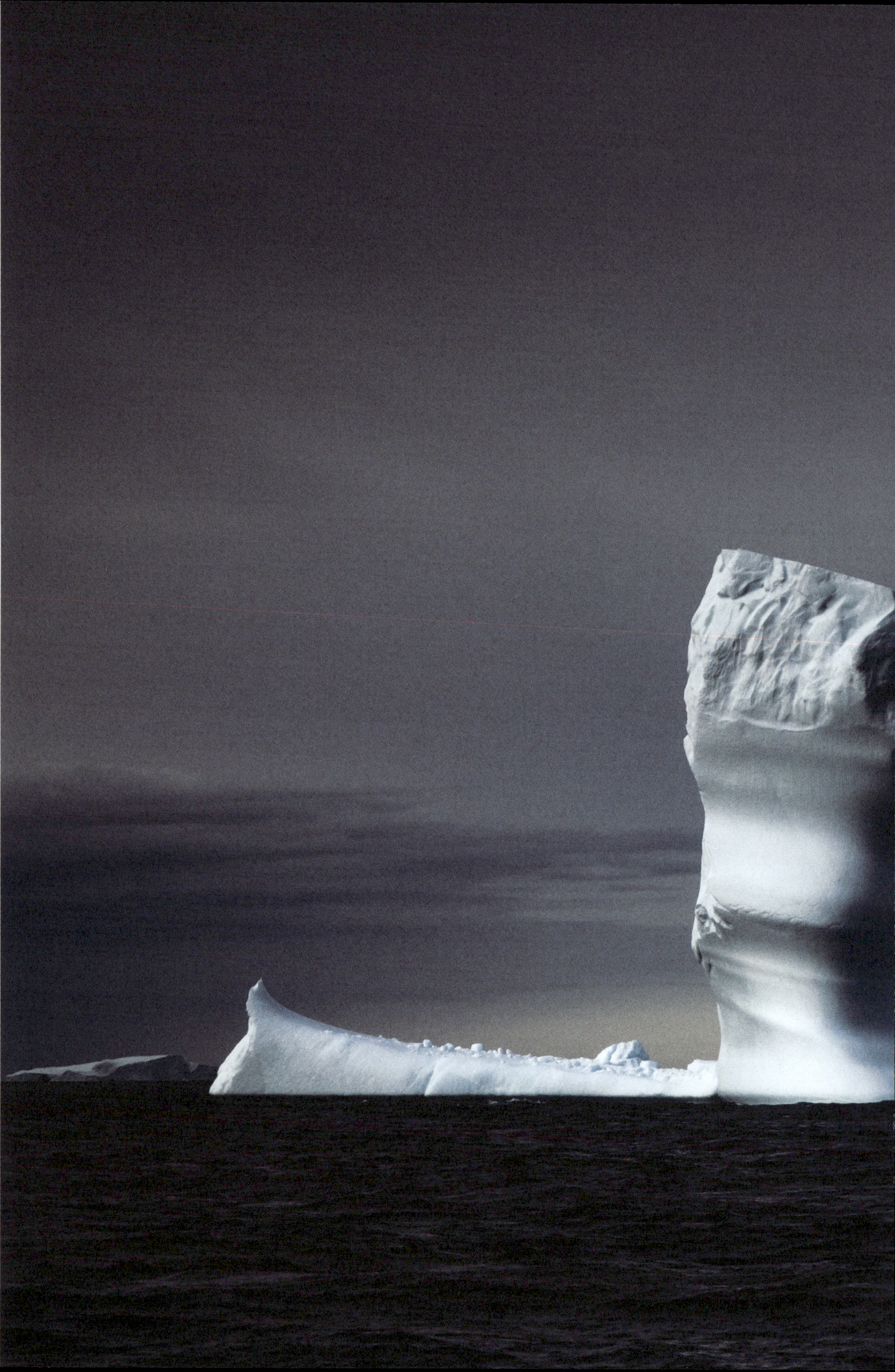

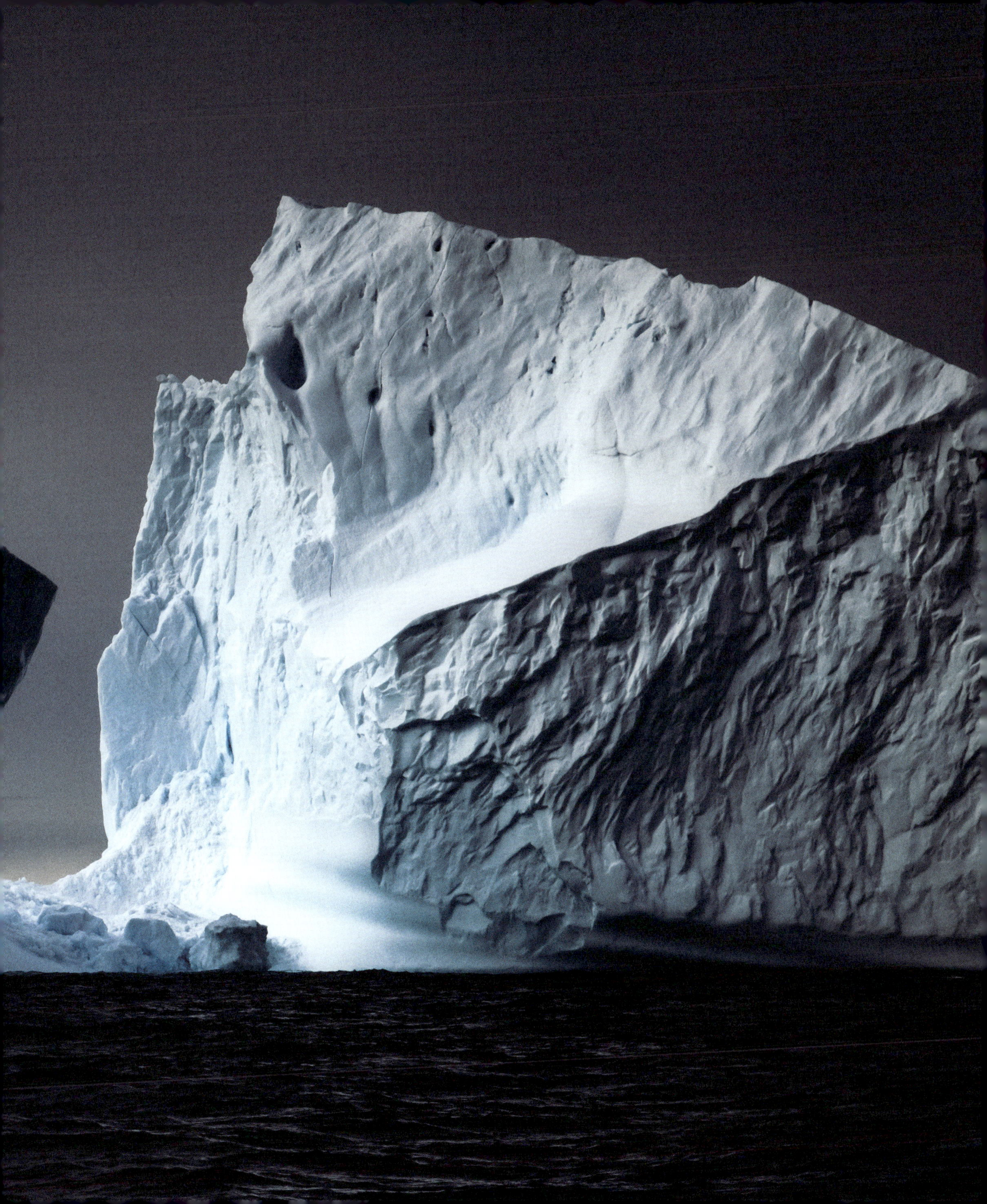

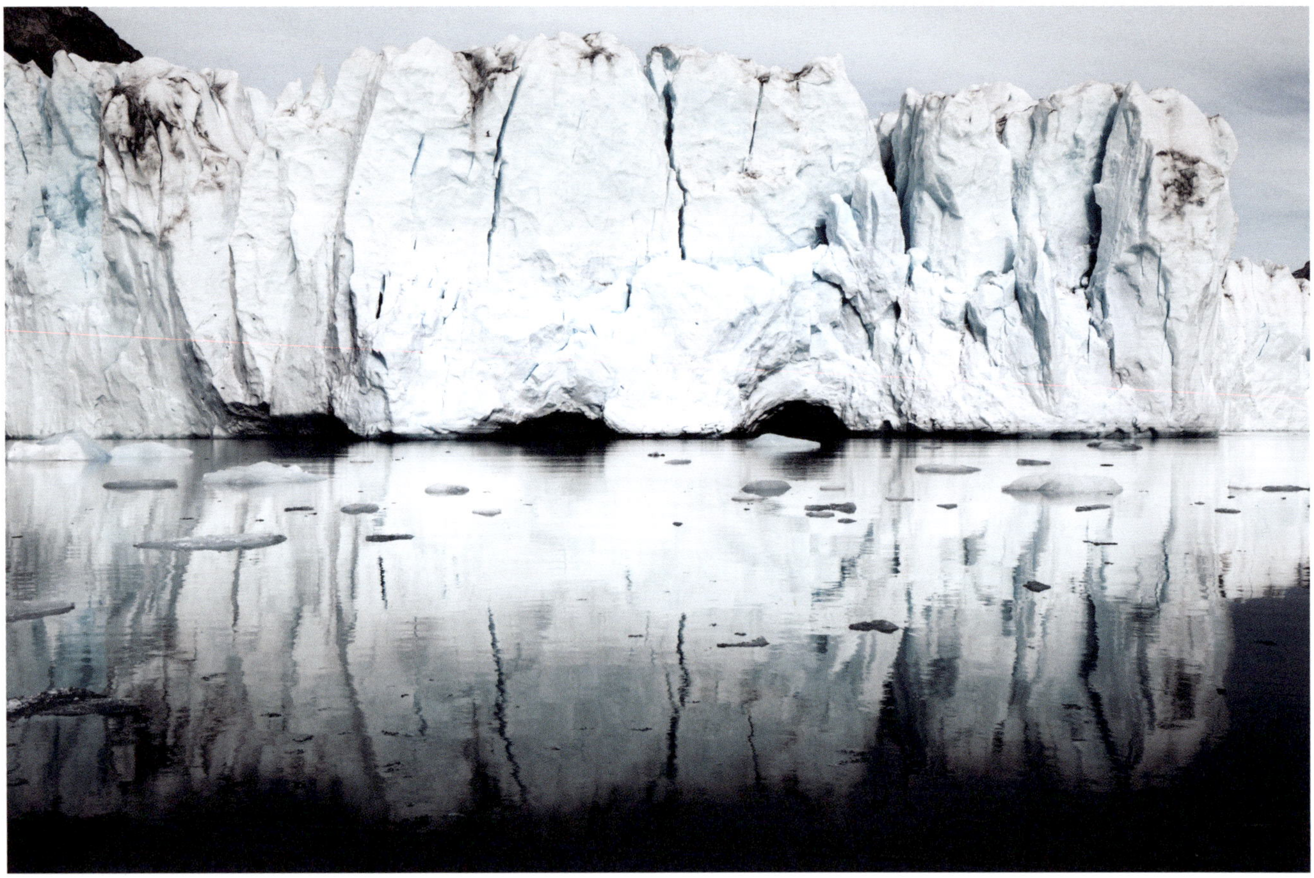

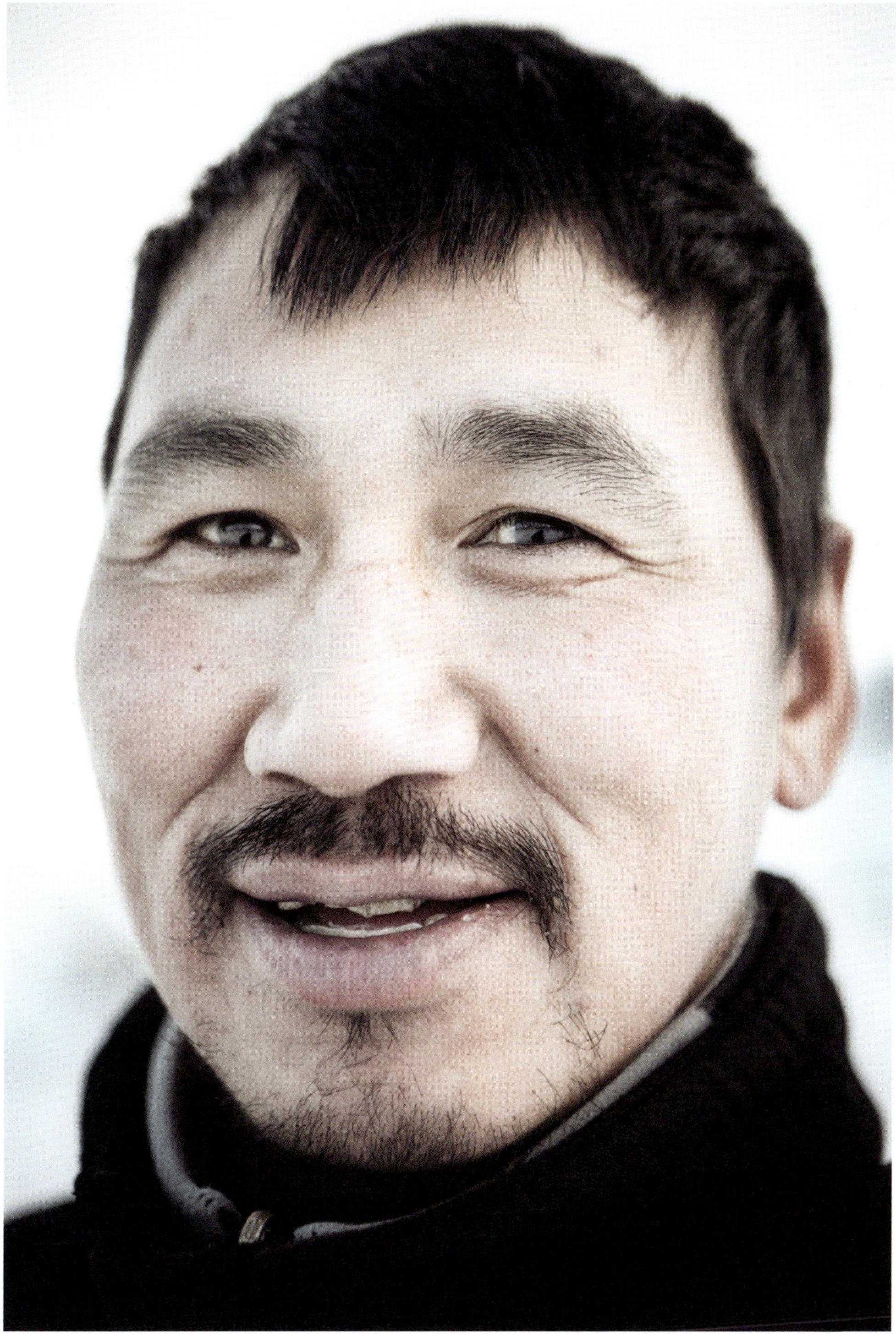

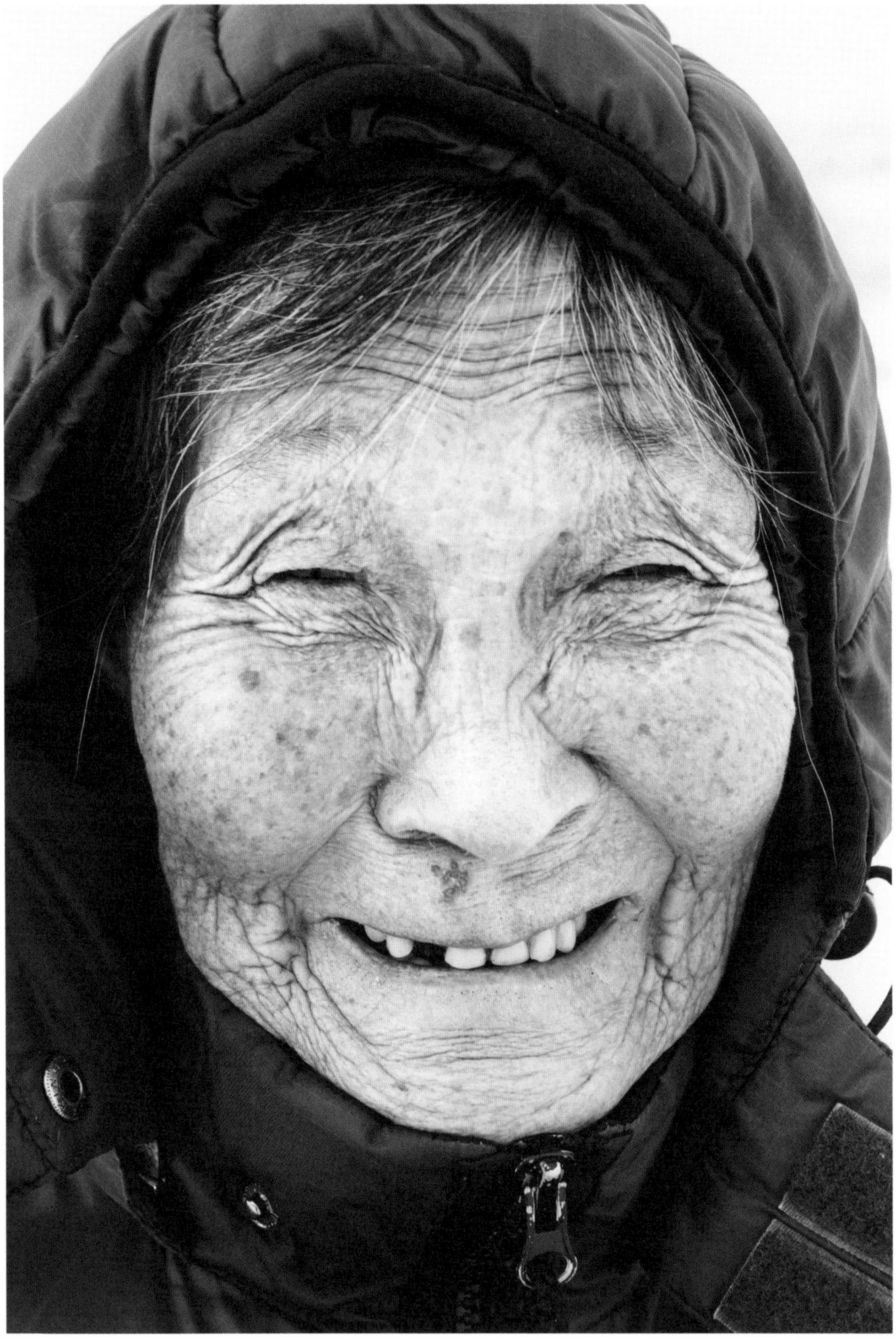

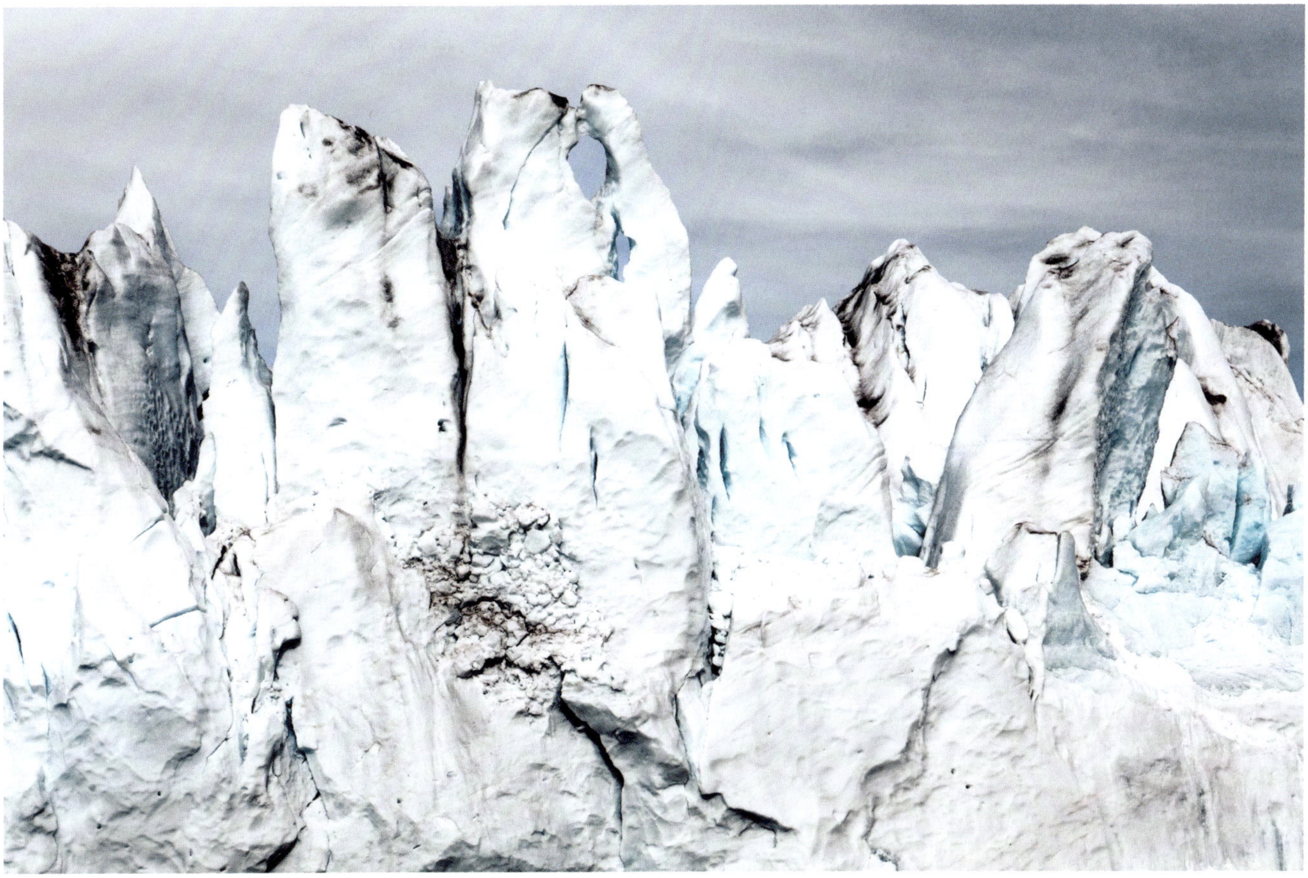

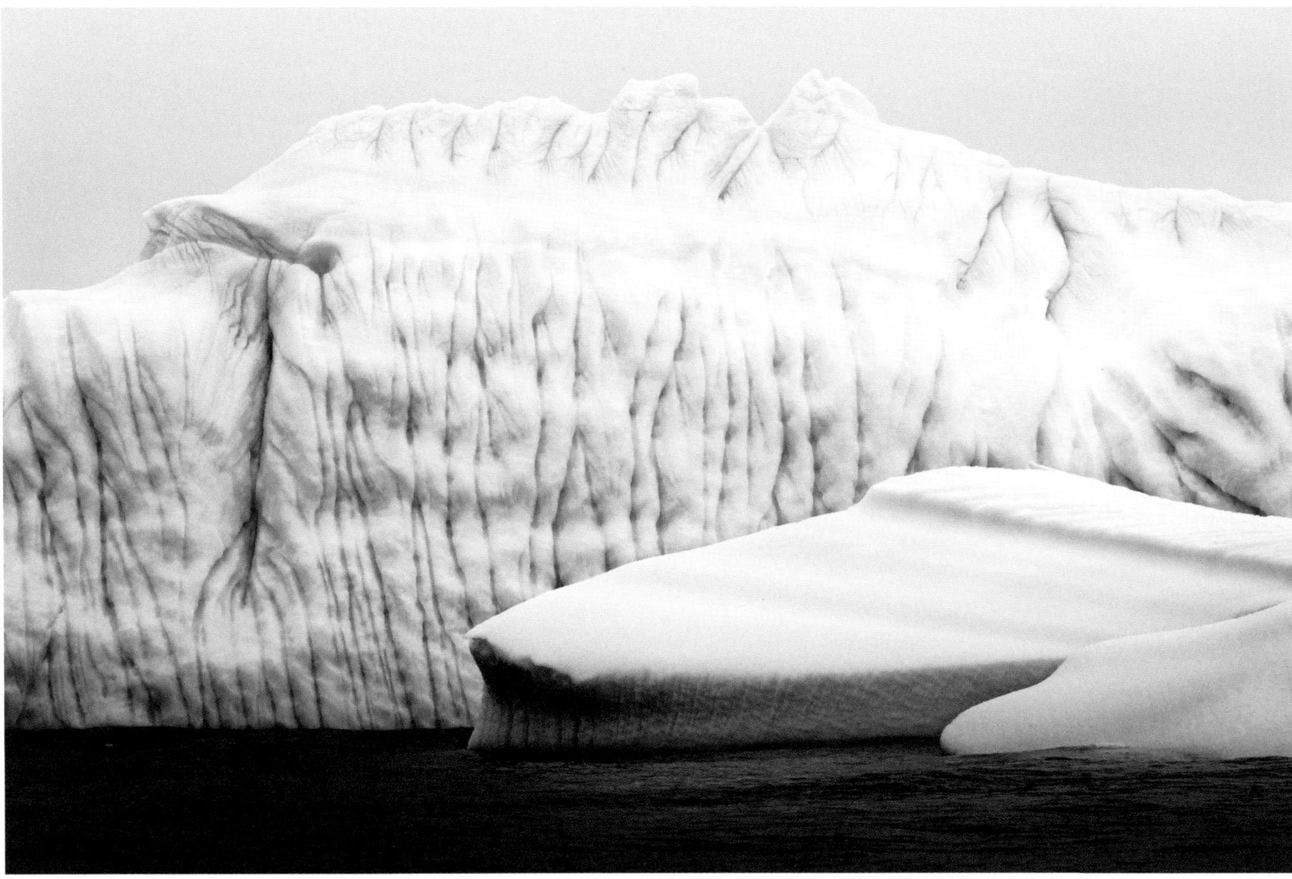

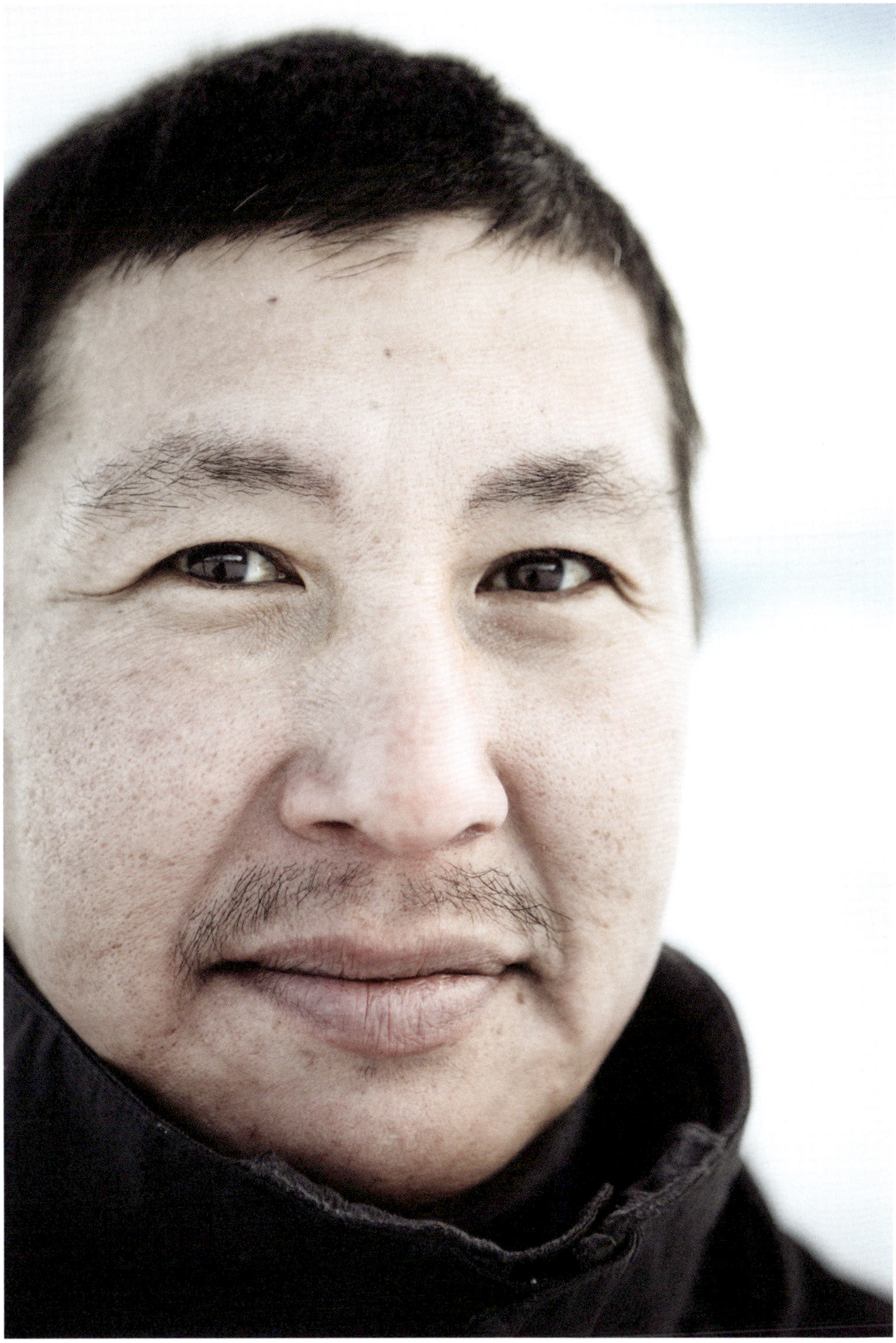

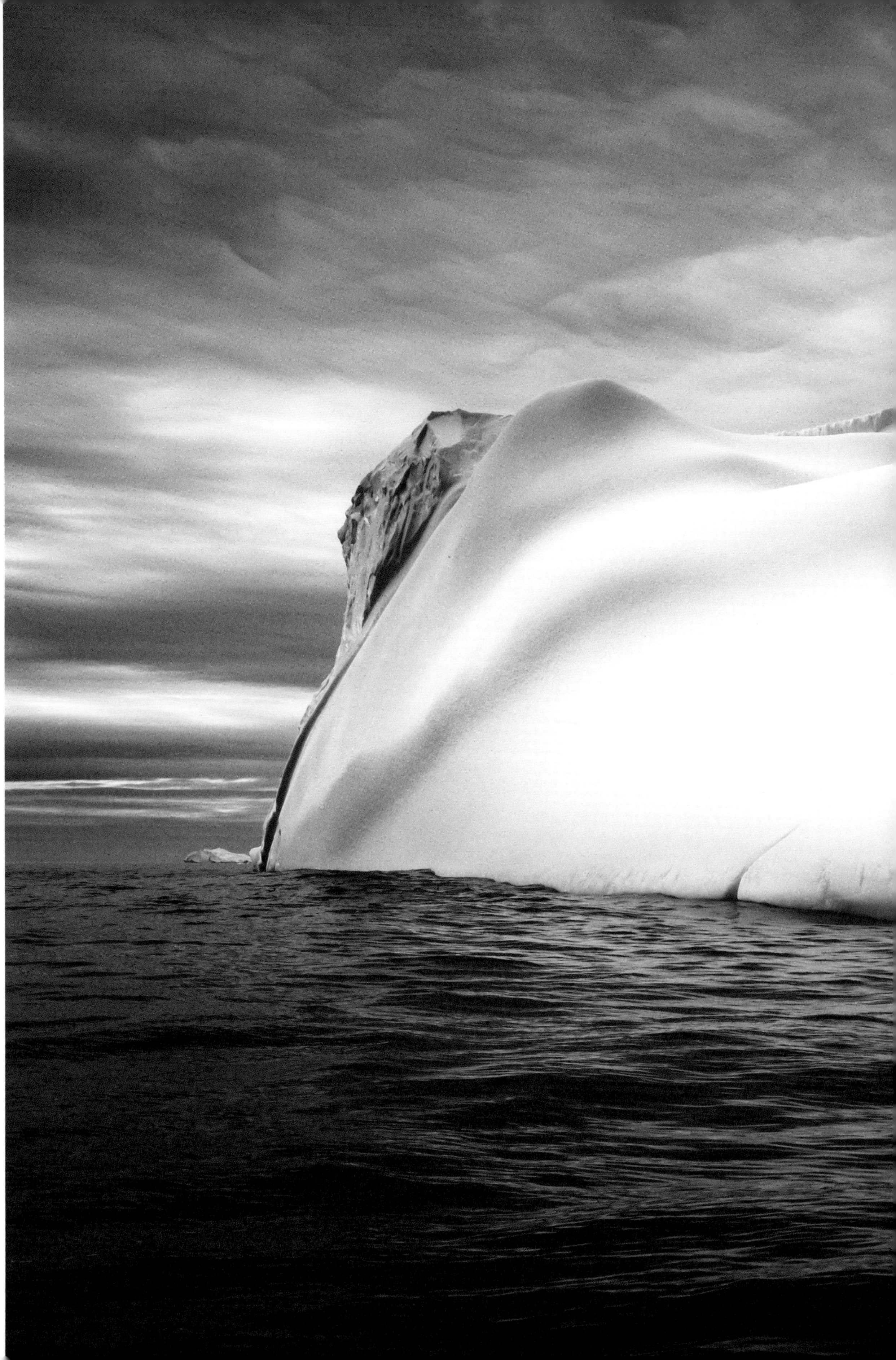

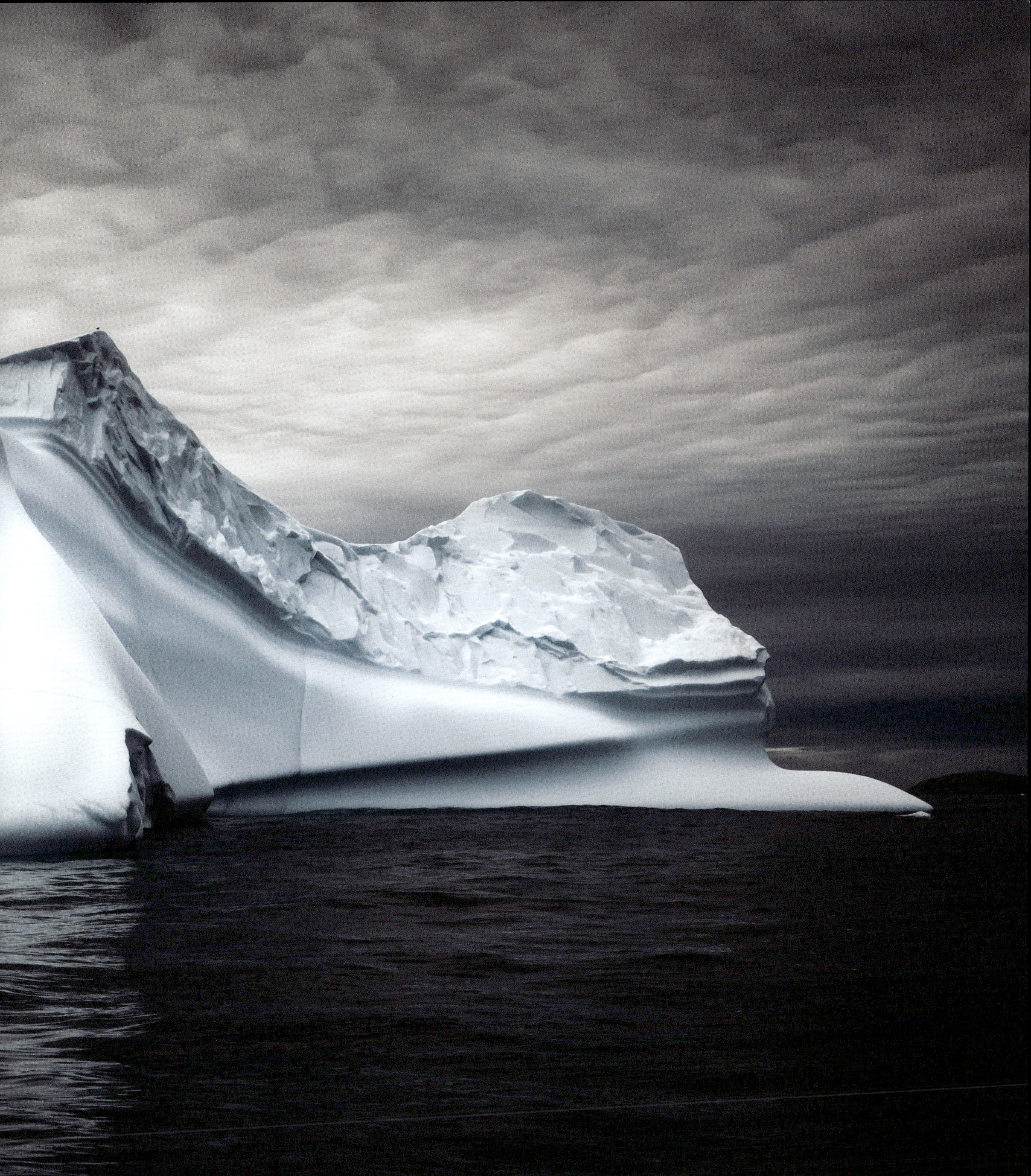

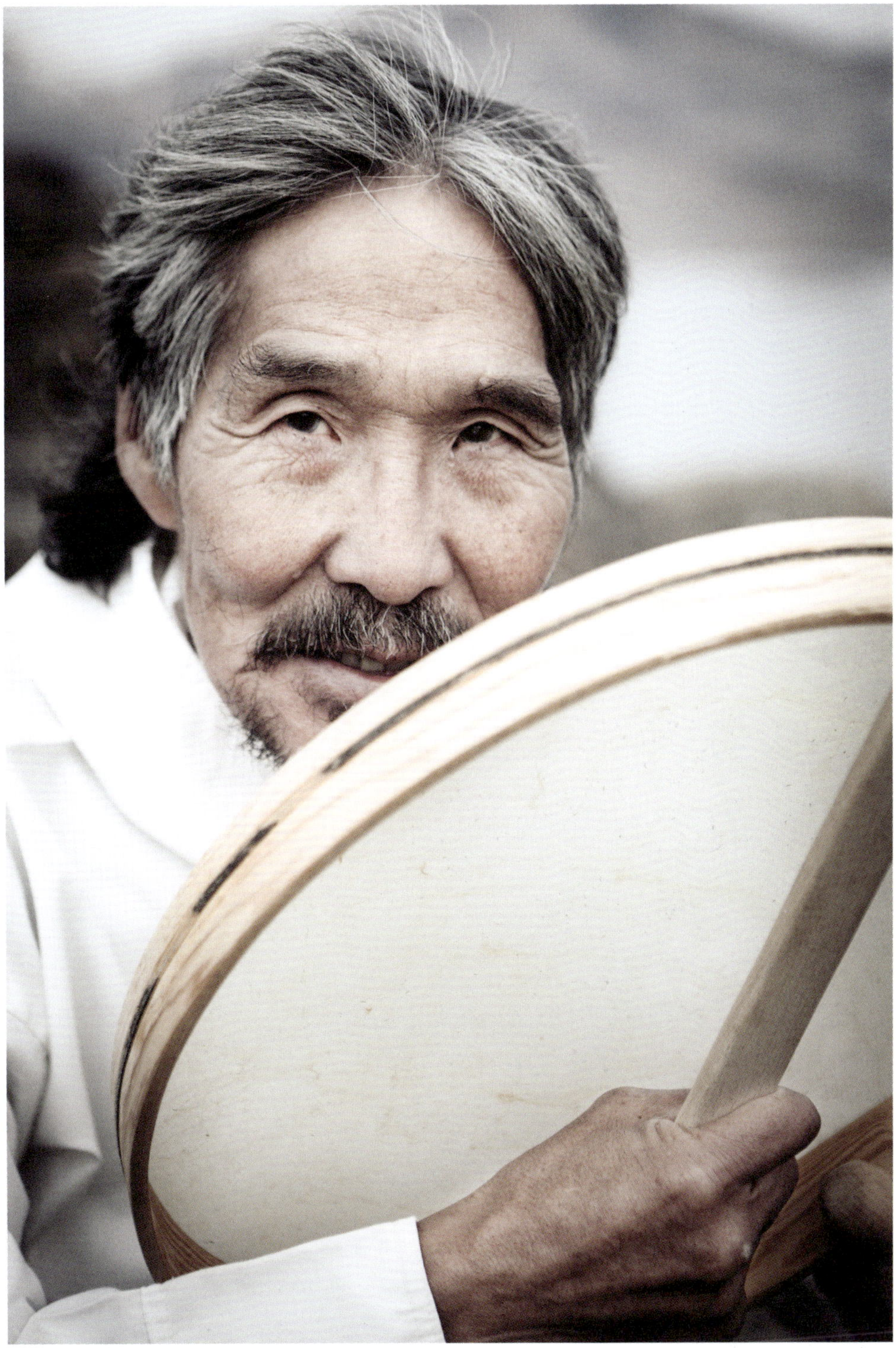

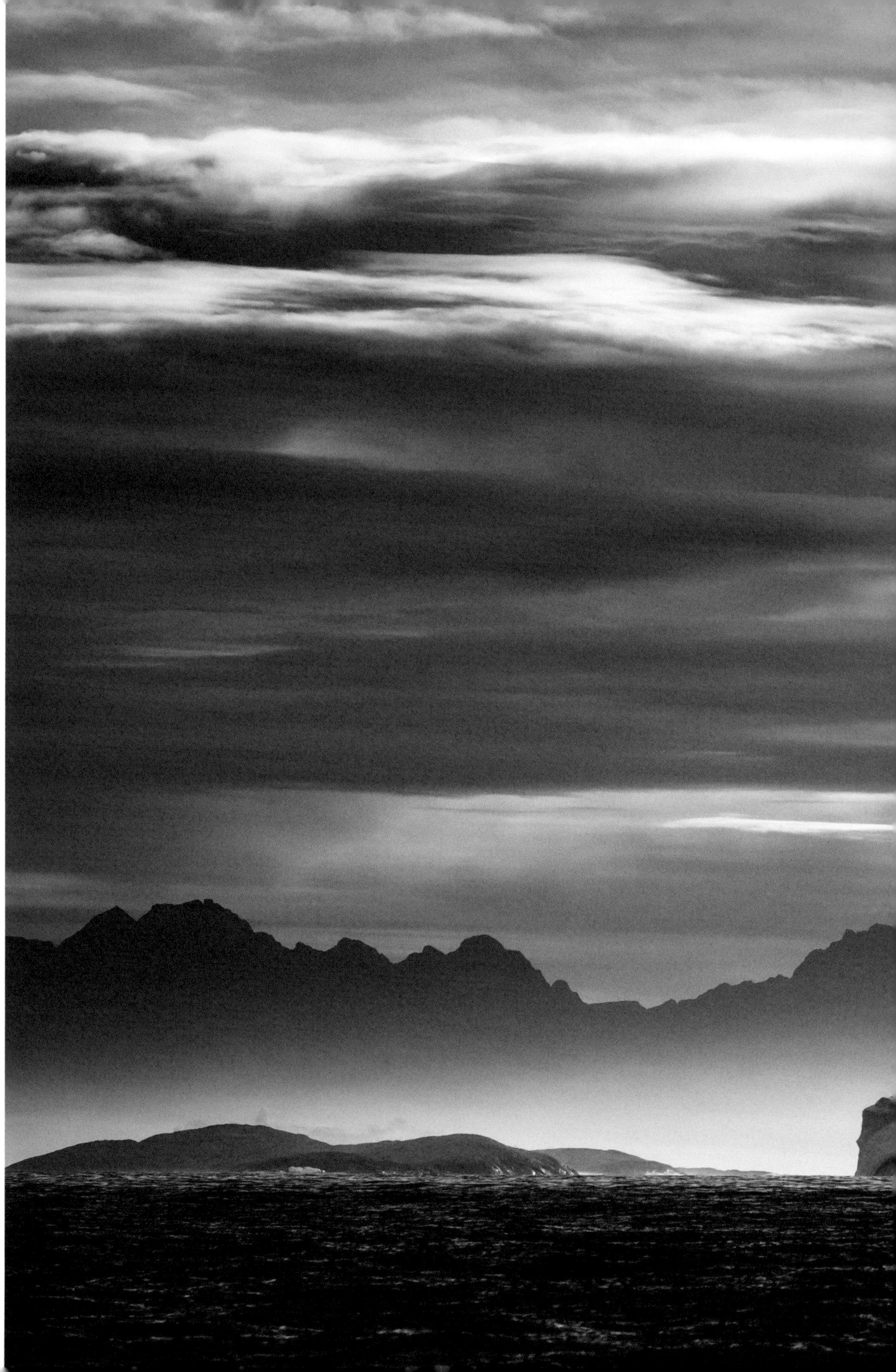

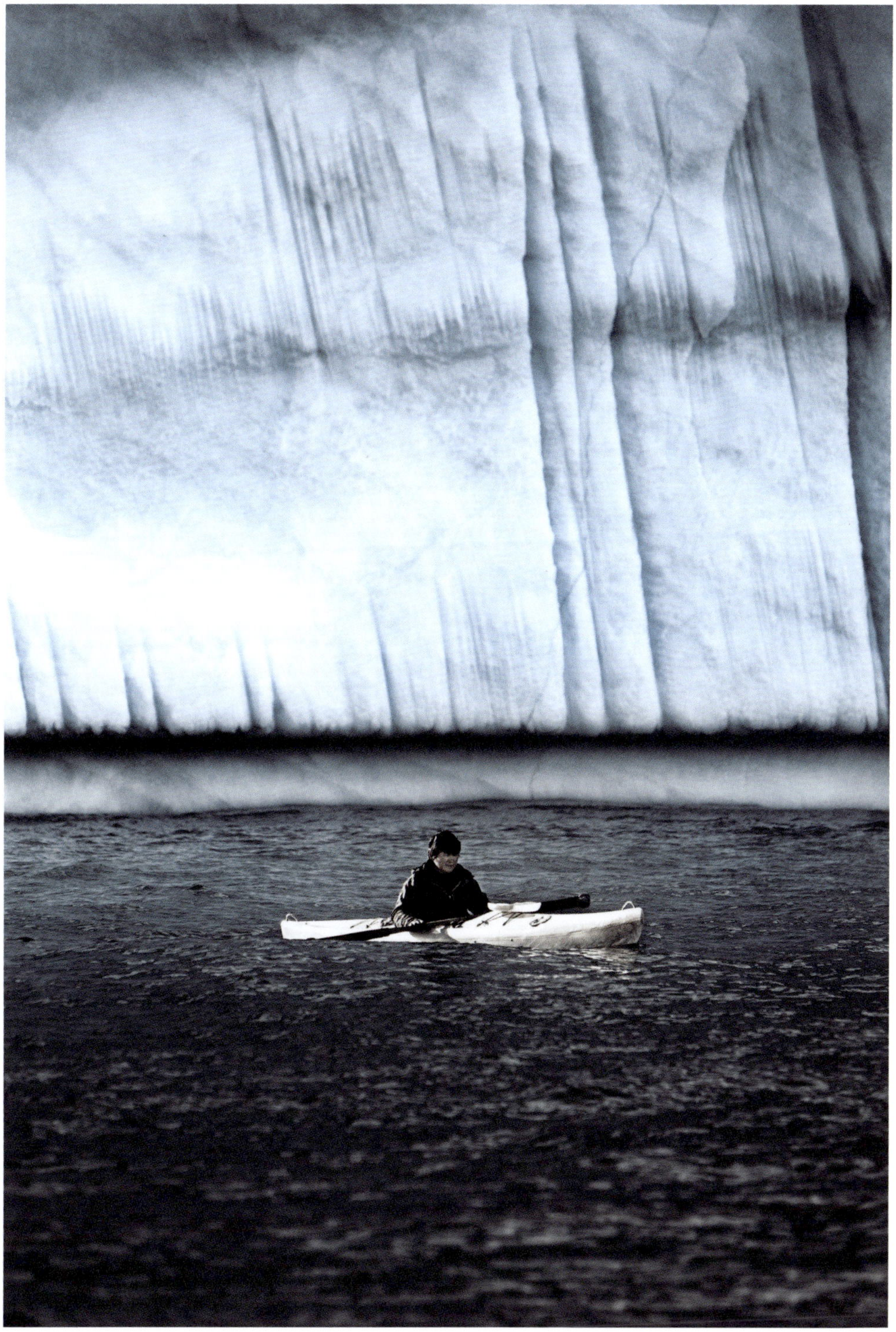

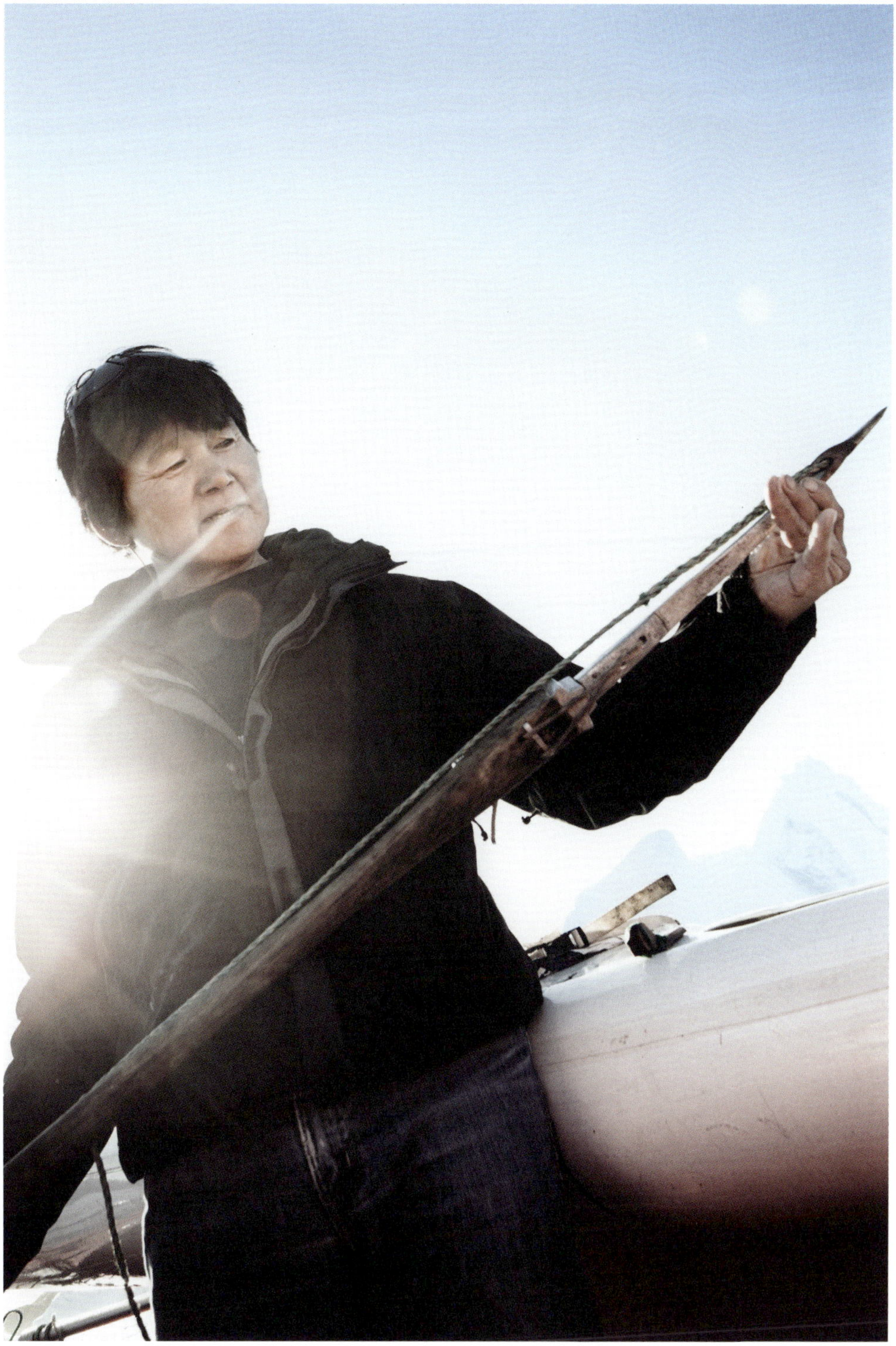

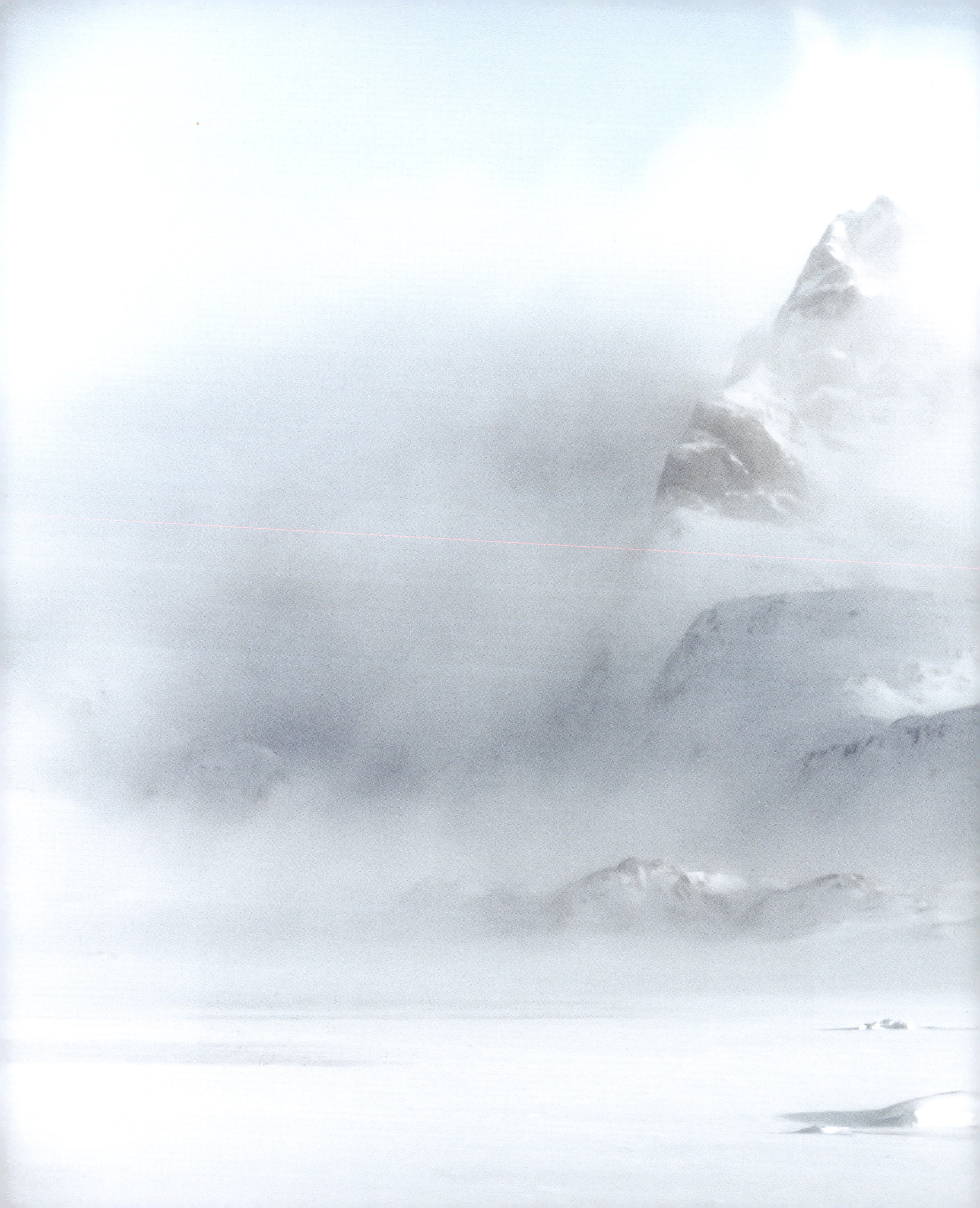

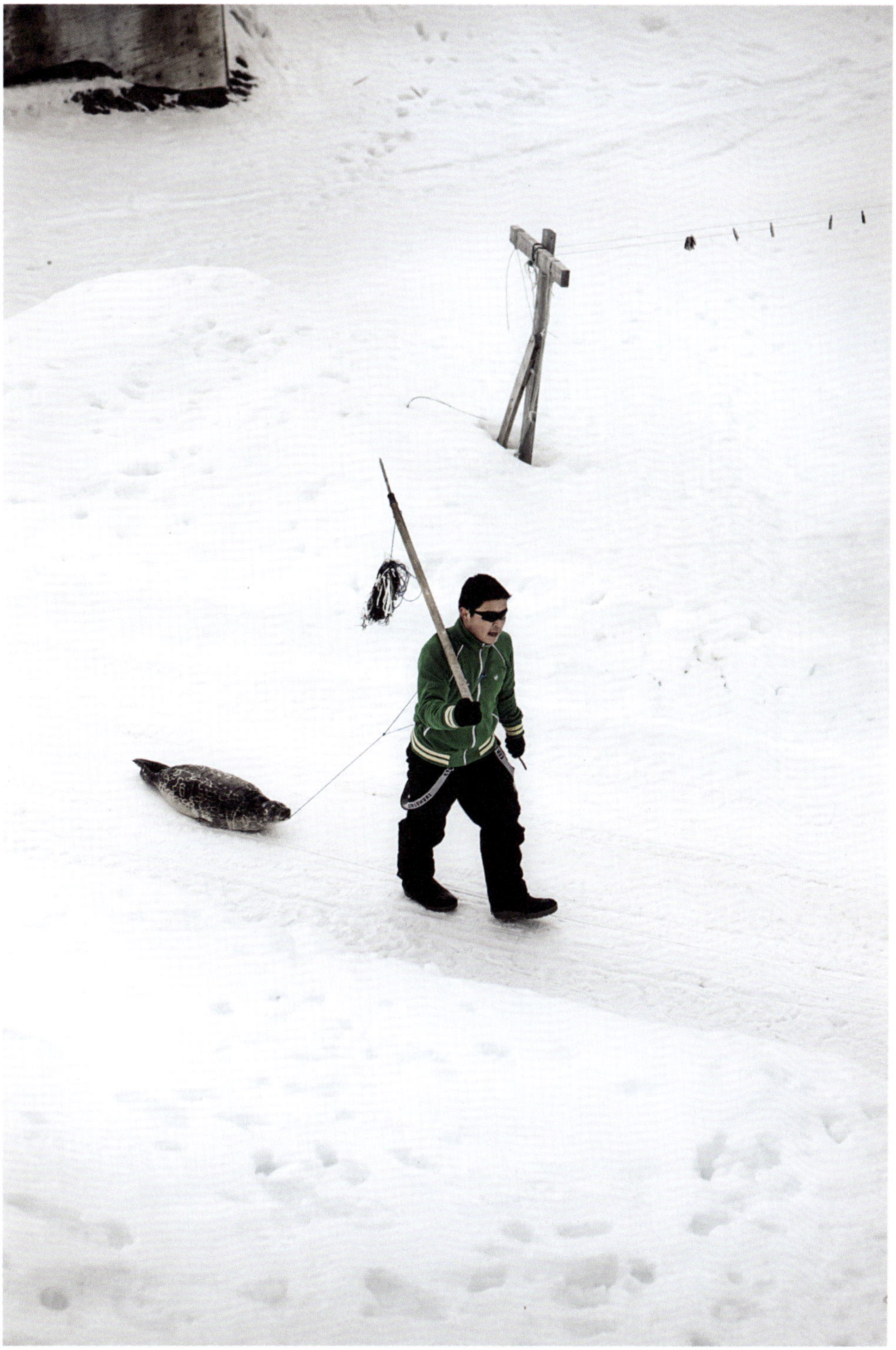

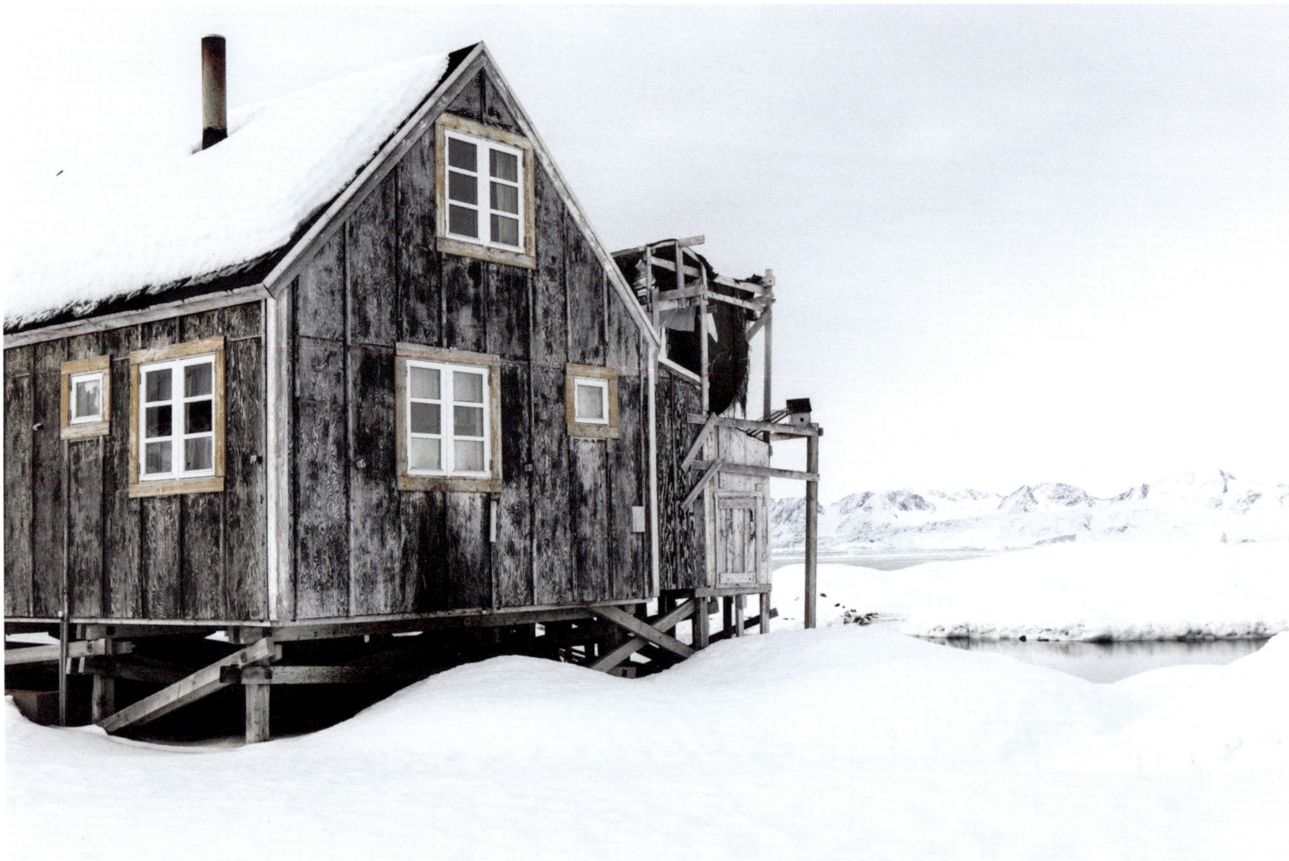

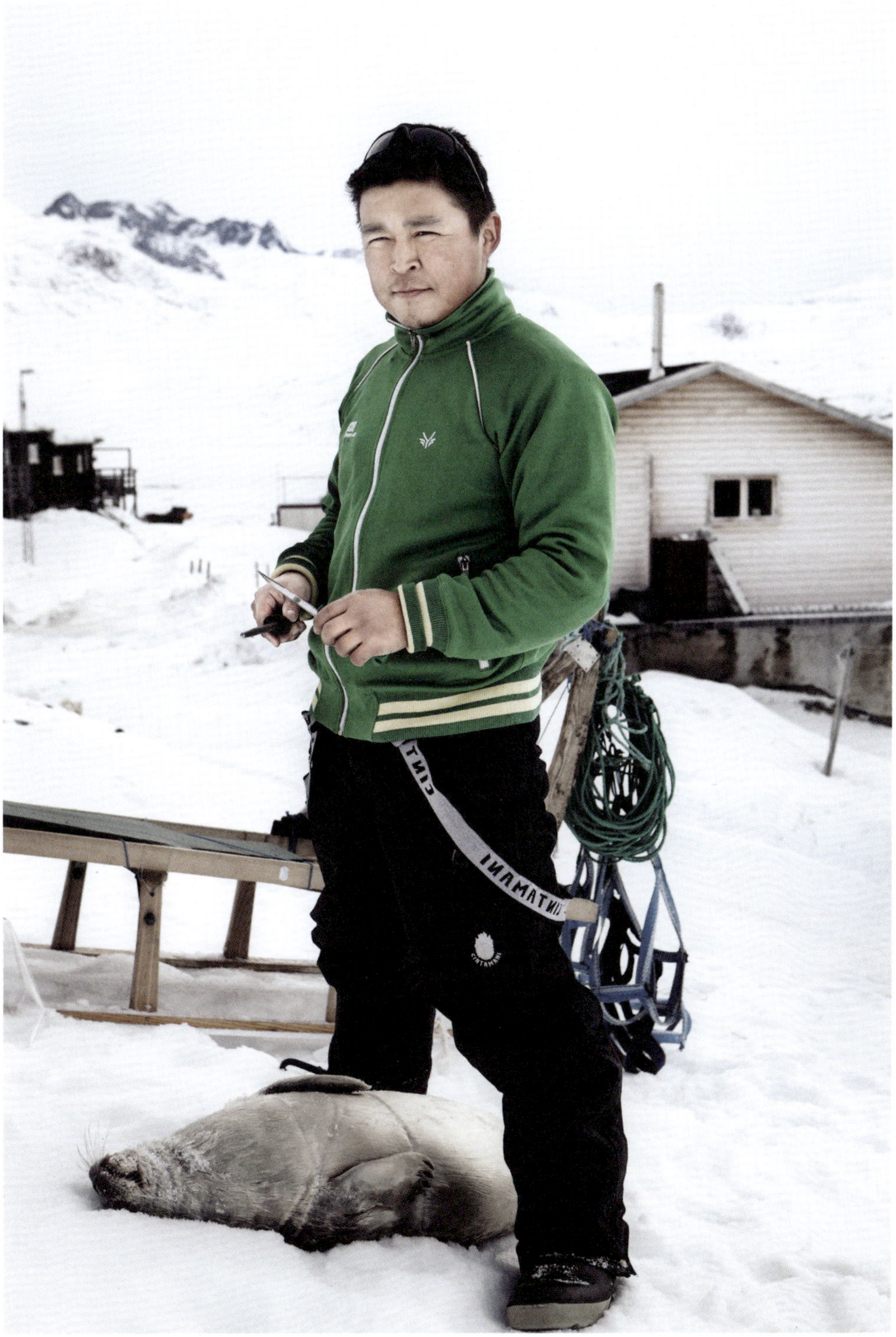

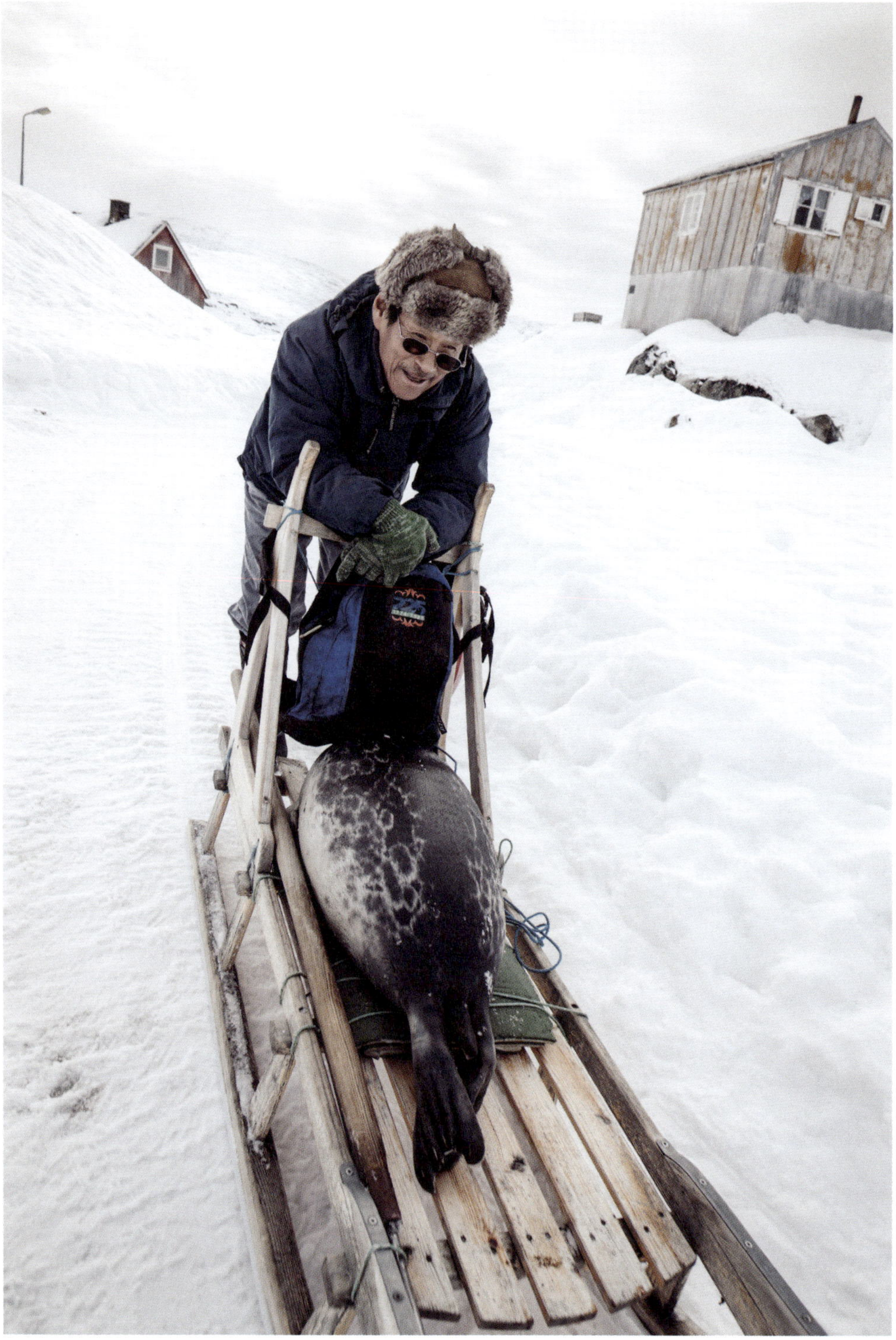

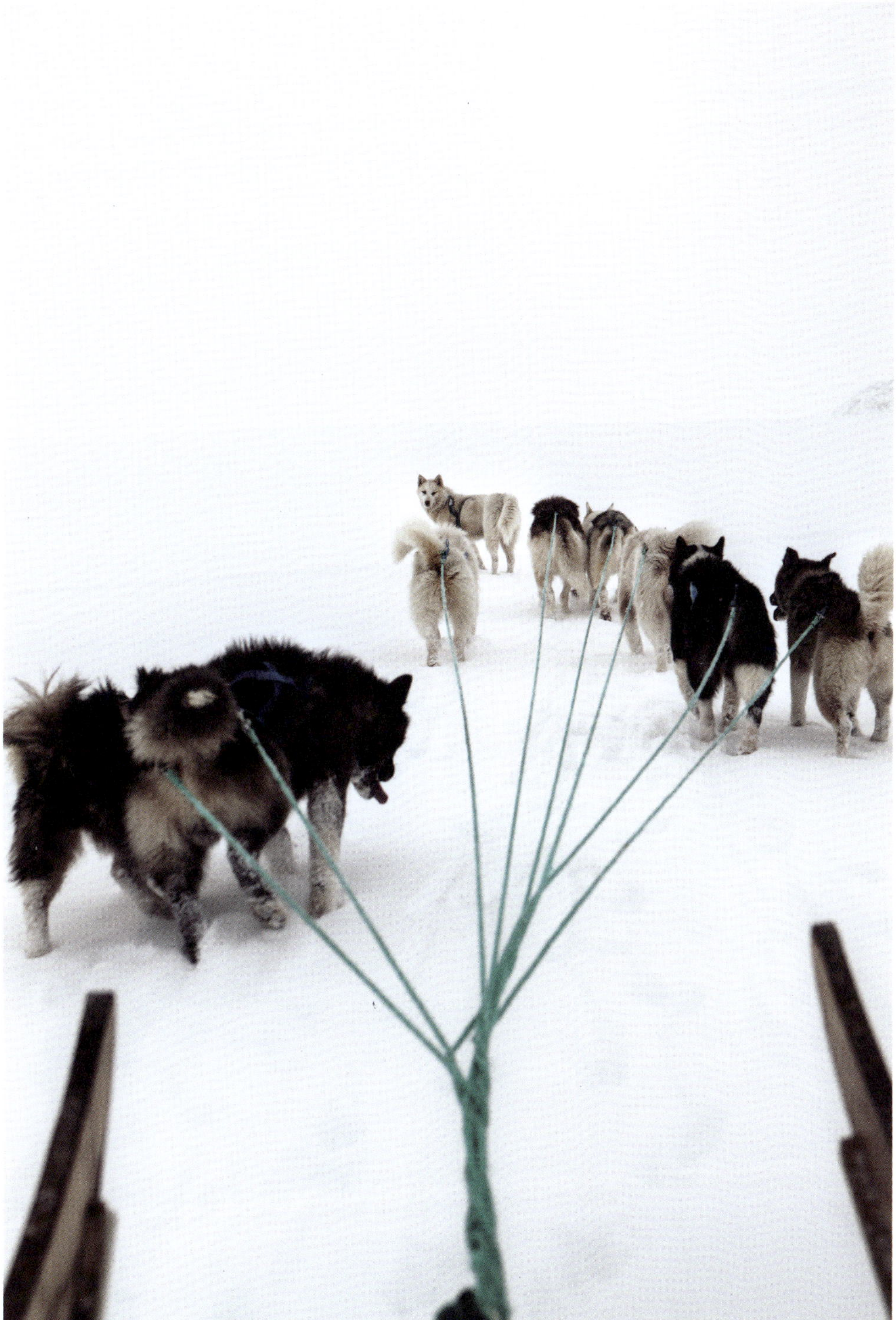

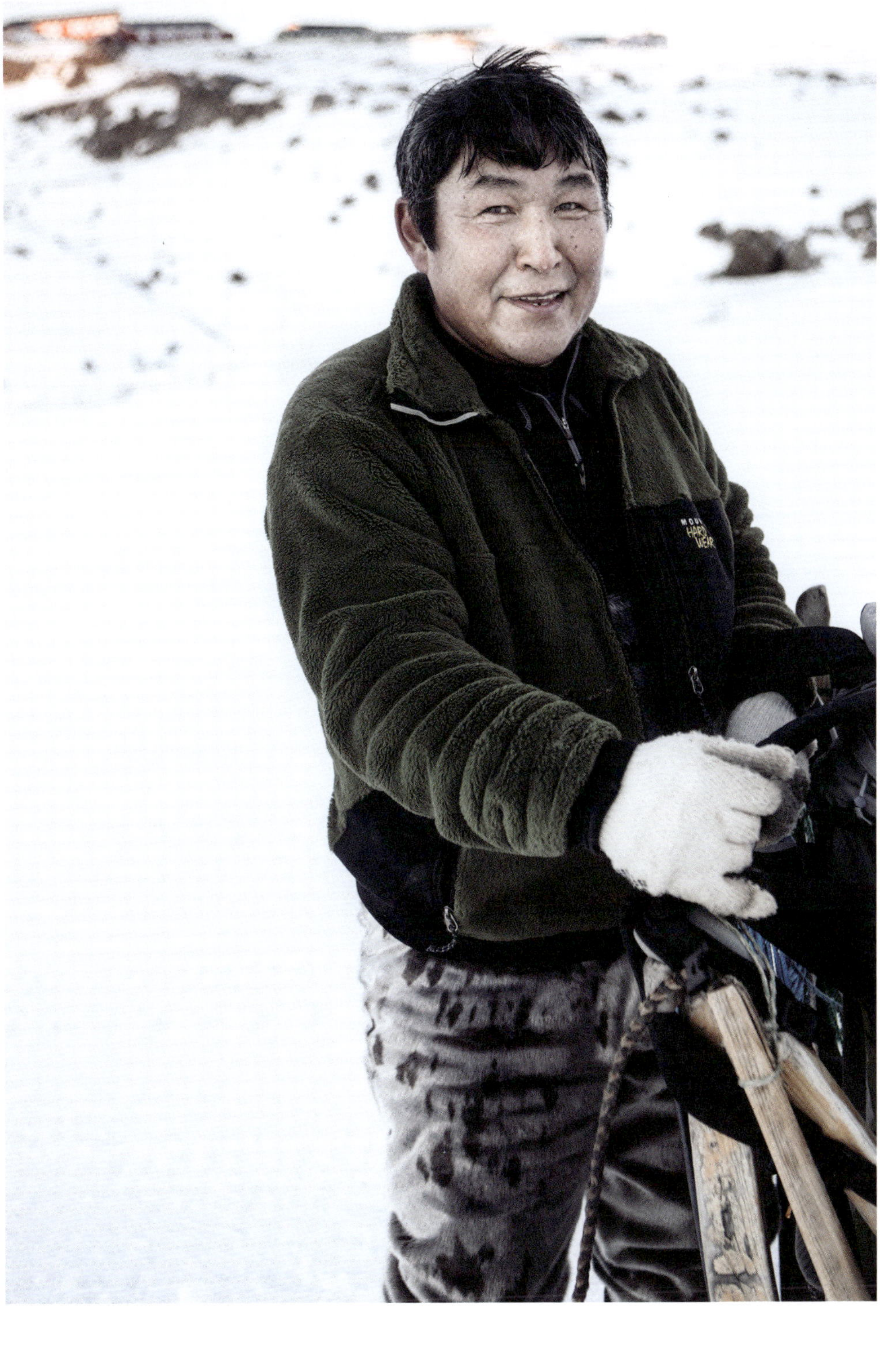

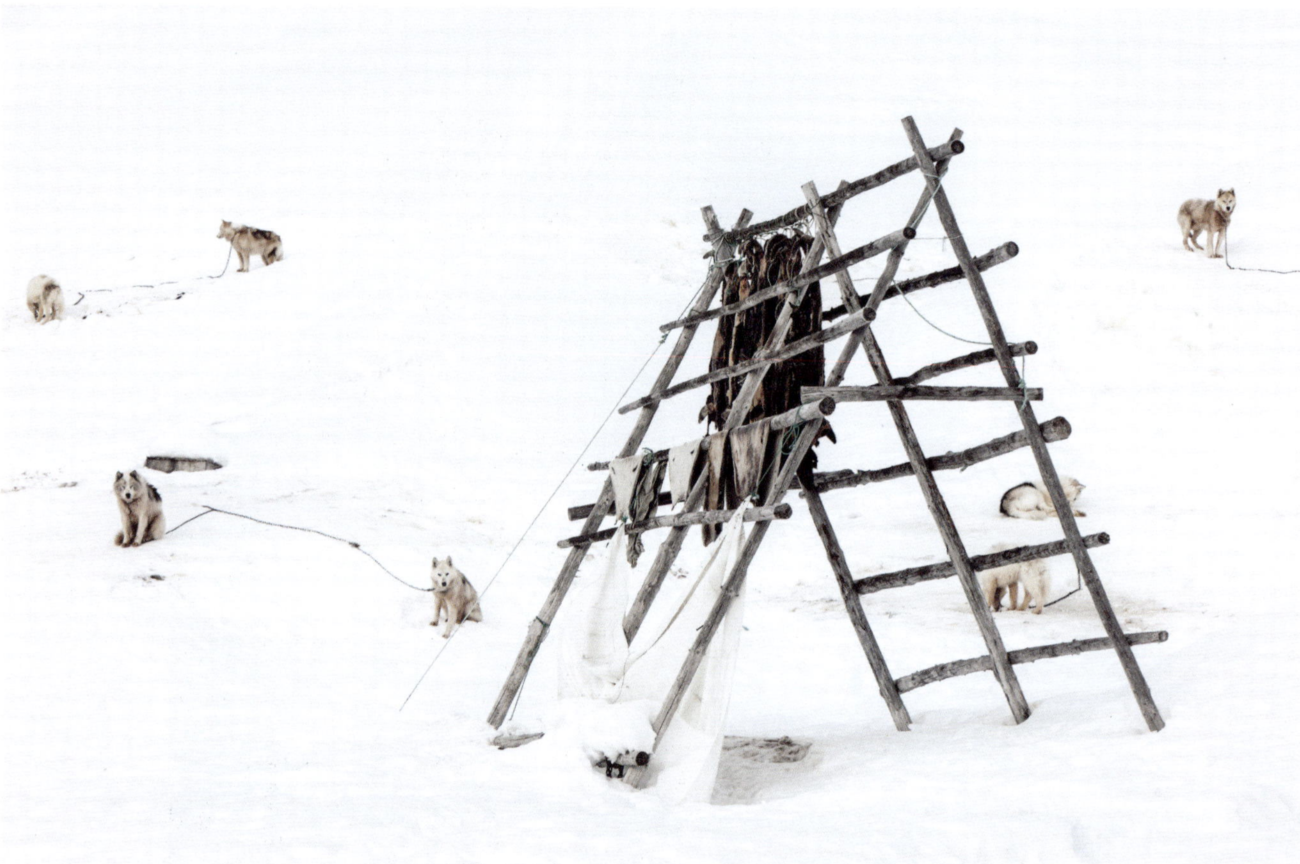

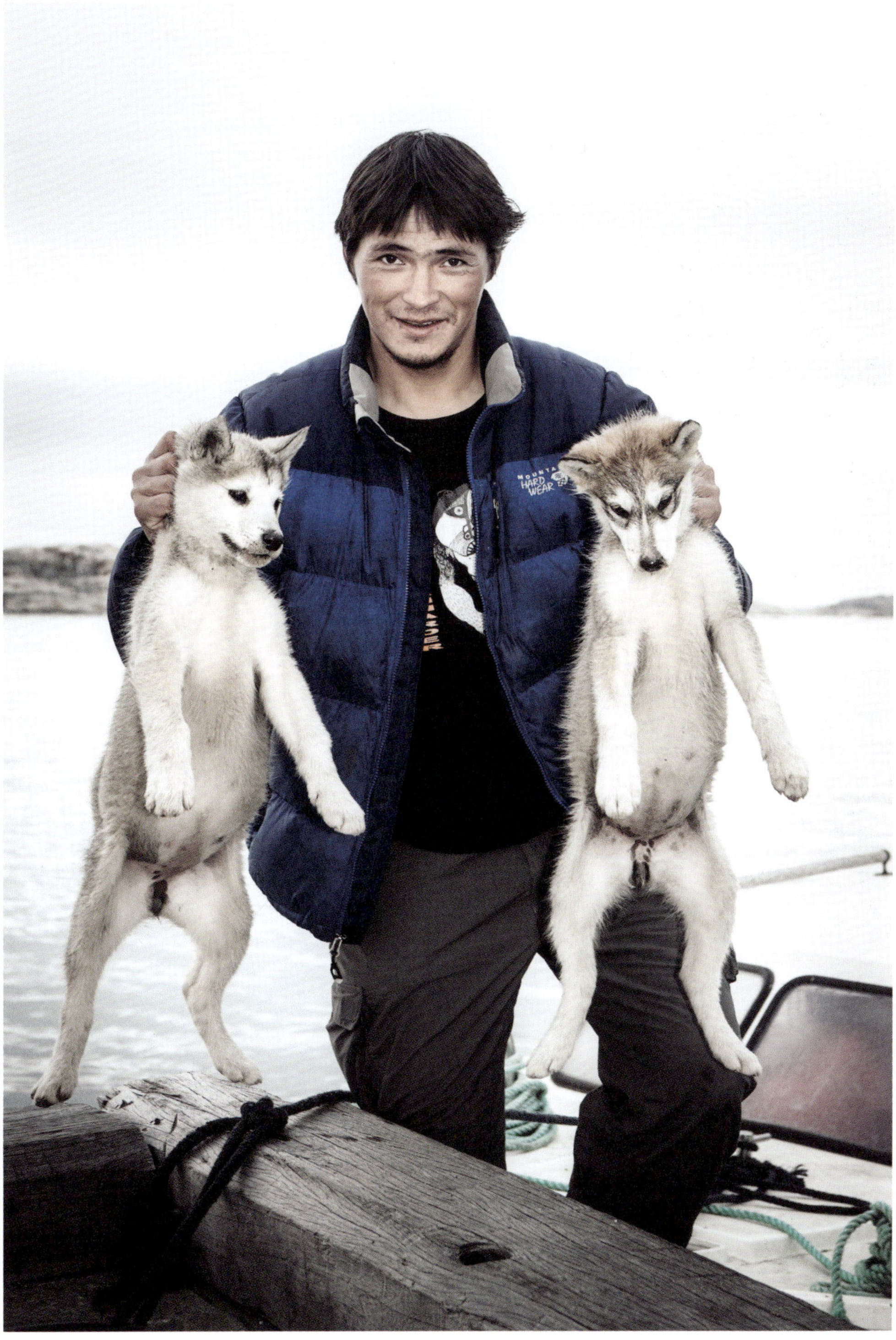
HARD
WEAR

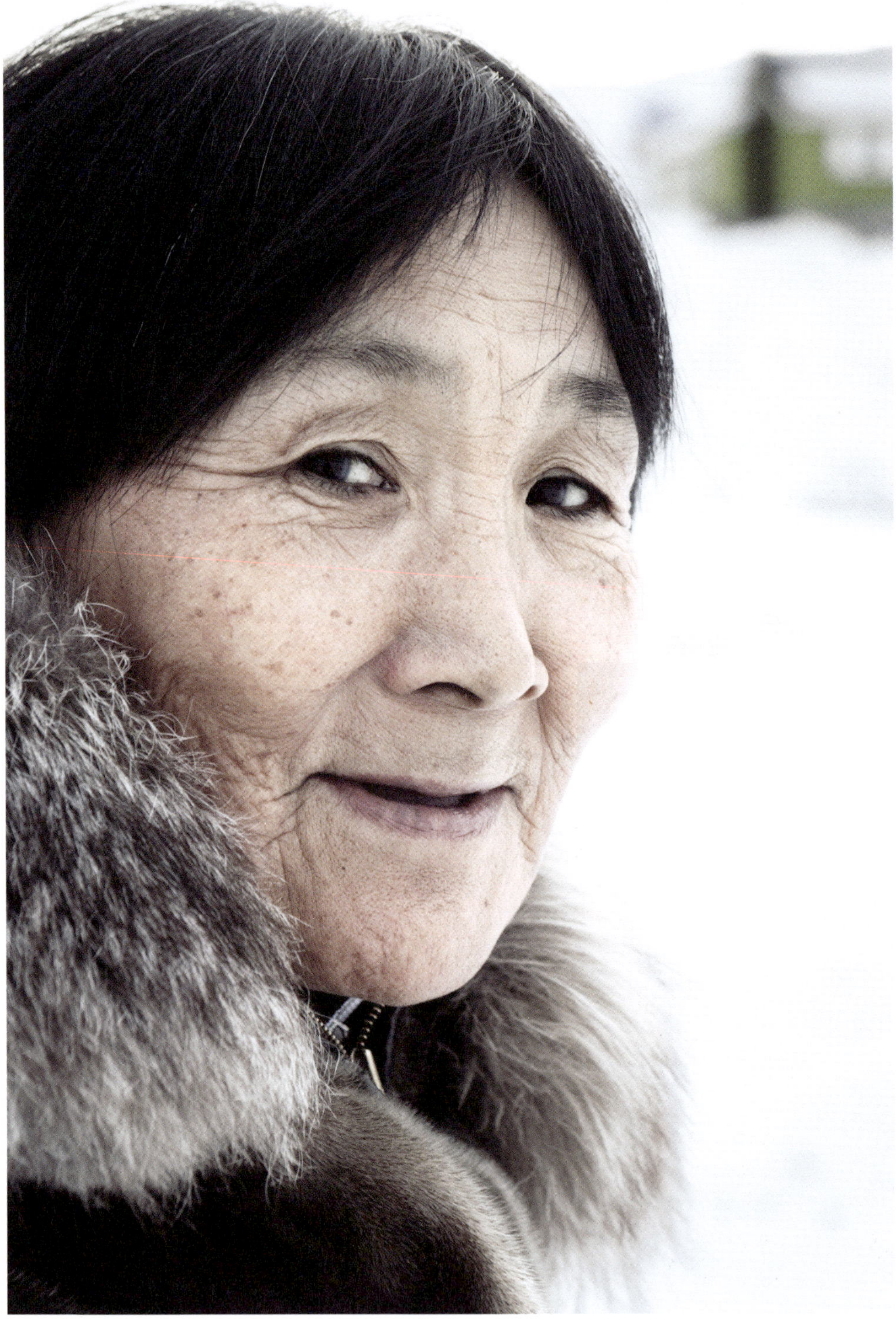

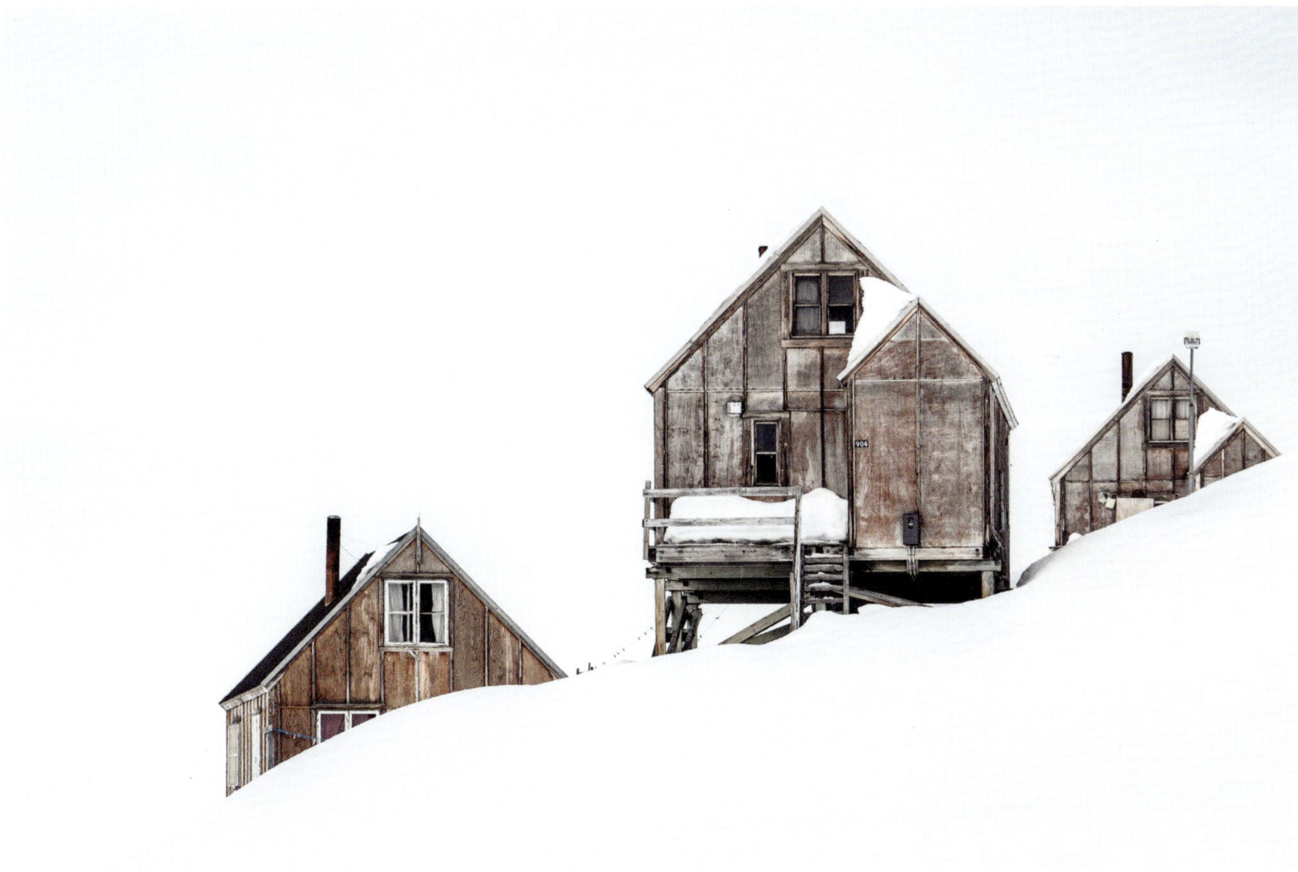

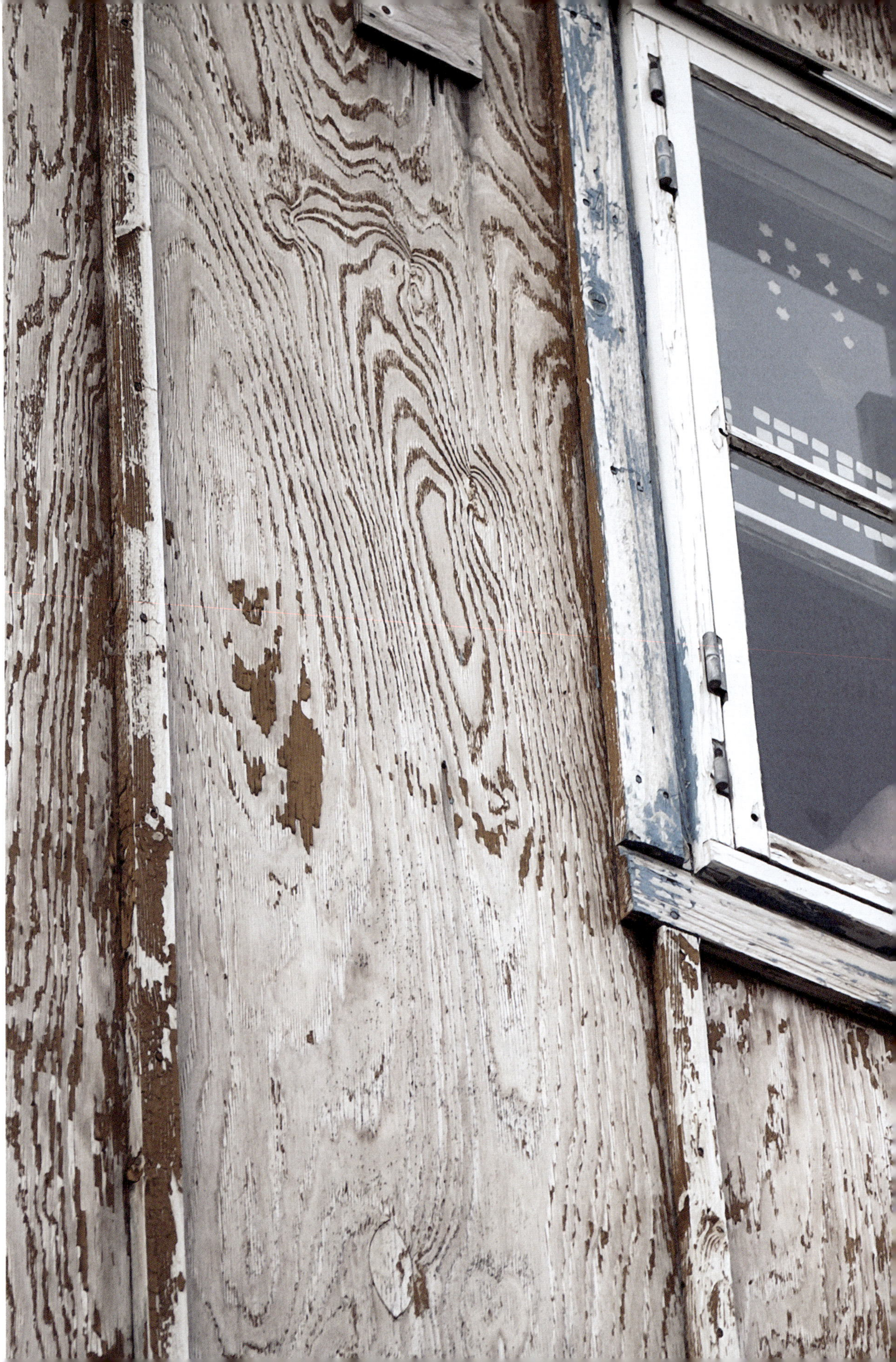

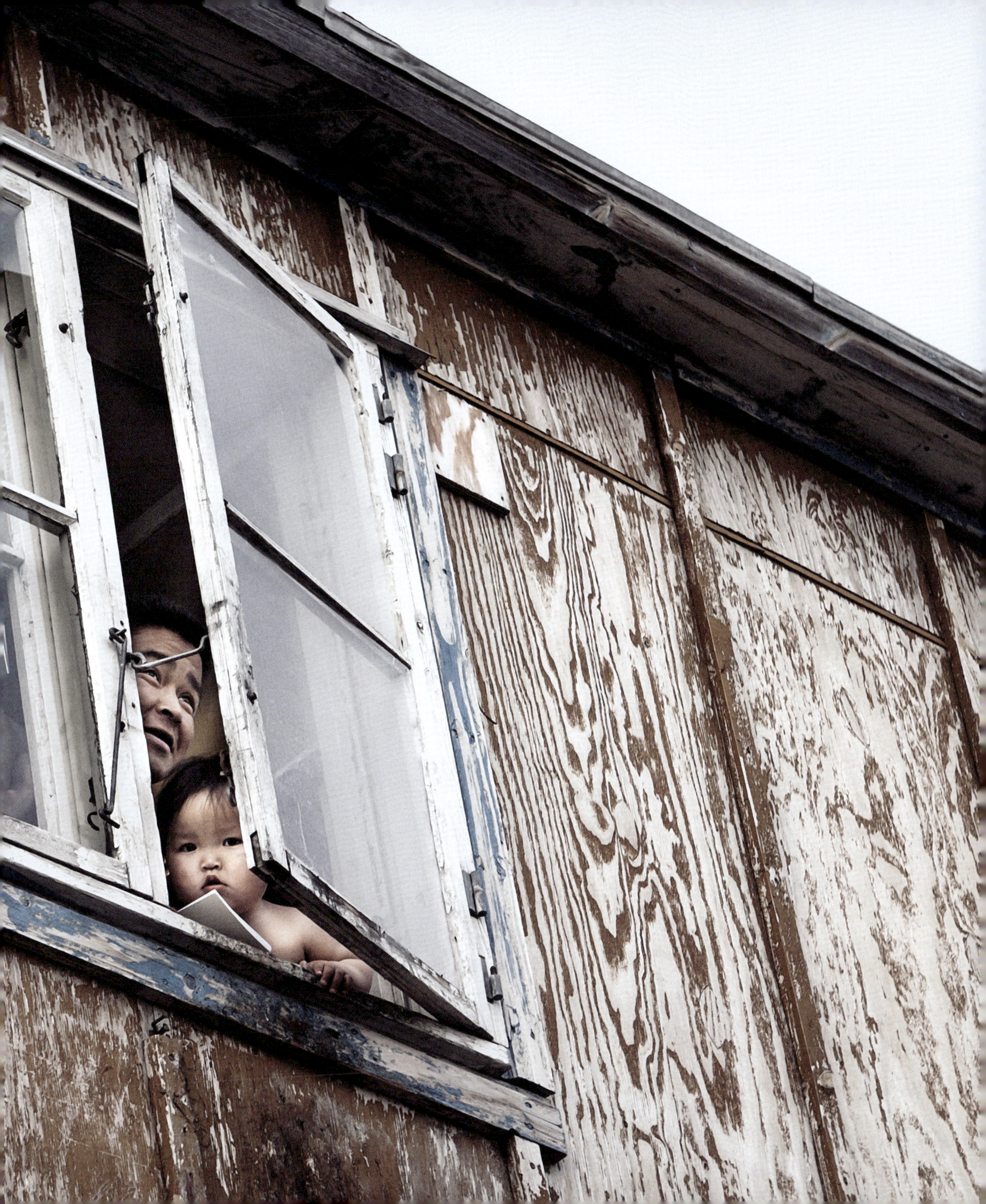

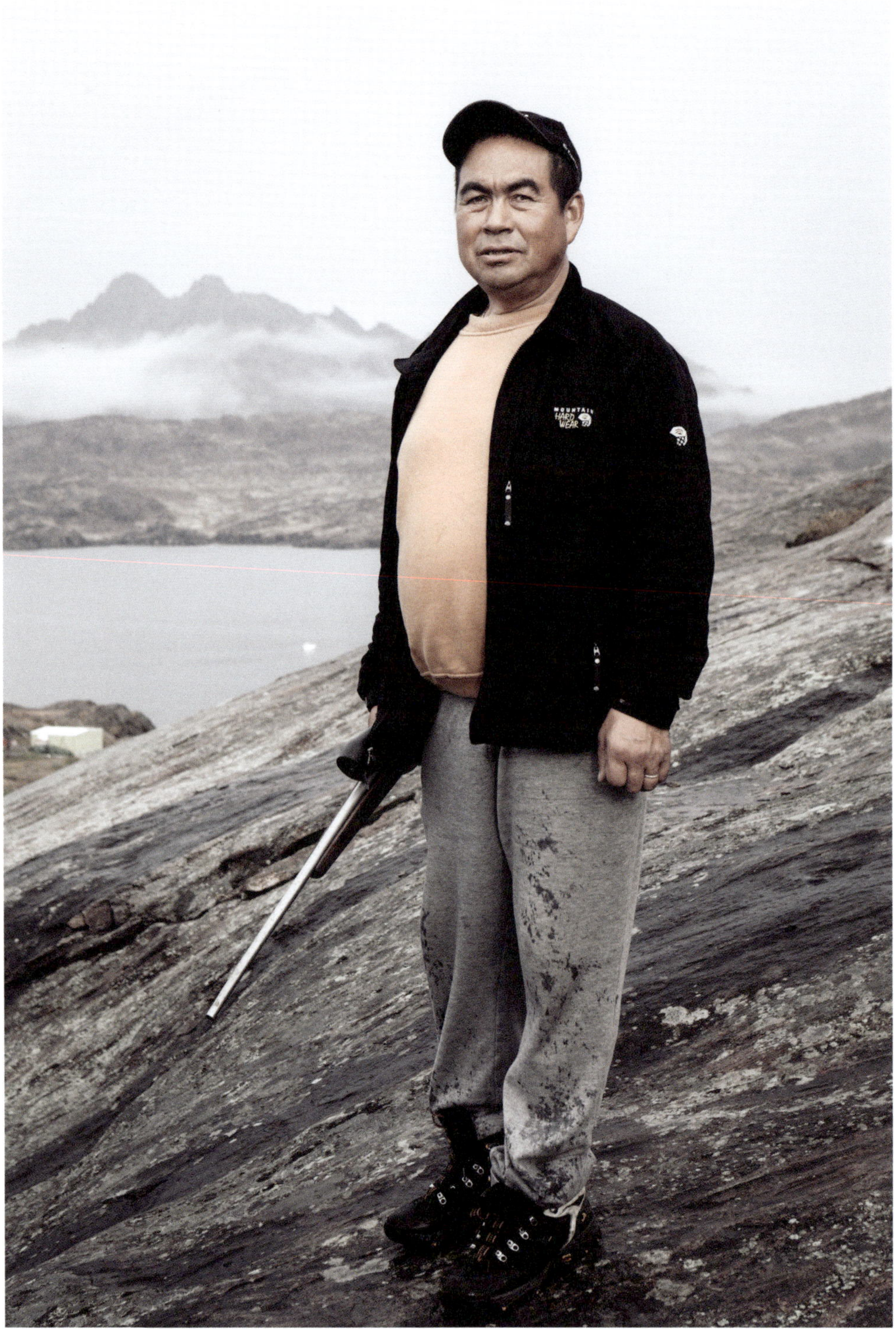
MOUNTAIN HARD WEAR

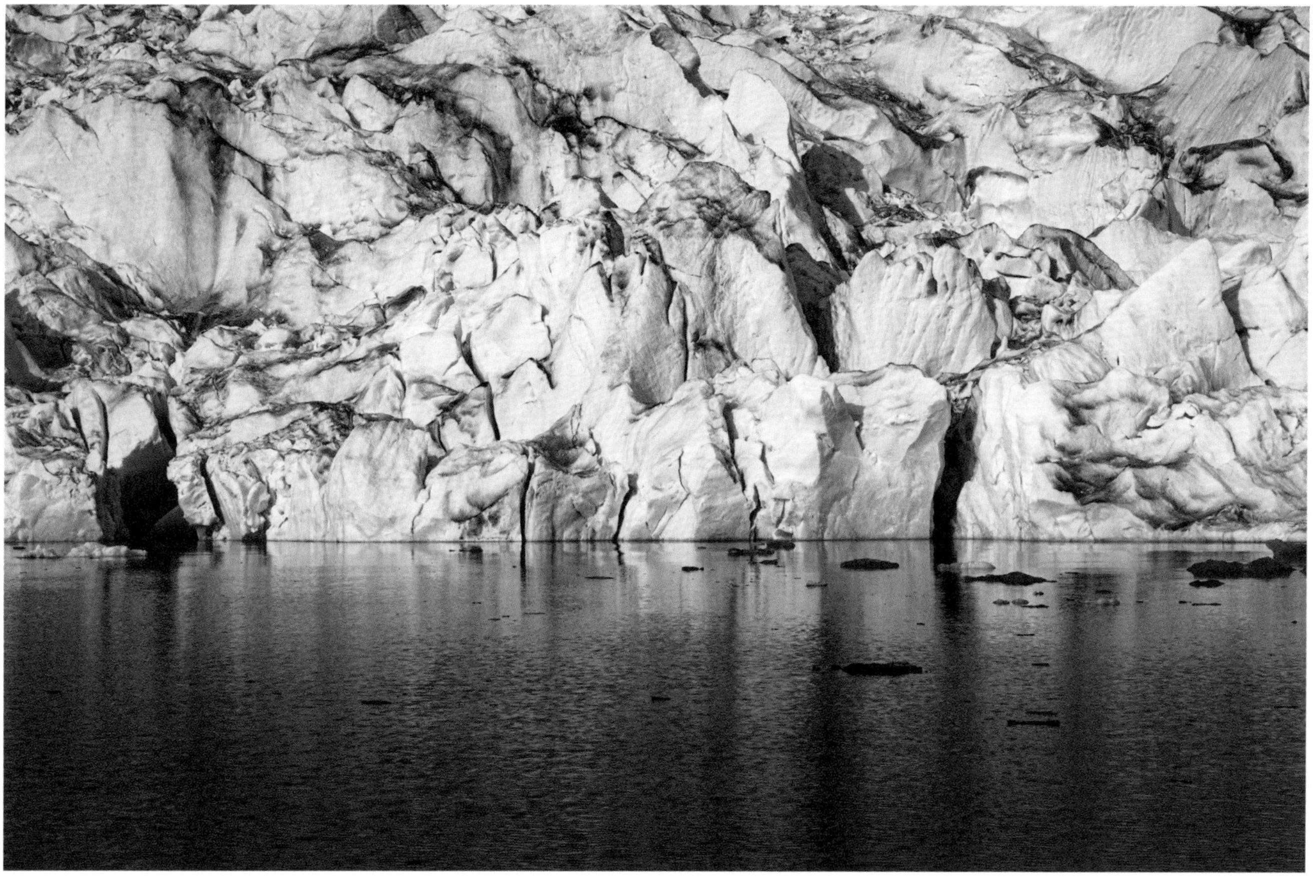

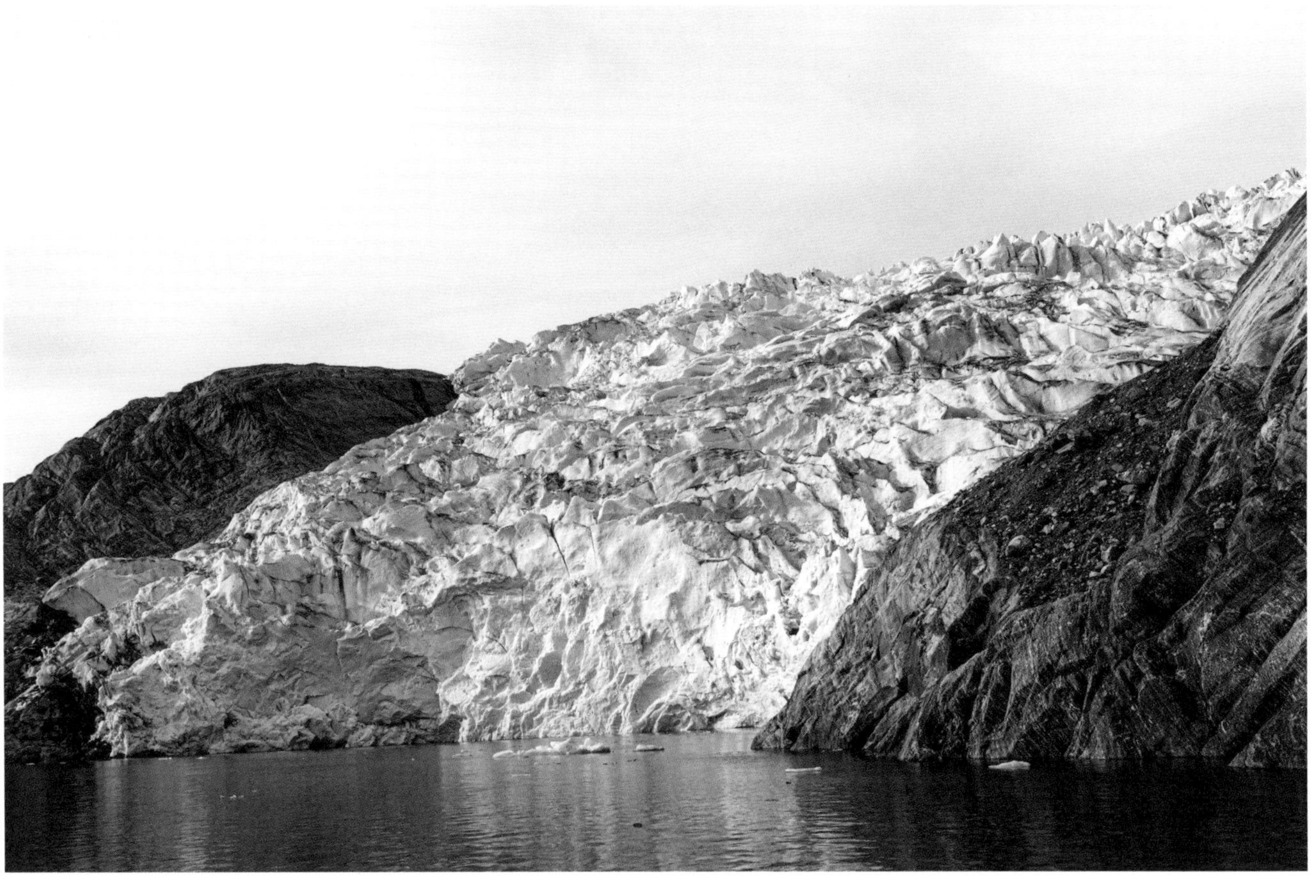

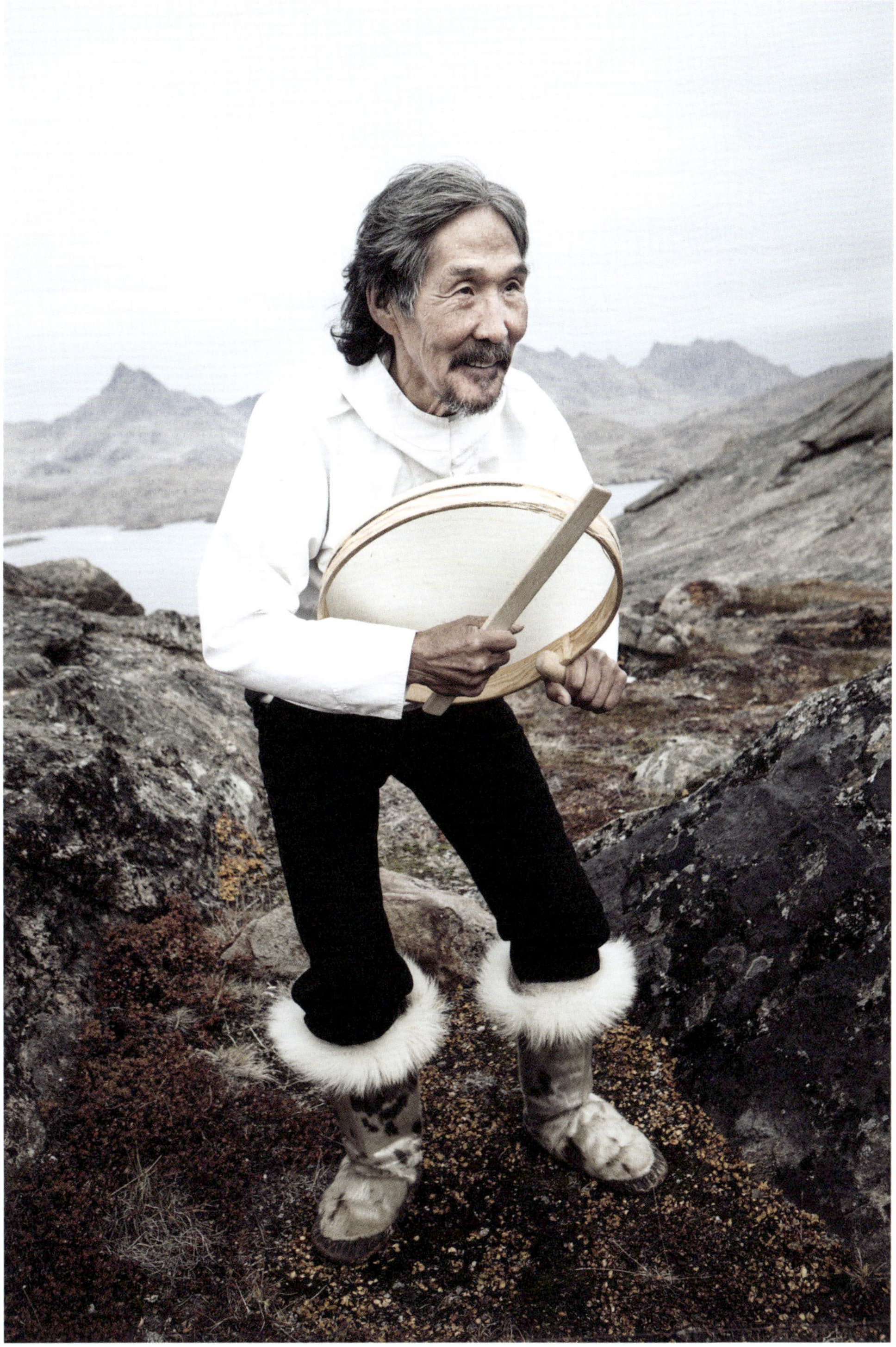

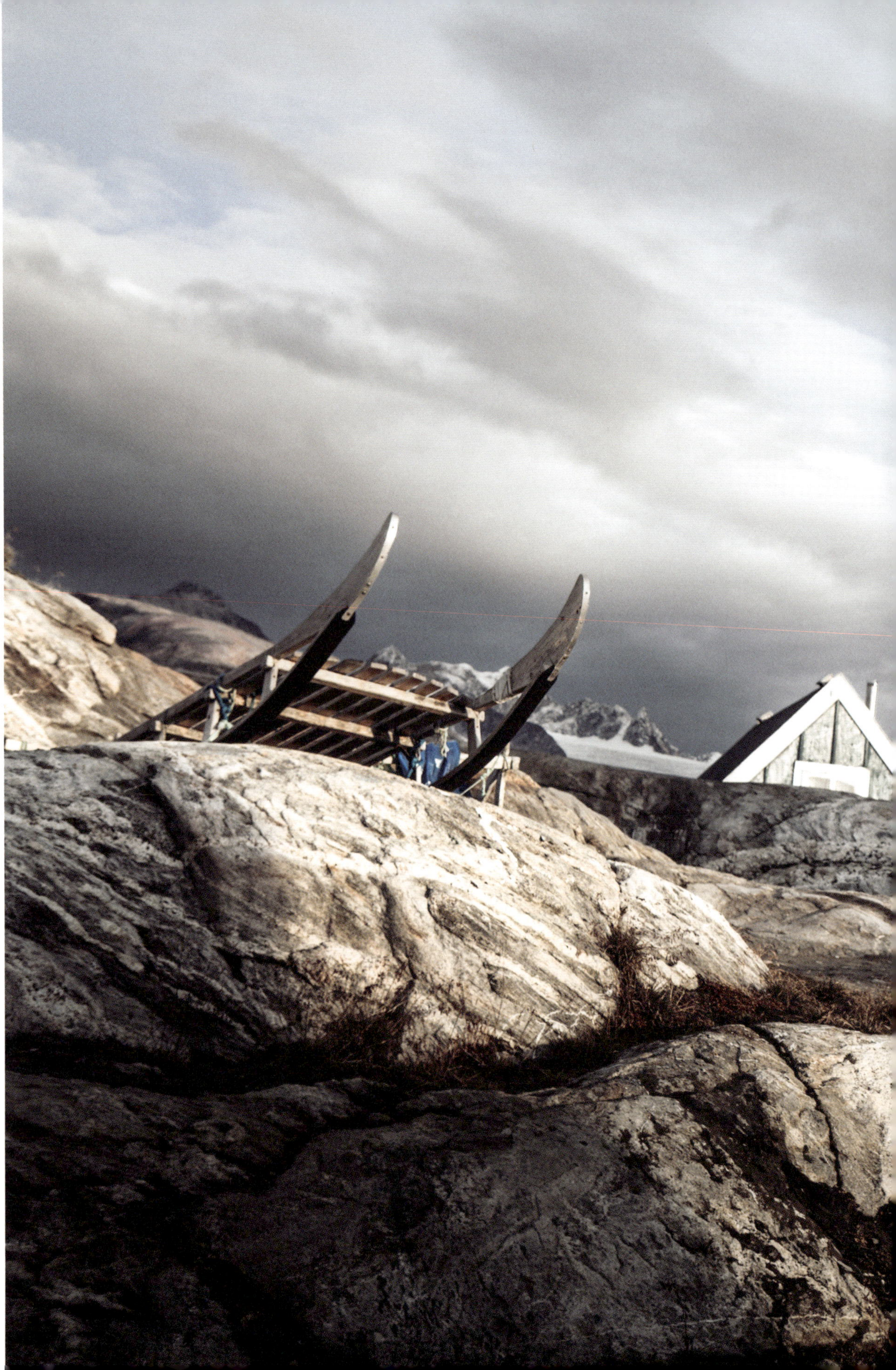

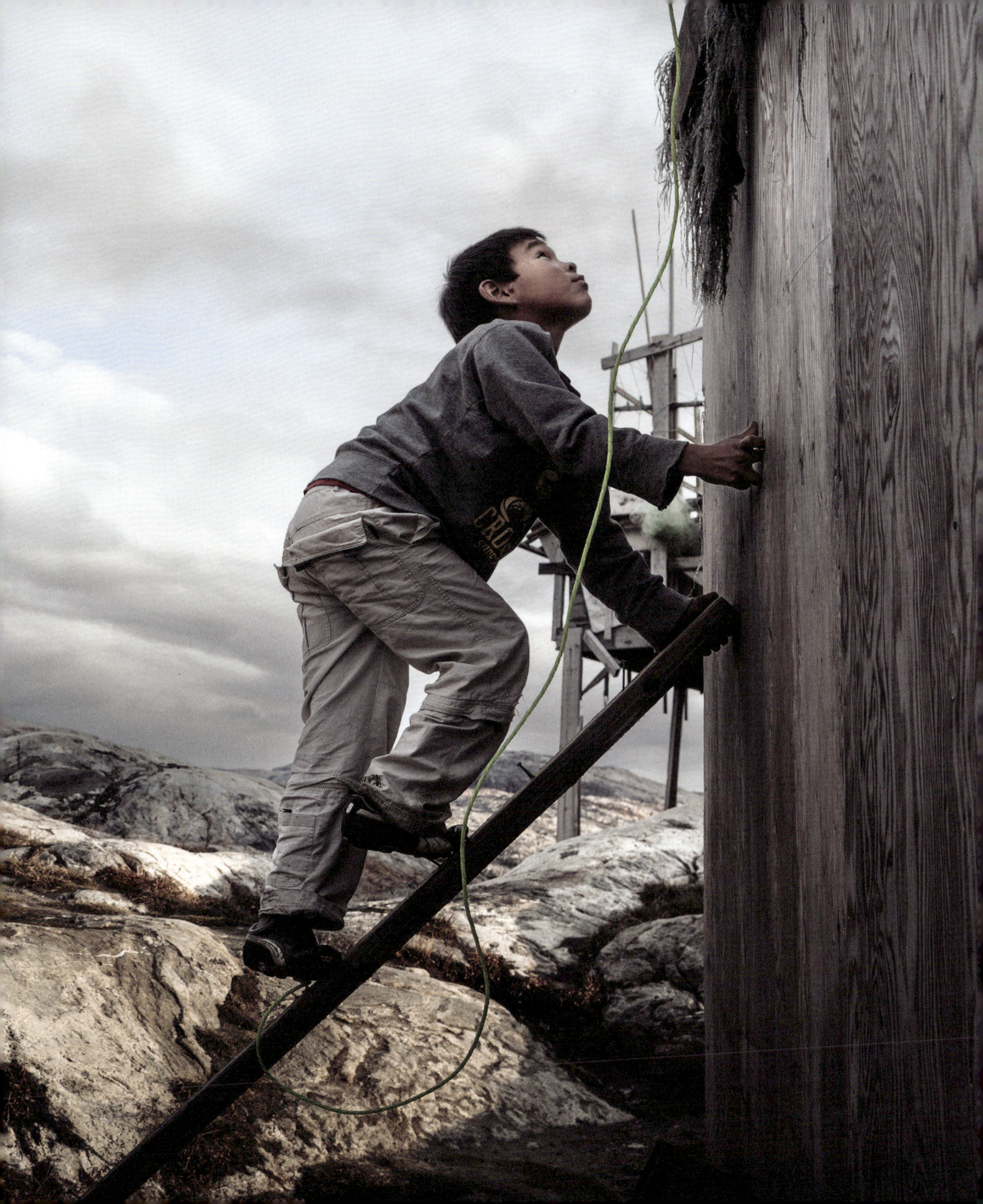

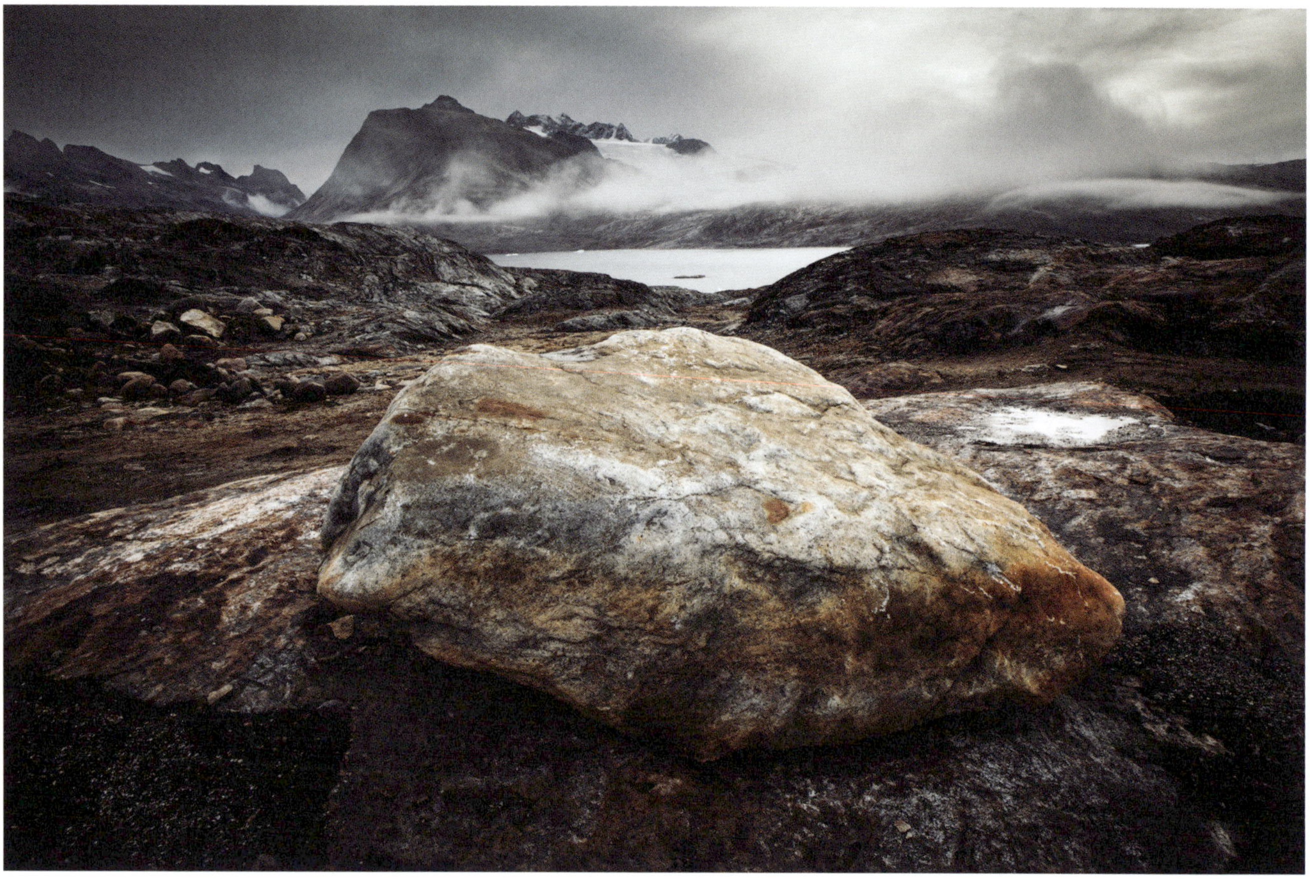

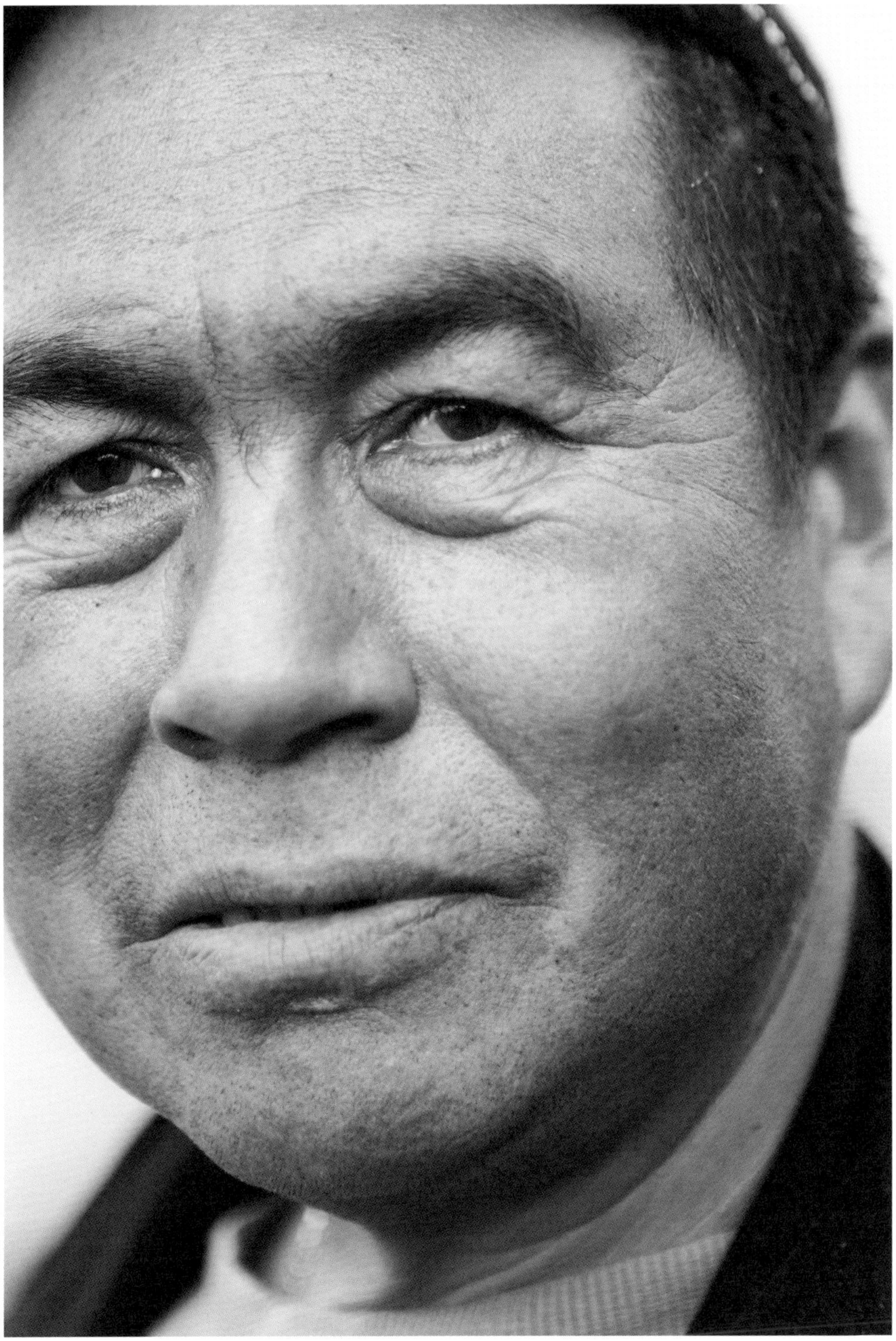

AMERICA

AMAZONIA

2° 27′ 46.35″ s 76° 59′ 34.62″ w

AMAZONIA

I visited the Achuar in the village of Sharamentsa and the Shuar in the village of Yuwientsa in Ecuador.

The Amazon basin stretches across nine countries in South America—Brazil, French Guyana, Guyana, Suriname, Venezuela, Colombia, Ecuador, Peru and Bolivia. Brazil accounts for around 65 percent of the area, with the other eight countries making up the remaining 35 percent. The Amazon rainforest is the world's largest tropical rainforest and a place of unique biodiversity. It is home to around ten percent of all known animal and plant species. Last but not least, it provides a habitat for over 30 million people, including around 320 indigenous population groups.

The Achuar live across the border between Ecuador and Peru and comprise around 18,500 people. Until the early 1940s, they were one of the last population groups that had hardly come into contact with the outside world. As warlike hunters, they originally lived scattered in their family clans in the forest. Today, they live in small village communities in the humid, deep part of the Peruvian and Ecuadorian Amazon basin, along the Pastaza River in the primary forest. The Shuar, on the other hand, live in the higher part of the Ecuadorian Amazon lowlands in the Andean foothills at an altitude of around 500 to 1000 meters and number around 40,000 people. The areas of the two groups border on each other.

The traditional culture of the Achuar and Shuar is characterized by the jungle of their settlement region. Their mythology is closely linked to nature and the laws of the universe. They believe in the spirituality of their surroundings, in spirits that live in waterfalls and on the banks of rivers. Christian missionaries have been able to gain a foothold in the area since the 1940s. Despite this missionary work, shamanism and the animistic beliefs associated with it are still firmly anchored in the Achuar and Shuar way of life. A hunter can only be successful if he lives in harmony with the "game mothers", kuntiniu nukuri, and treats the game he hunts with respect.

AMAZONIEN

Ich besuchte die Achuar im Dorf Sharamentsa und die Shuar im Dorf Yuwientsa in Ecuador.

Das Amazonasbecken erstreckt sich über neun Staaten Südamerikas – Brasilien, Französisch-Guyana, Guyana, Suriname, Venezuela, Kolumbien, Ecuador, Peru und Bolivien. Etwa 65 Prozent der Fläche entfällt auf Brasilien, die restlichen 35 Prozent verteilen sich auf die anderen acht Länder. Der Amazonas-Regenwald als weltweit größter tropischer Regenwald besitzt eine einzigartige Biodiversität. Er beherbergt rund zehn Prozent aller bekannten Tier- und Pflanzenarten. Nicht zuletzt bietet er einen Lebensraum für über 30 Millionen Menschen, davon rund 320 indigene Bevölkerungsgruppen.

Die Achuar siedeln grenzübergreifend in Ecuador und Peru und umfassen etwa 18.500 Menschen. Bis in die frühen 1940er Jahre waren sie eine der letzten Bevölkerungsgruppen, die noch kaum mit der Außenwelt in Berührung gekommen waren. Als kriegerische Jäger lebten sie ursprünglich in ihren Familienclans verstreut im Wald. Heute siedeln sie in kleinen Dorfgemeinschaften im feuchten, tiefen Teil des peruanischen und ecuadorianischen Amazonasbeckens, unter anderem am Fluss Pastaza im Primärwald. Die Shuar hingegen leben im höhergelegenen Teil des ecuadorianischen Amazonastieflandes im Bereich der Andenvorgebirge auf einer Höhe von etwa 500 bis 1000 Metern und zählen ungefähr 40.000 Menschen. Die Gebiete der beiden Gruppen grenzen aneinander.

Die traditionelle Kultur der Achuar und Shuar ist vom Urwald ihres Siedlungsgebiets geprägt. Ihre Mythologie ist der Natur und den Gesetzen des Universums eng verbunden. Sie glauben an die Beseeltheit ihrer Umgebung, an Geistwesen, die in Wasserfällen und an den Flussufern wohnen. Seit den 1940er Jahren des letzten Jahrhunderts konnten christliche Missionare in der Gegend Fuß fassen. Trotz dieser Missionierung ist der Schamanismus und der damit einhergehende animistische Glaube in der Lebensweise der Achuar und Shuar noch immer verankert. Ein Jäger kann nur dann erfolgreich sein, wenn er in Harmonie mit den „Wildmüttern", Kuntiniu Nukuri, lebt und dem Wild, das er jagt, mit Respekt begegnet.

The habitats of the indigenous peoples of the Amazon basin are threatened by slash-and-burn agriculture to clear land for cattle pastures and monocultures such as soy, by illegal gold mining, deforestation and oil extraction. These activities not only destroy the habitat of the communities in the forest, but also have a major impact on the global climate. Satellite images from the Brazilian space agency Instituto Nacional de Pesquisas Espaciais (INPE) have shown that around 436,600 square kilometers have been cut down in the Brazilian part of the rainforest alone over the past 30 years. This is roughly equivalent to the combined area of Germany and Austria.

According to a study by the Intergovernmental Panel on Climate Change (IPCC), the forest is increasingly losing its ability to regenerate – in the years between 2010 and 2019, the Amazon rainforest has emitted more carbon dioxide than it absorbed.

Die Lebensräume der indigenen Völker des Amazonasbeckens werden durch Brandrodung zur Landgewinnung für Viehweiden und Monokulturen wie Soja, durch illegalen Goldabbau, Abholzung und Ölförderung bedroht. Diese Aktivitäten zerstören nicht nur den Lebensraum der Gemeinschaften im Wald, sondern haben auch starke Auswirkungen auf das Weltklima. Satellitenaufnahmen der brasilianischen Weltraumbehörde Instituto Nacional de Pesquisas Espaciais (INPE) haben ergeben, dass in den vergangenen 30 Jahren allein im brasilianischen Teil des Regenwalds etwa 436.600 Quadratkilometer abgeholzt wurden. Das entspricht etwa der Fläche Deutschlands und Österreichs zusammen.

Laut einer Studie des Intergovernmental Panel on Climate Change (IPCC) verliert der Wald zunehmend seine Fähigkeit zur Regeneration – in den Jahren zwischen 2010 und 2019 gab der Amazonas-Regenwald mehr Kohlendioxid ab, als er aufnahm.

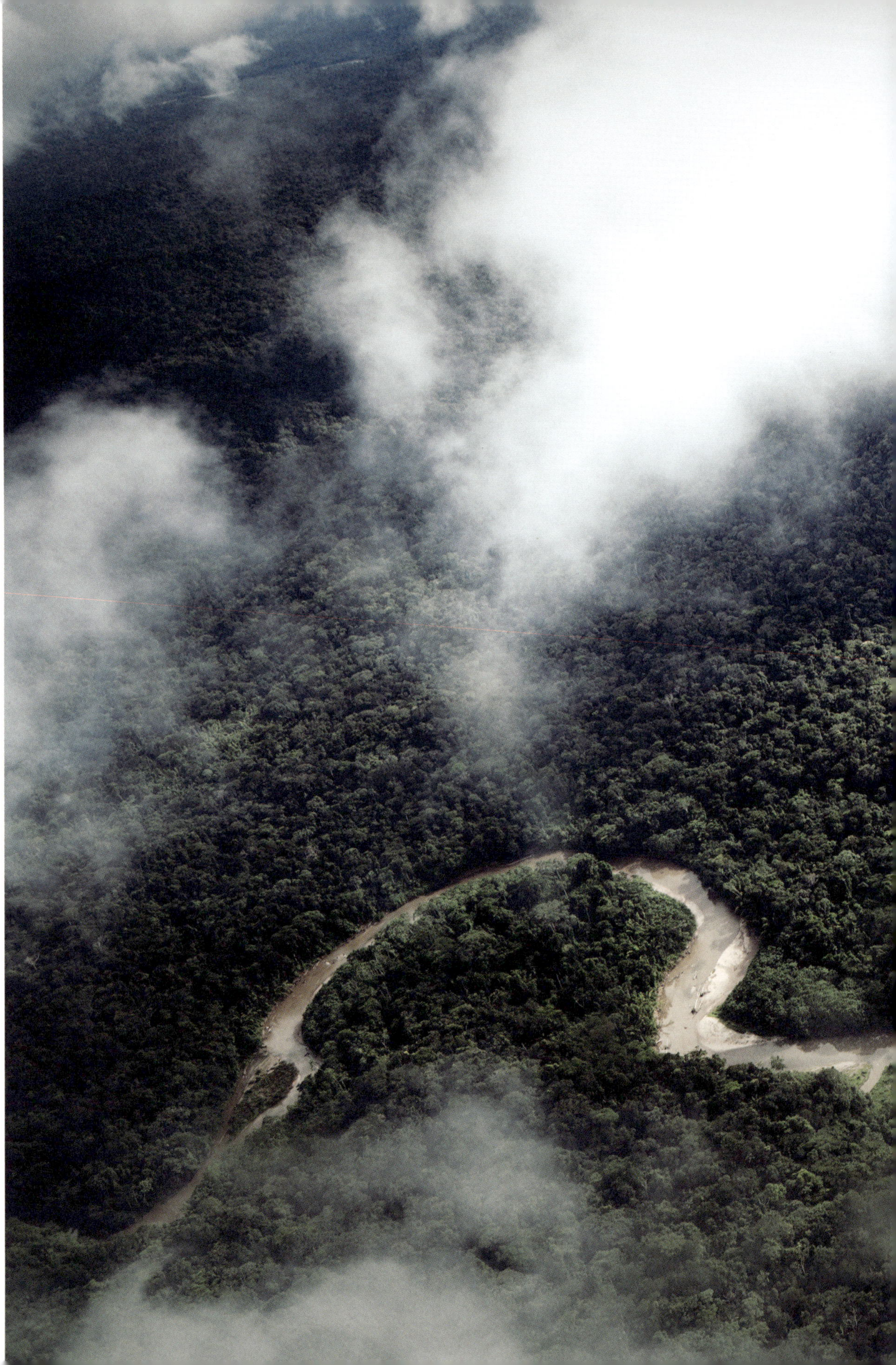

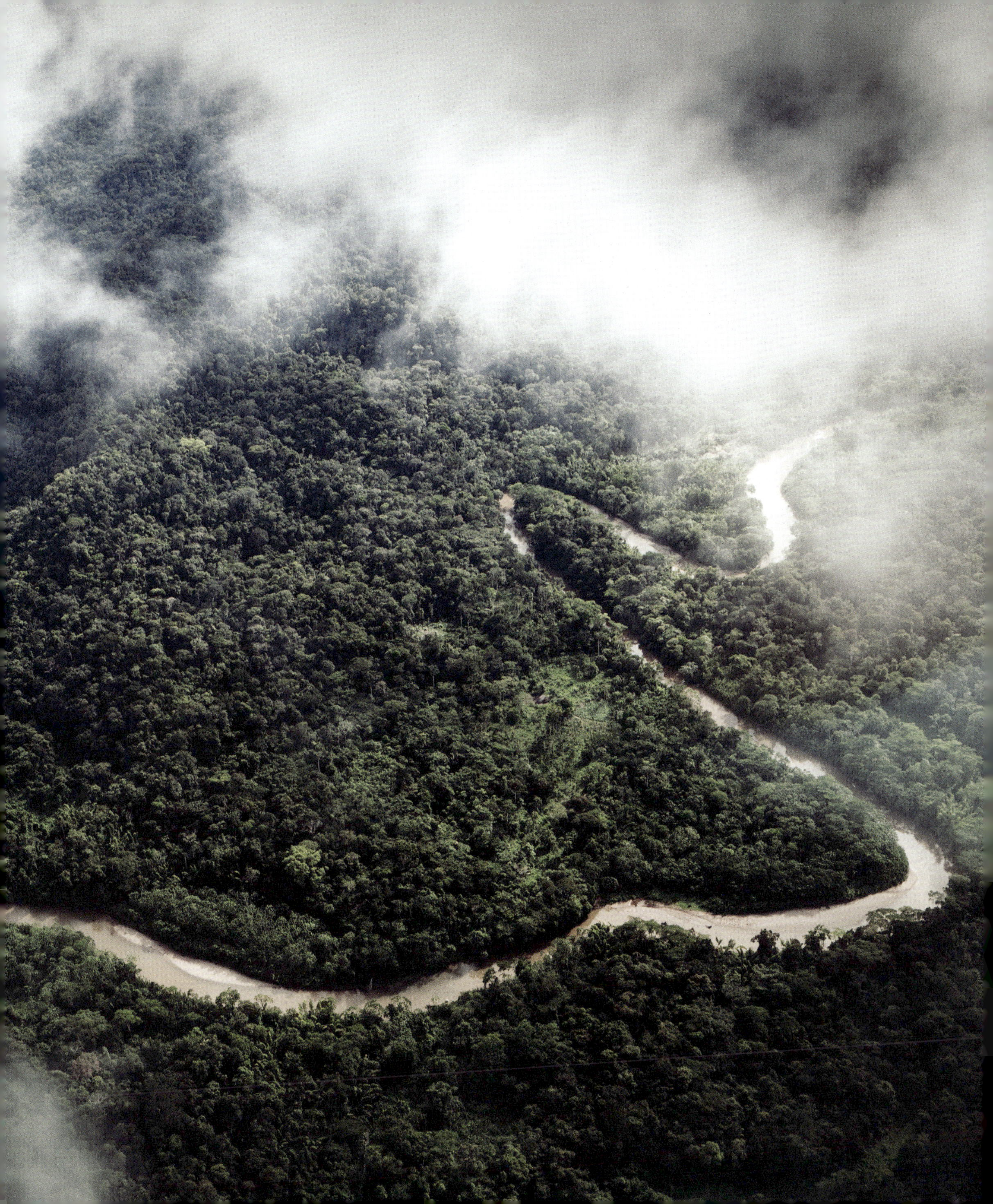

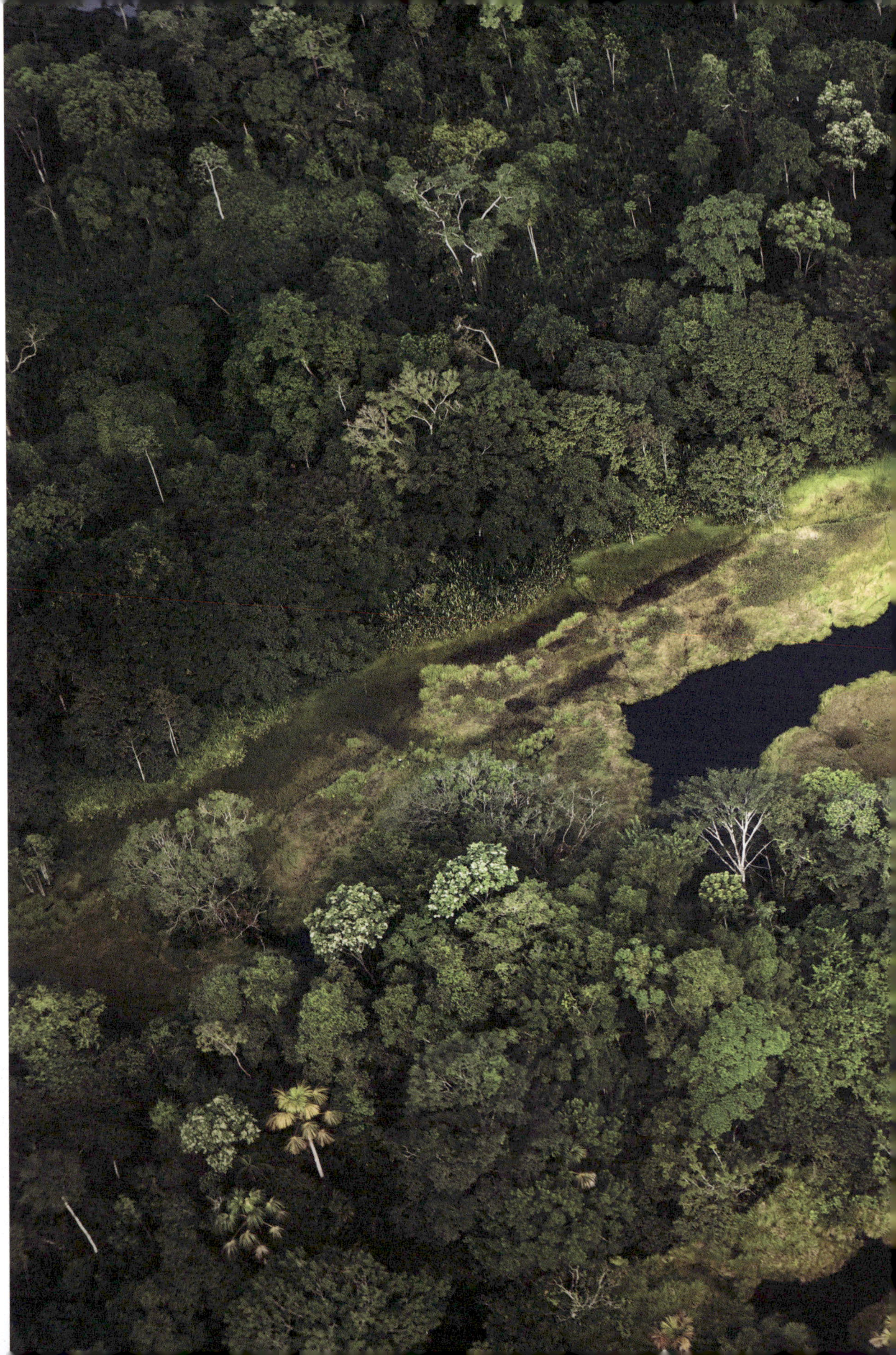

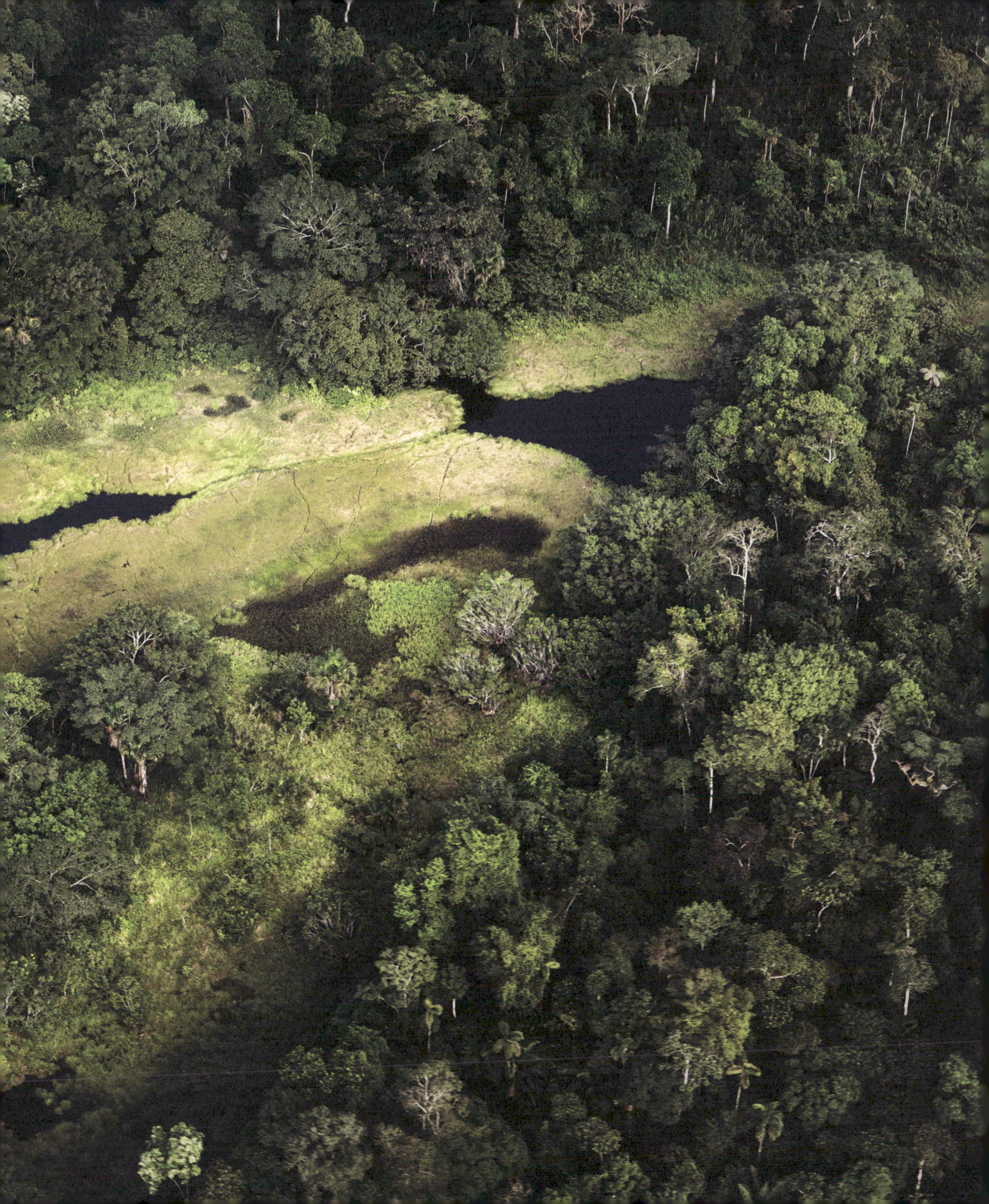

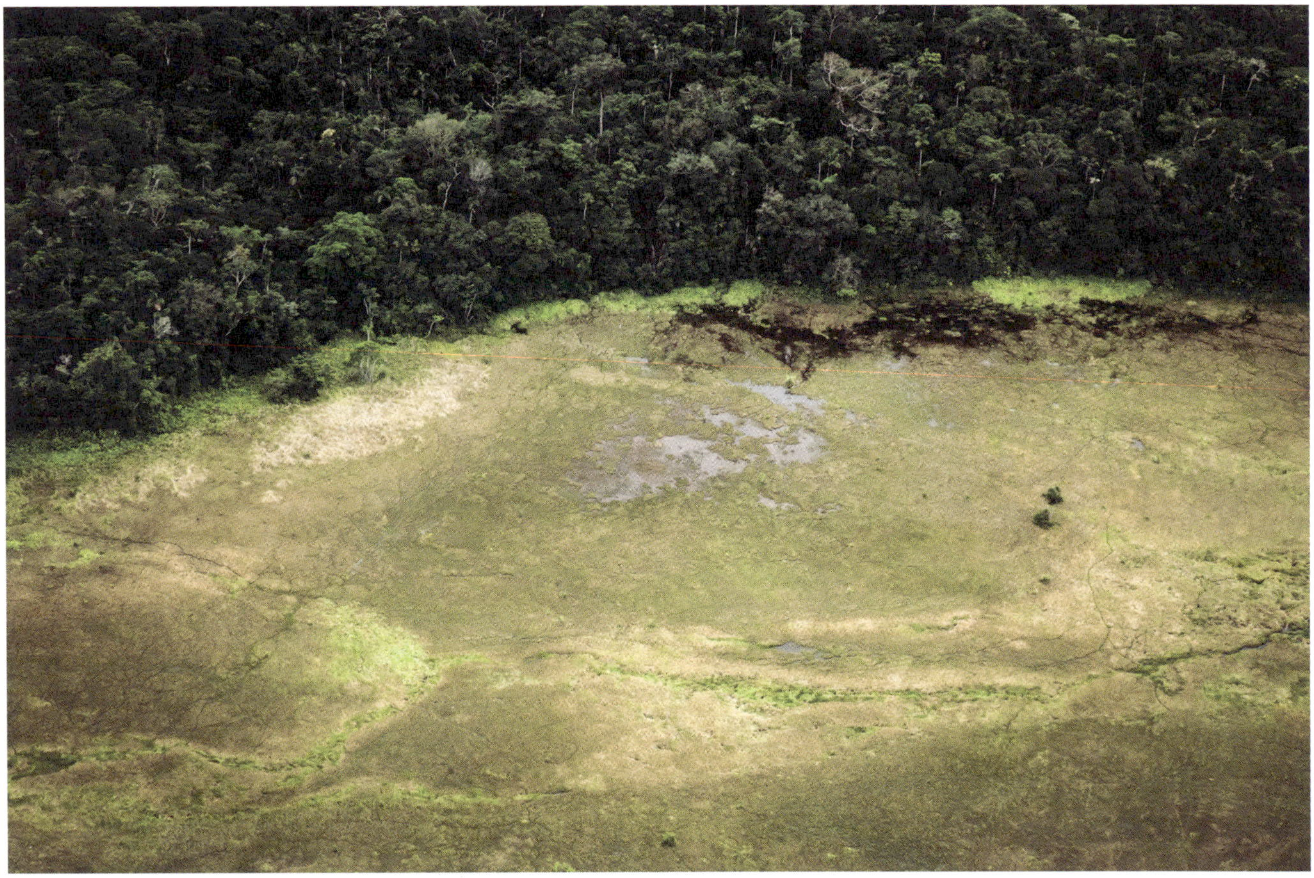

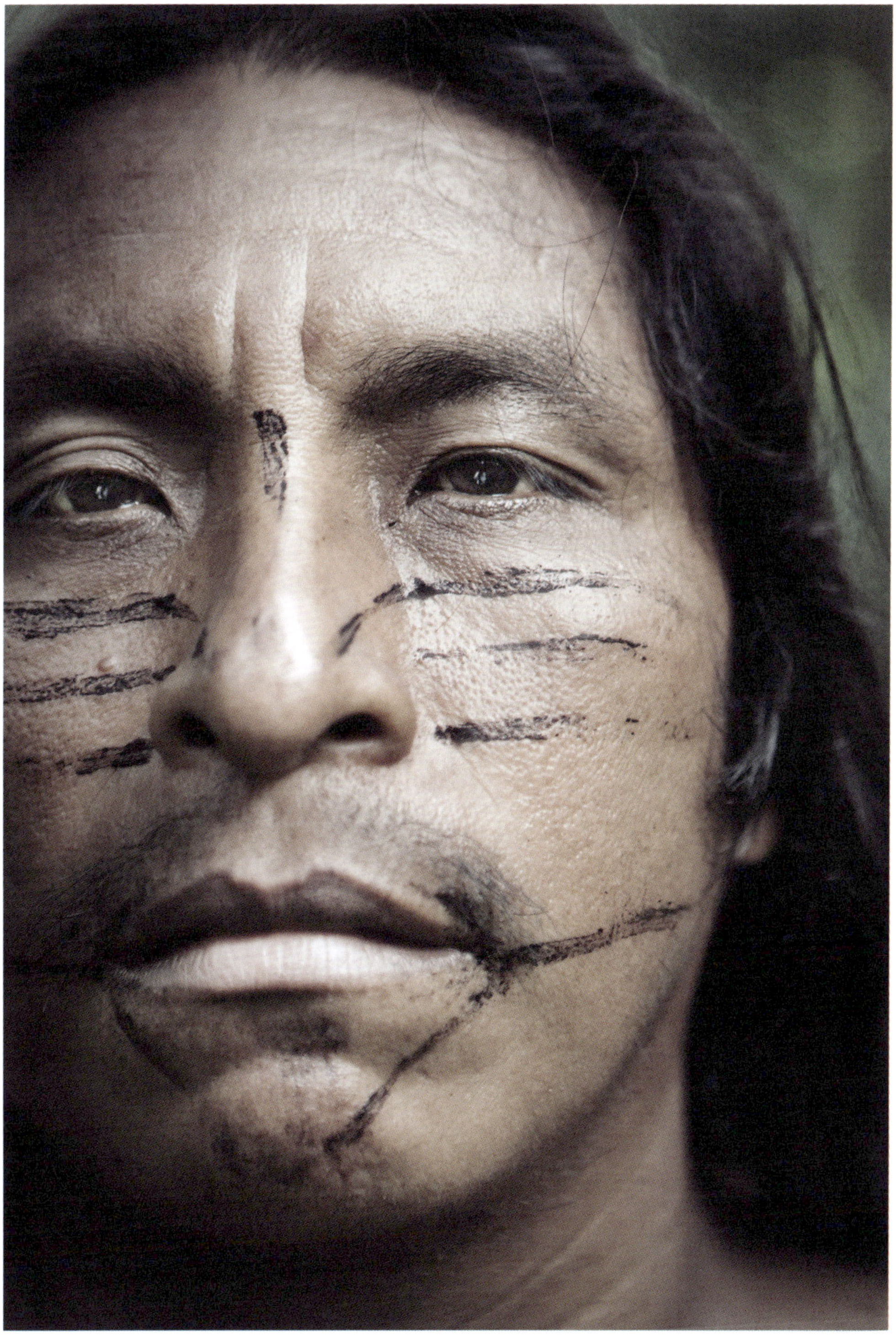

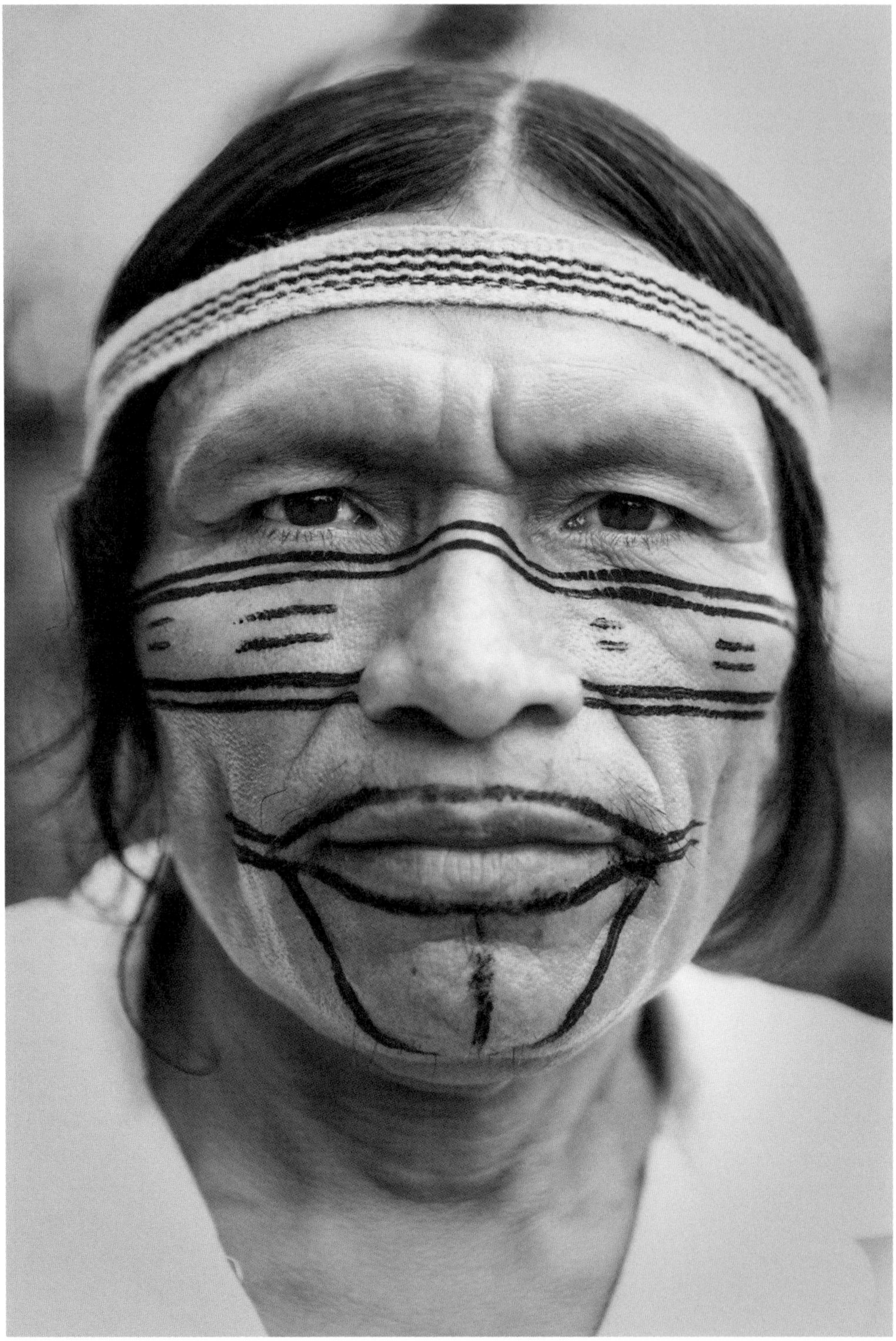

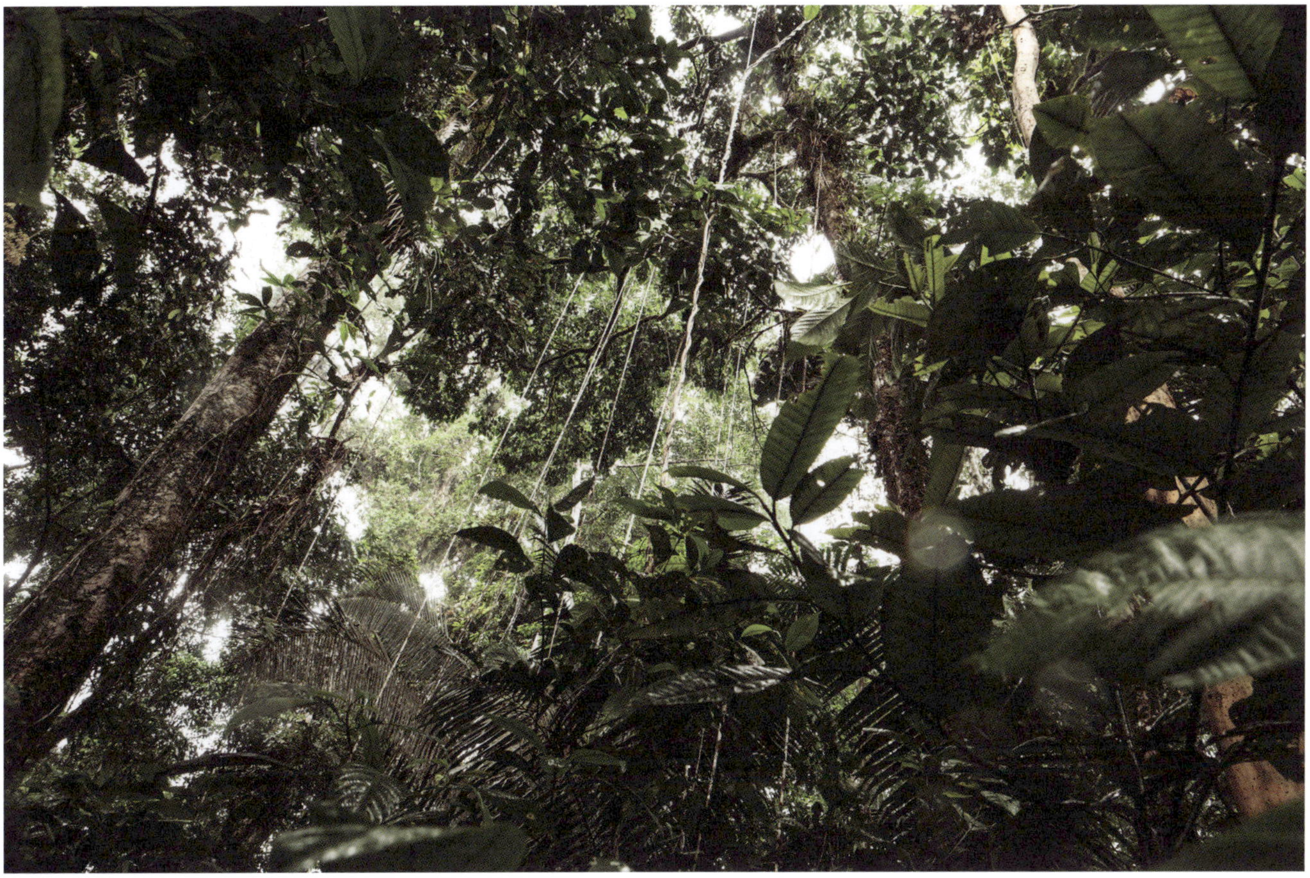

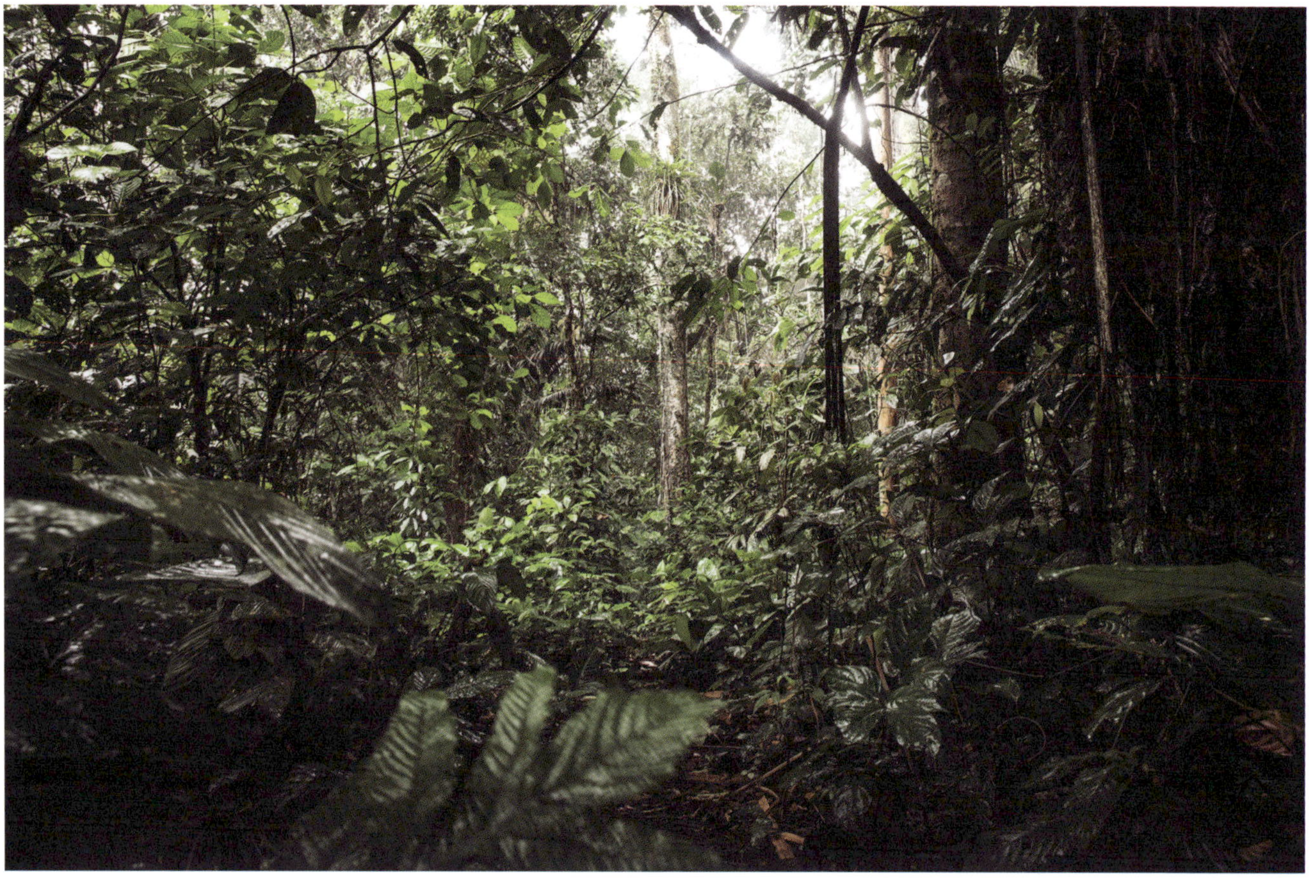

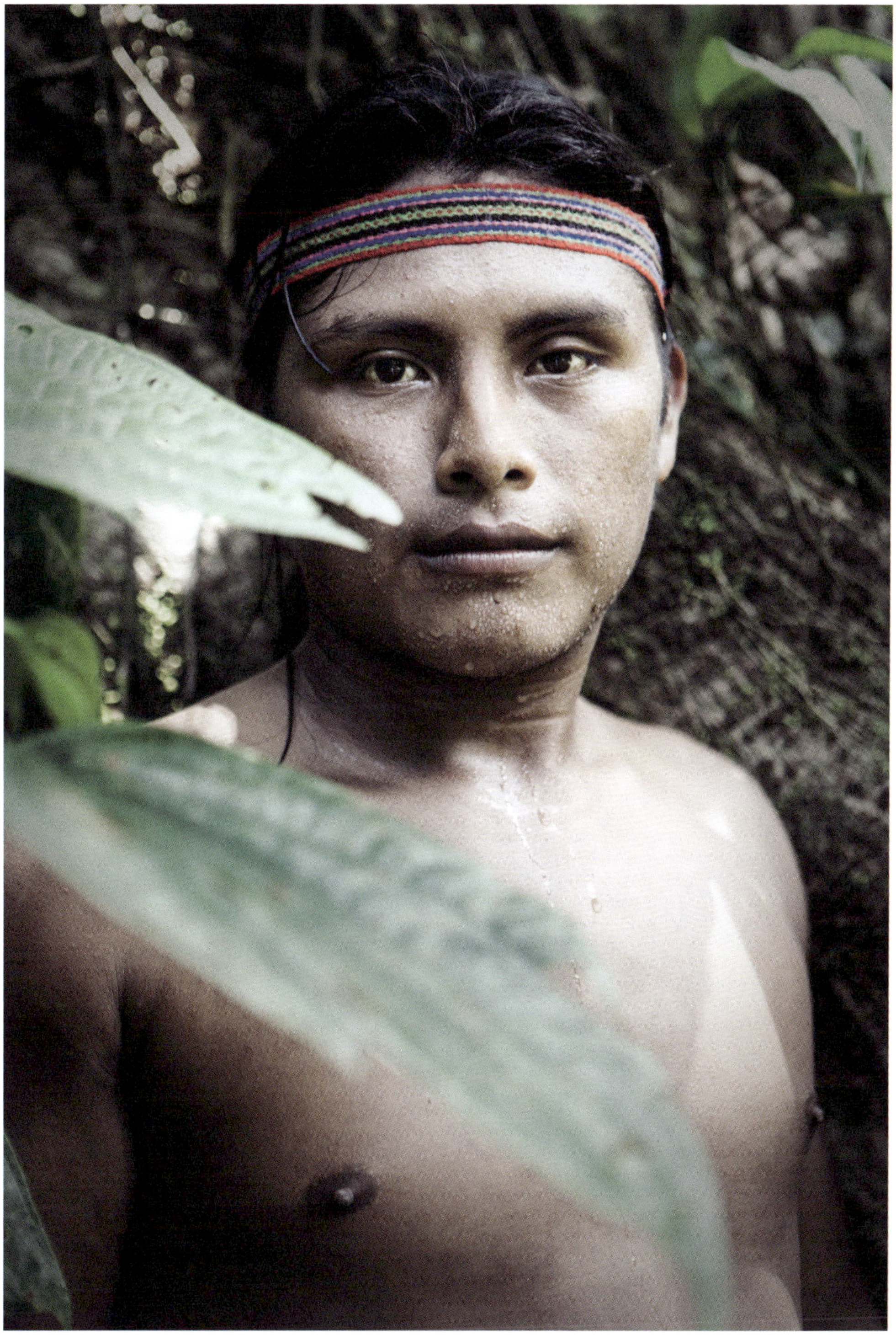

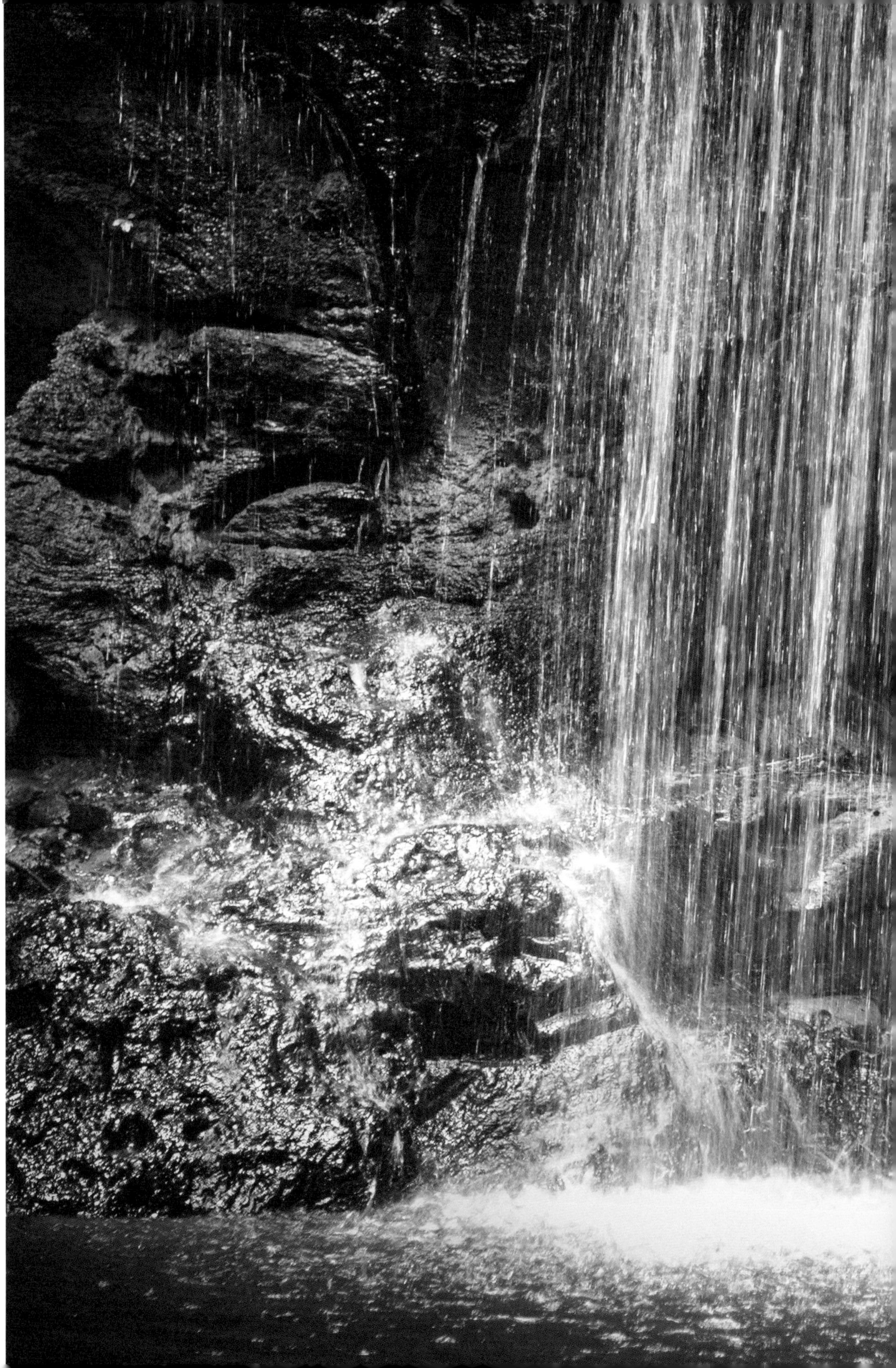

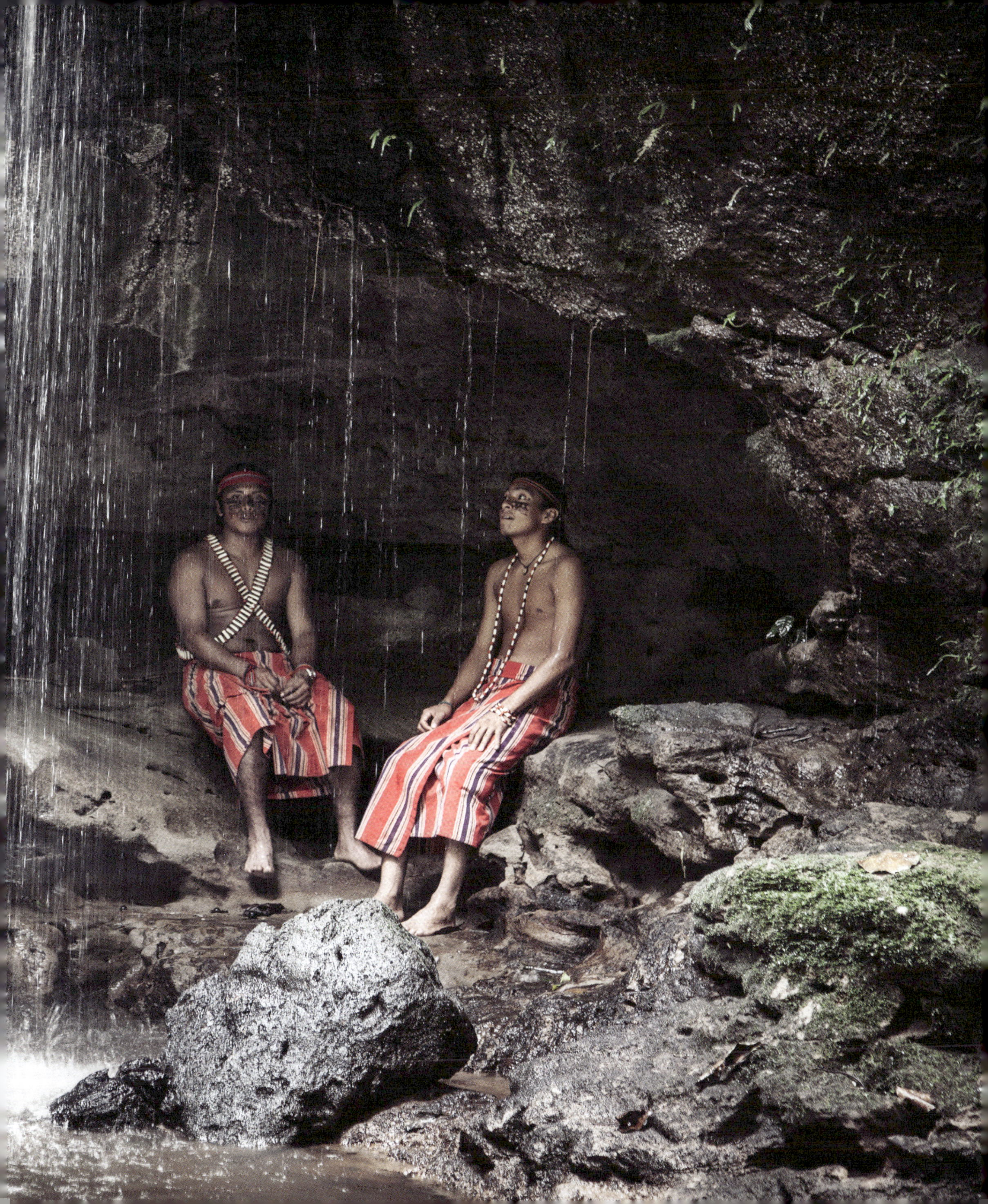

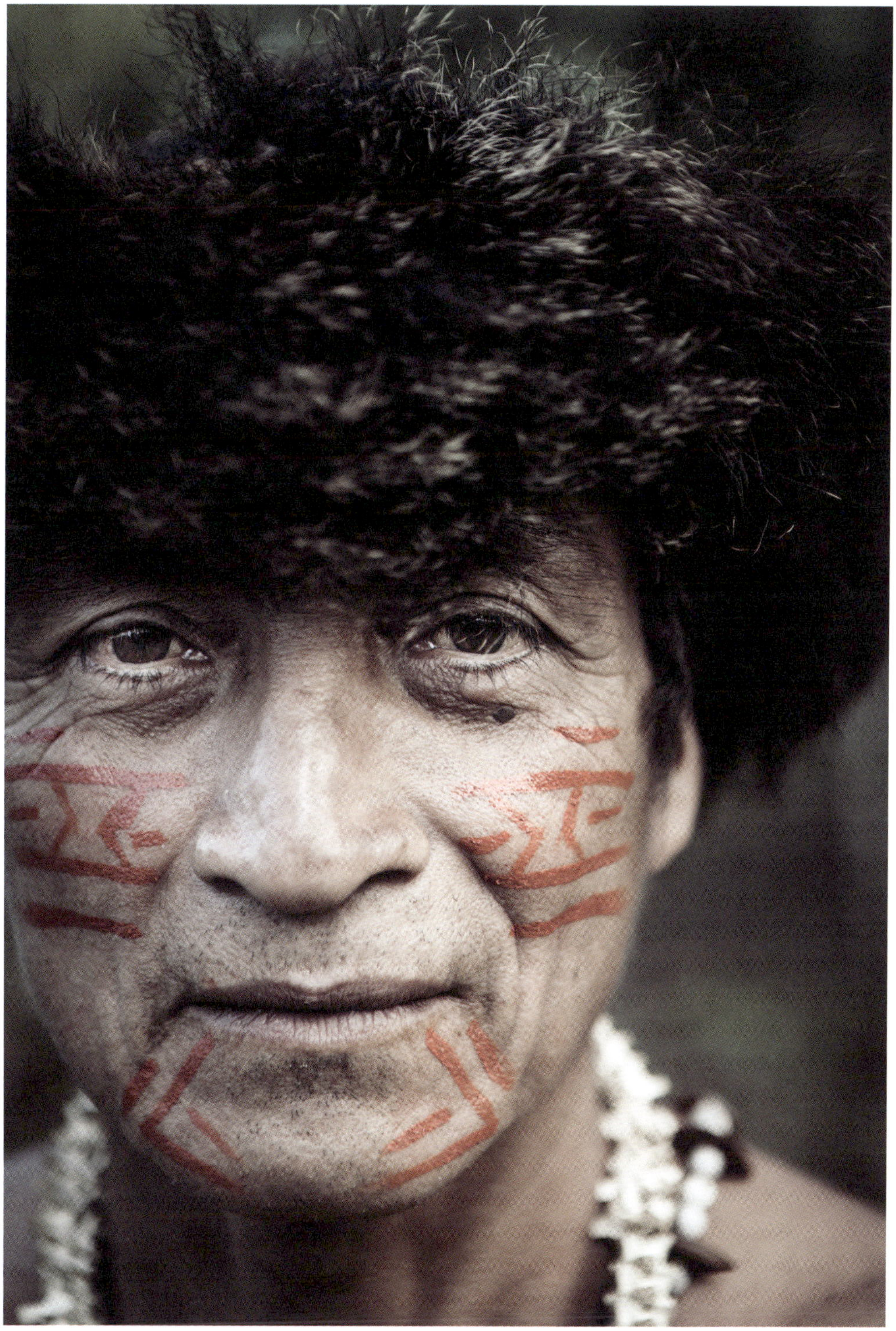

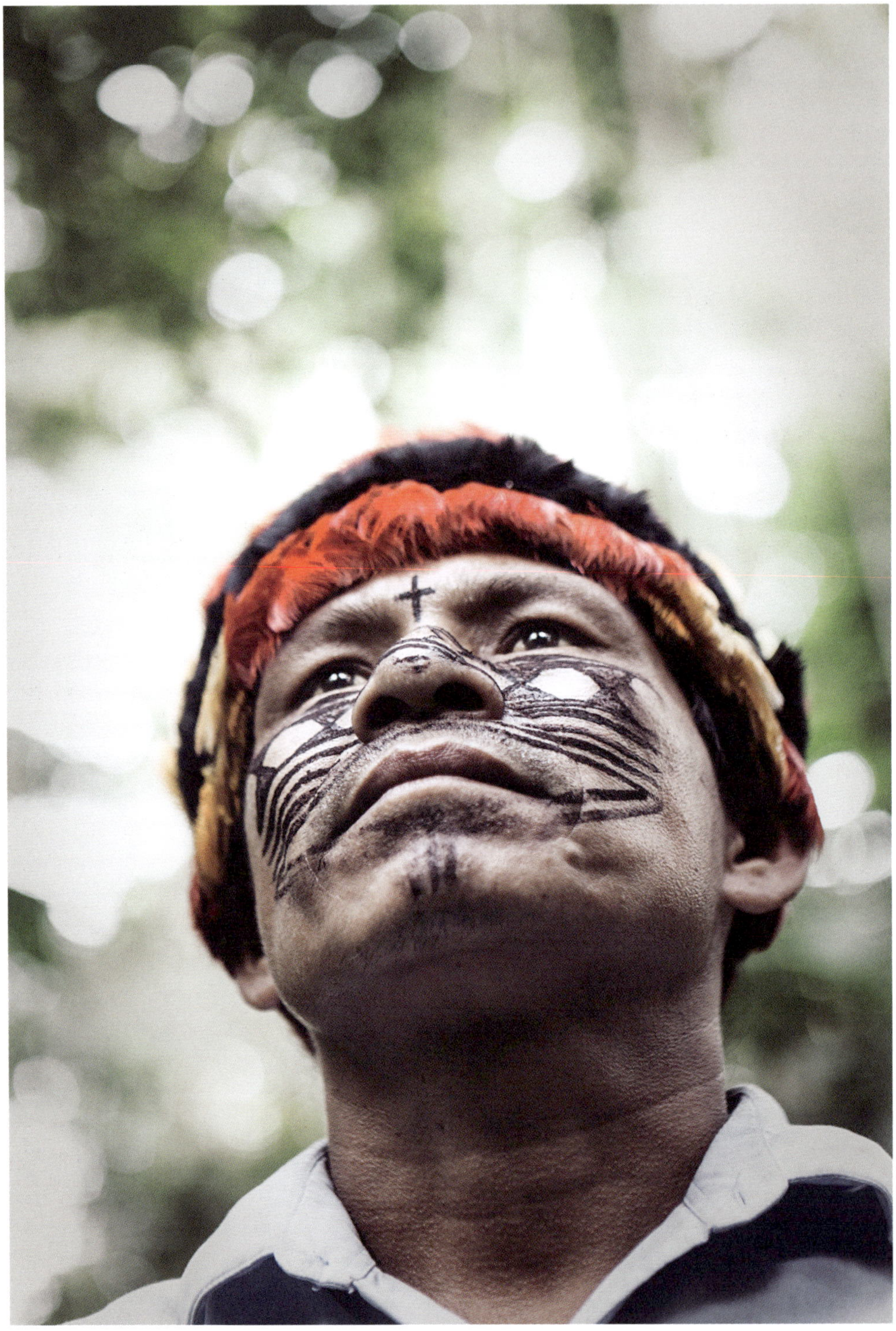

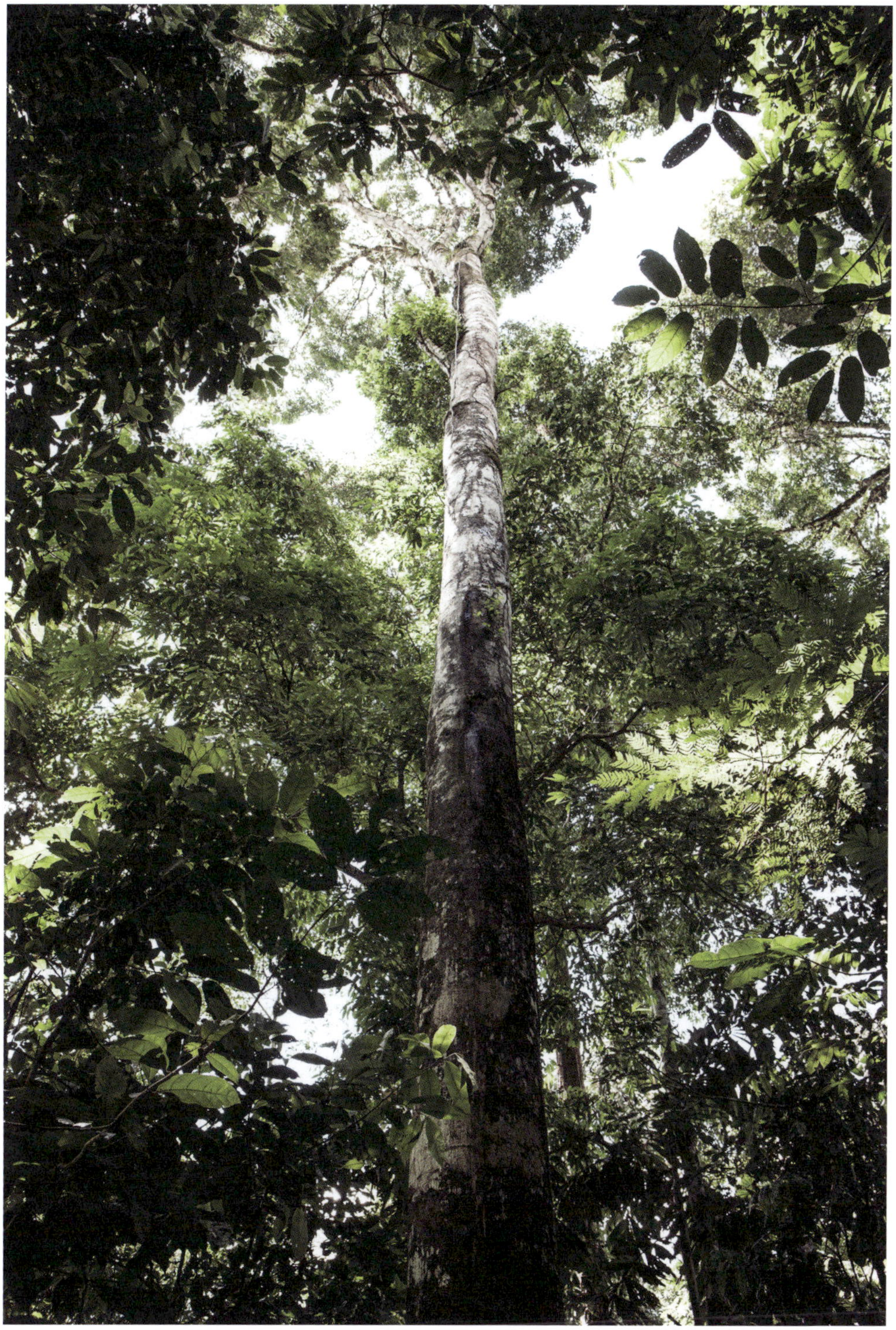

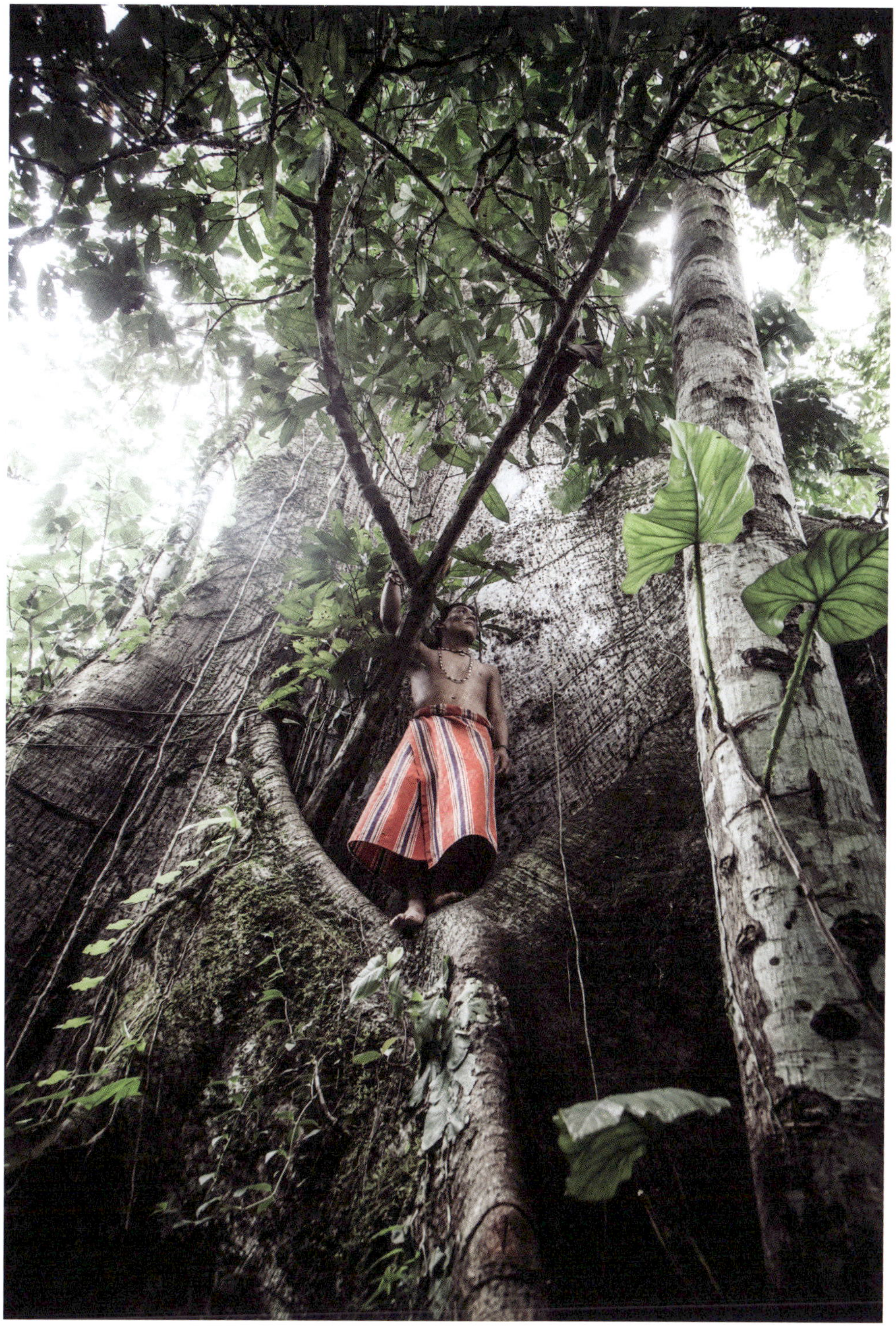

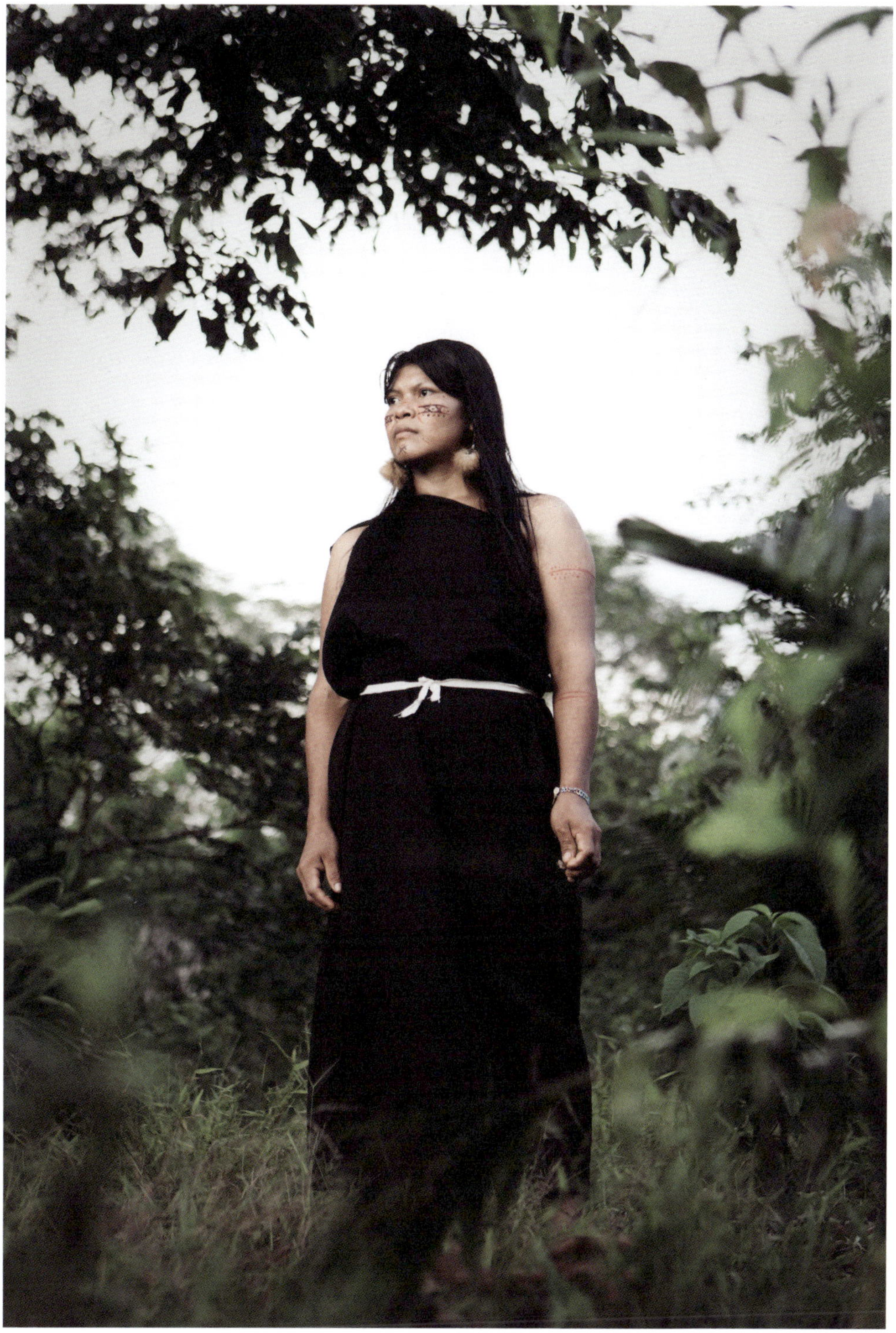

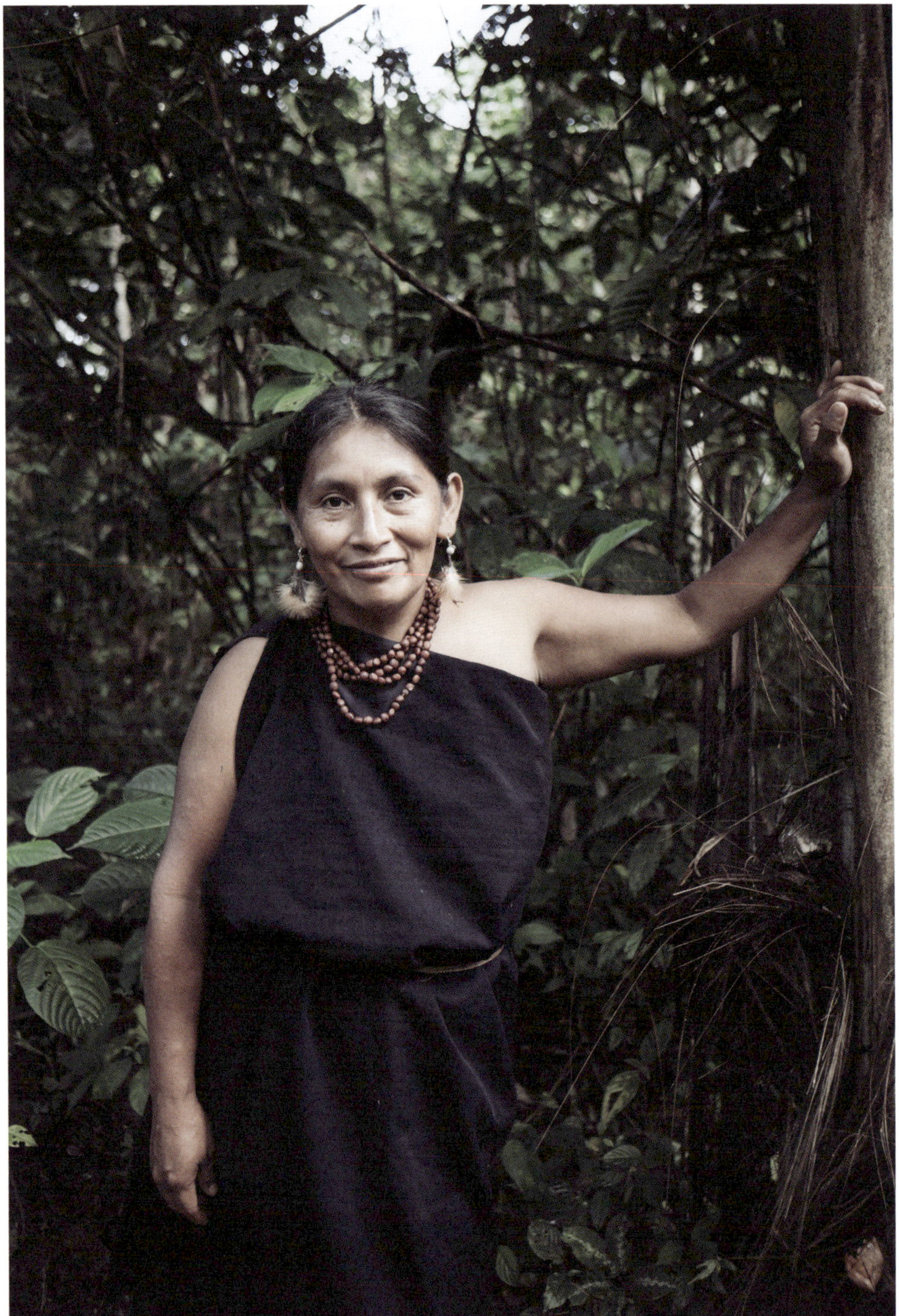

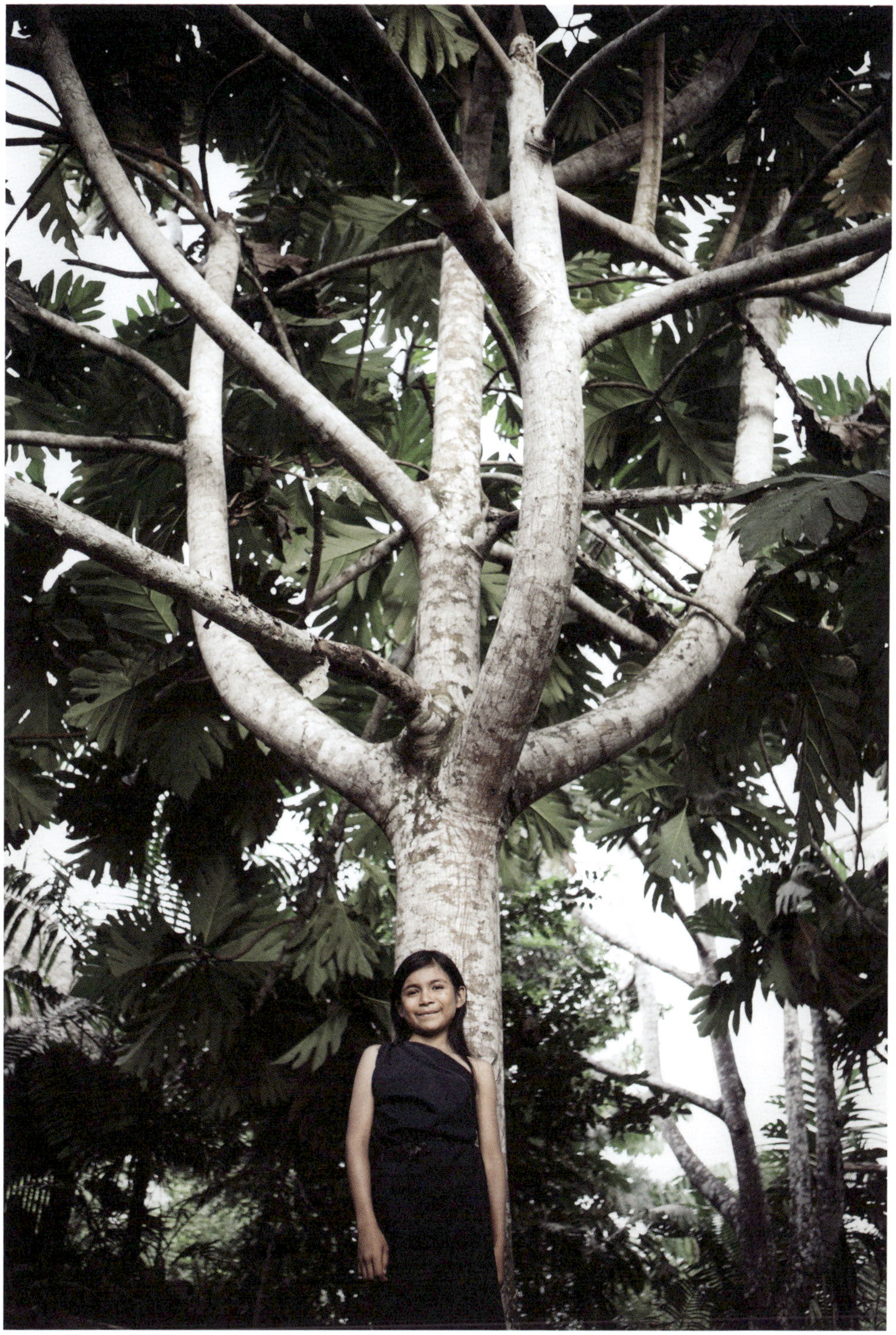

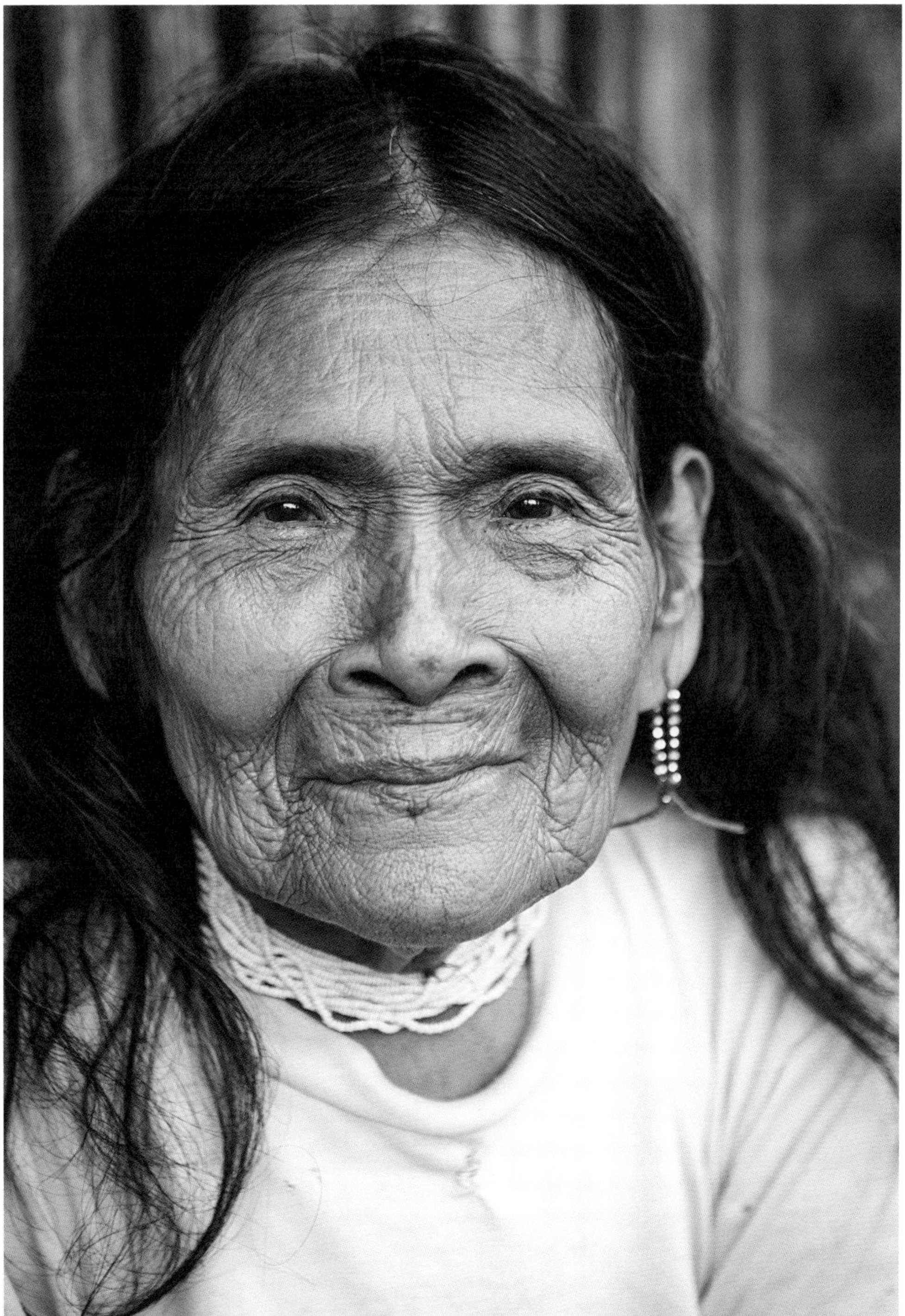

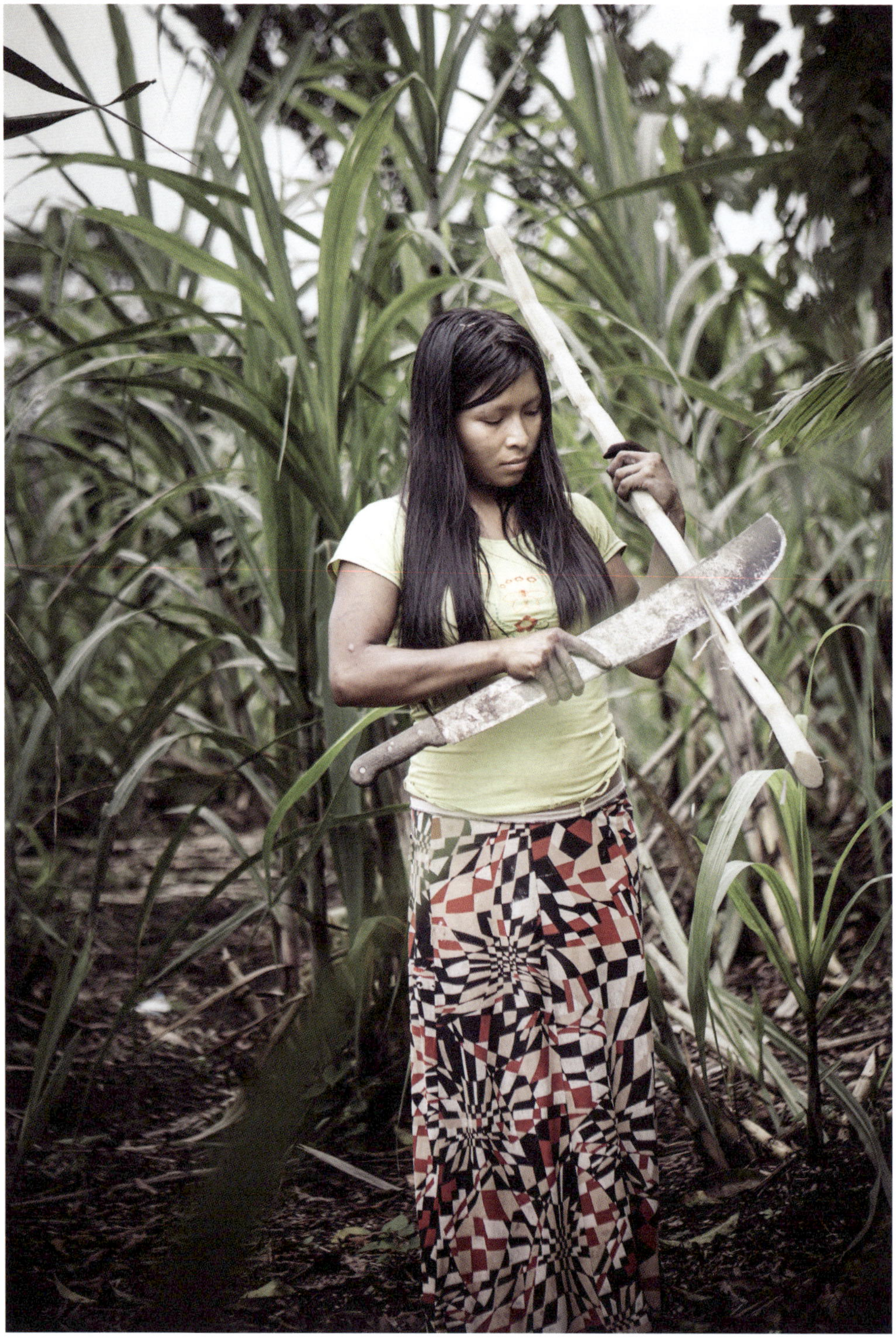

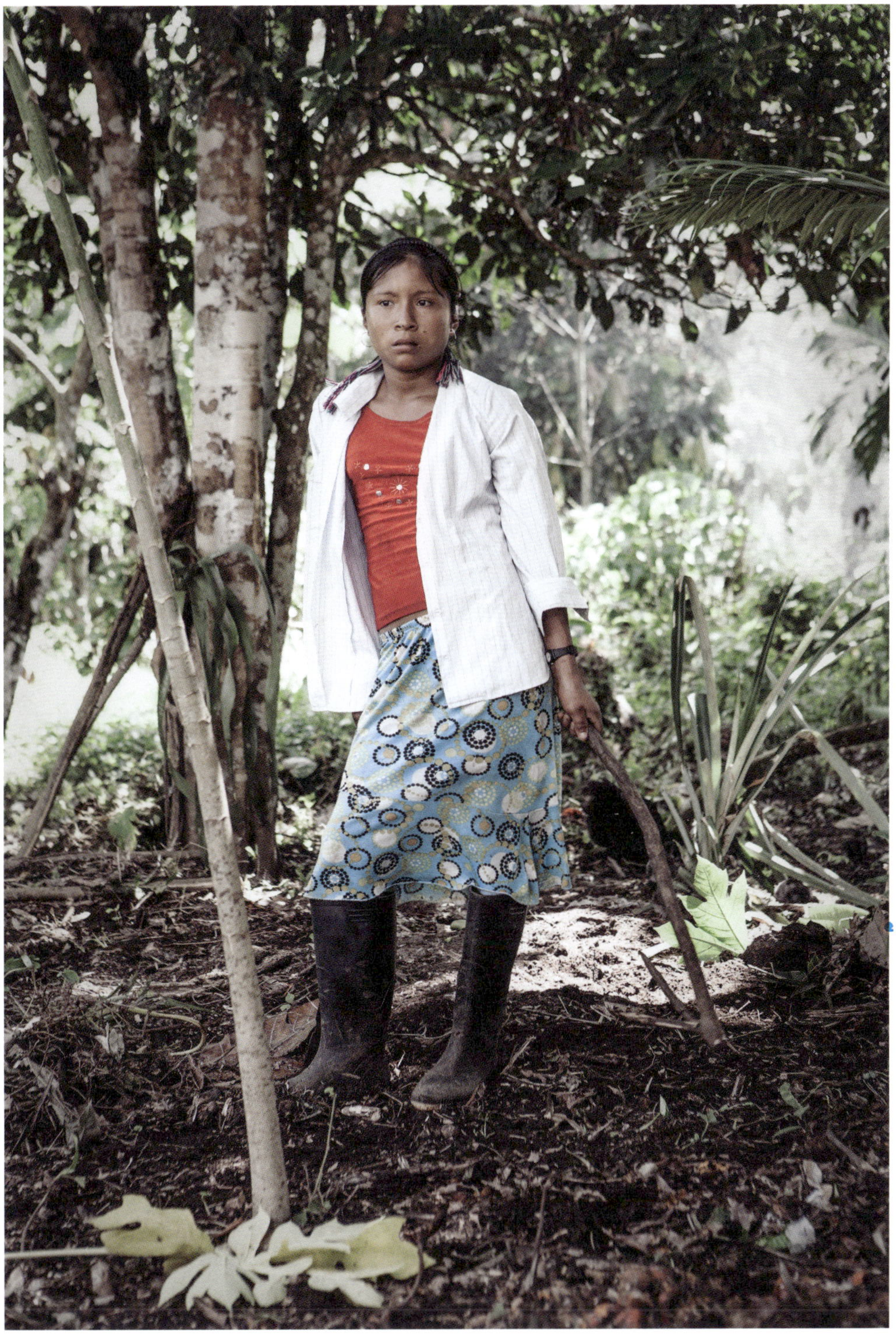

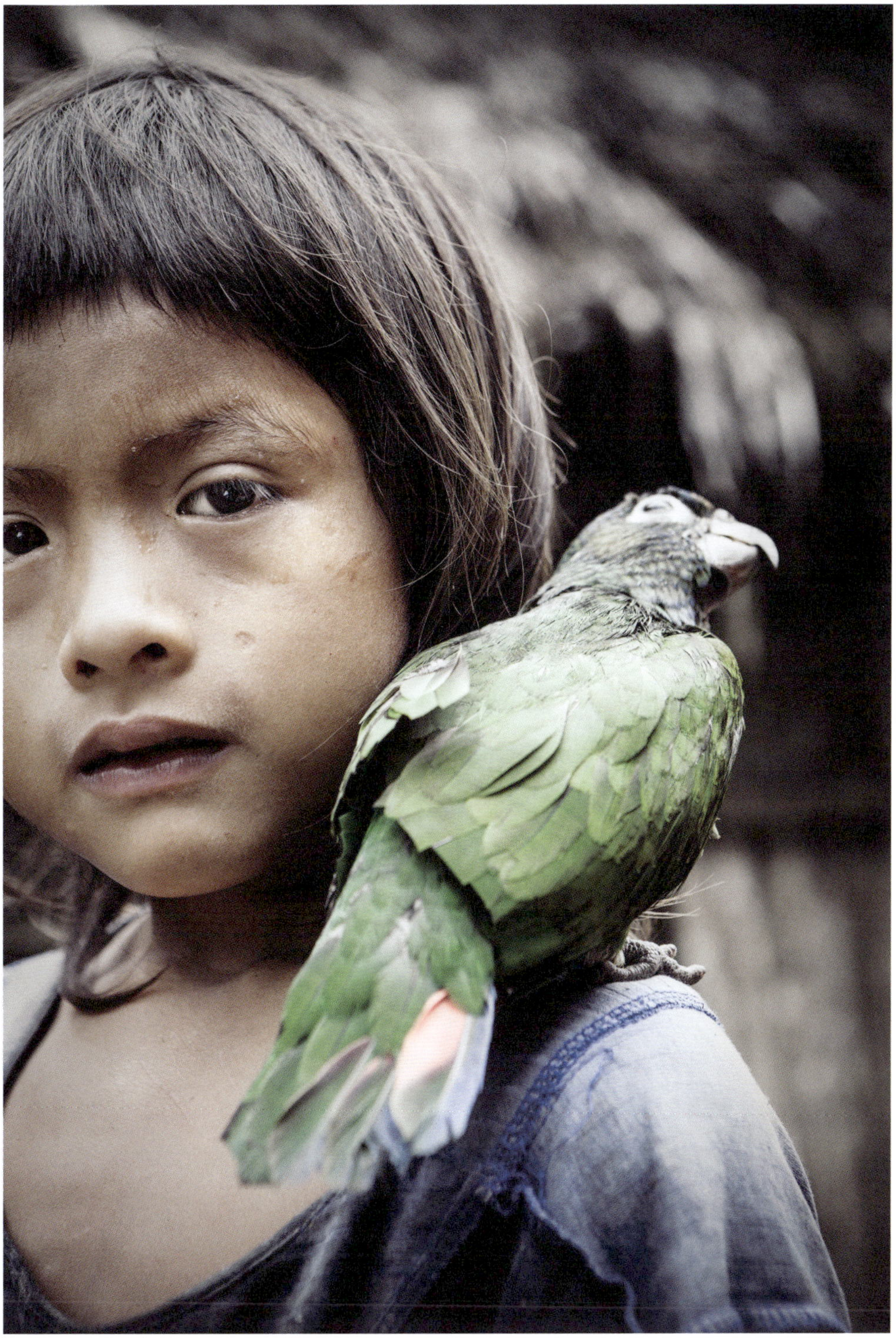

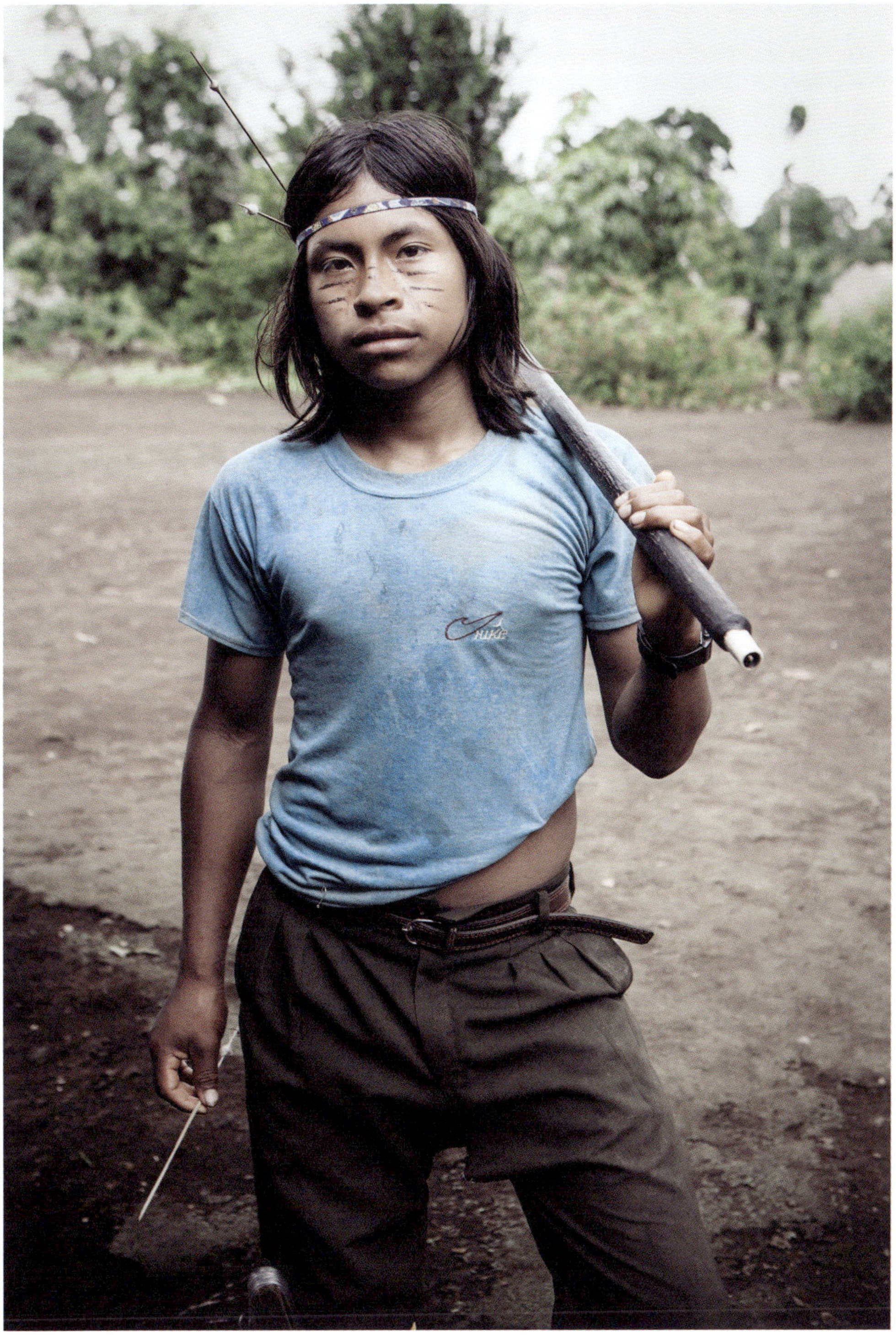

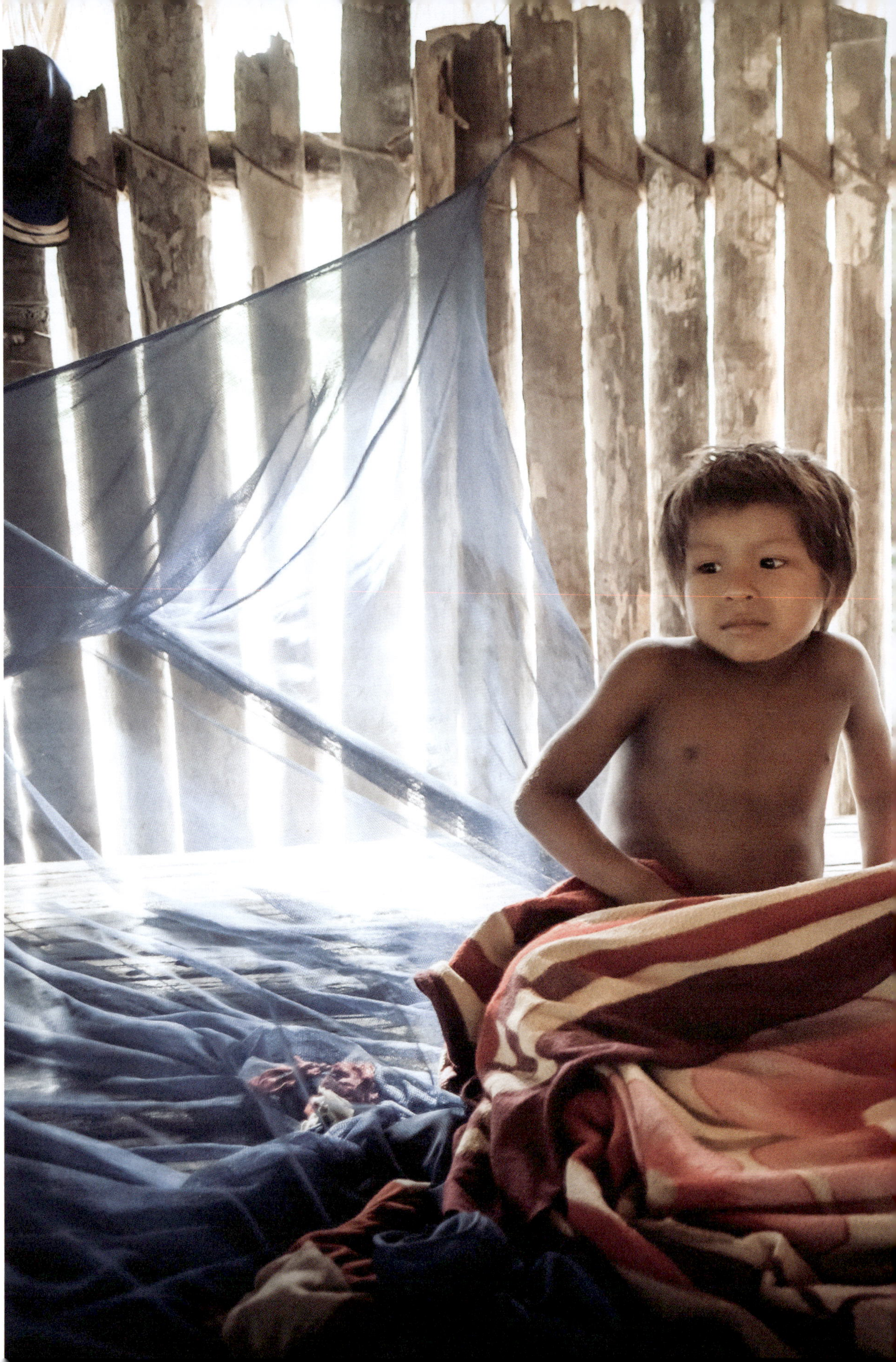

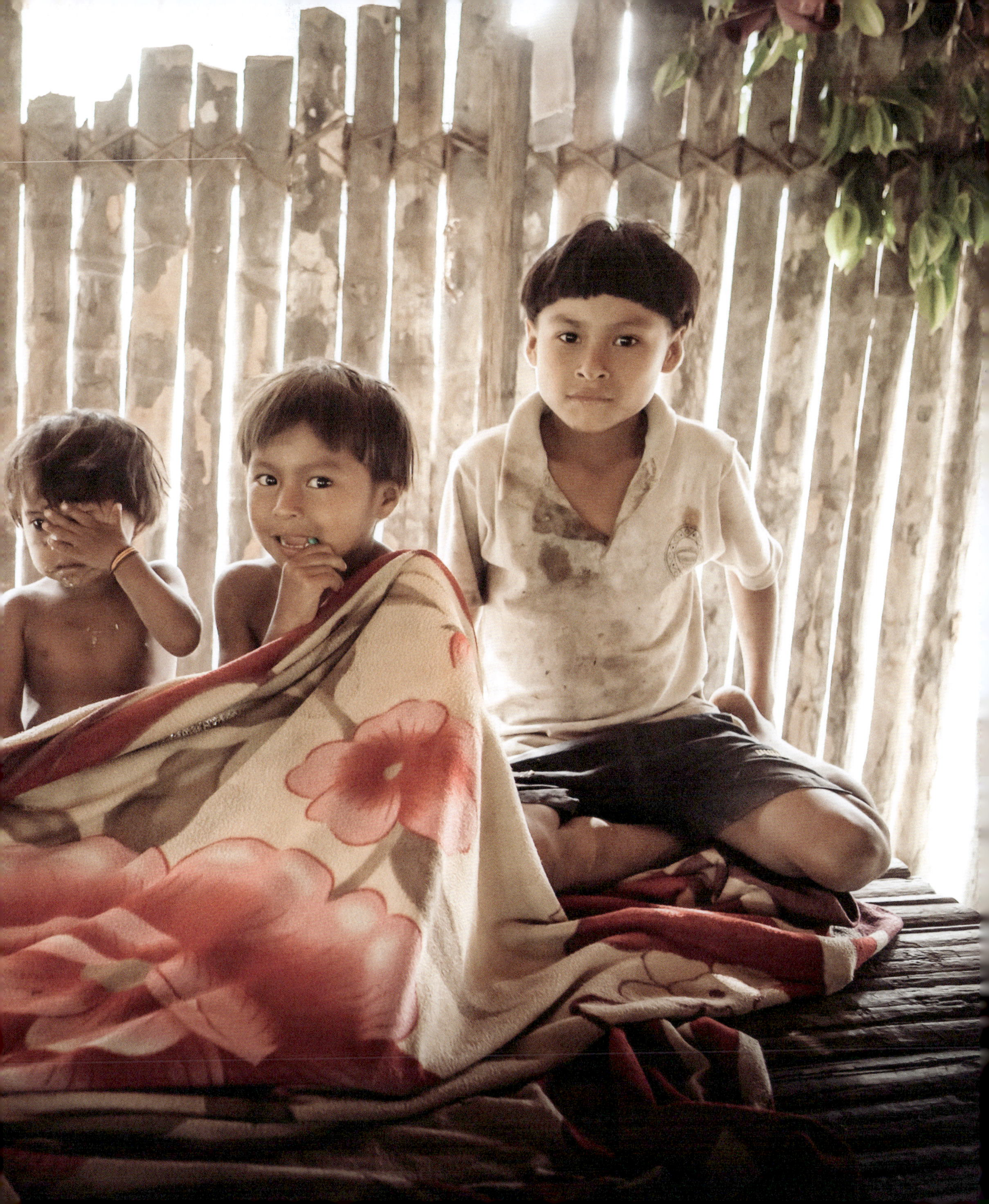

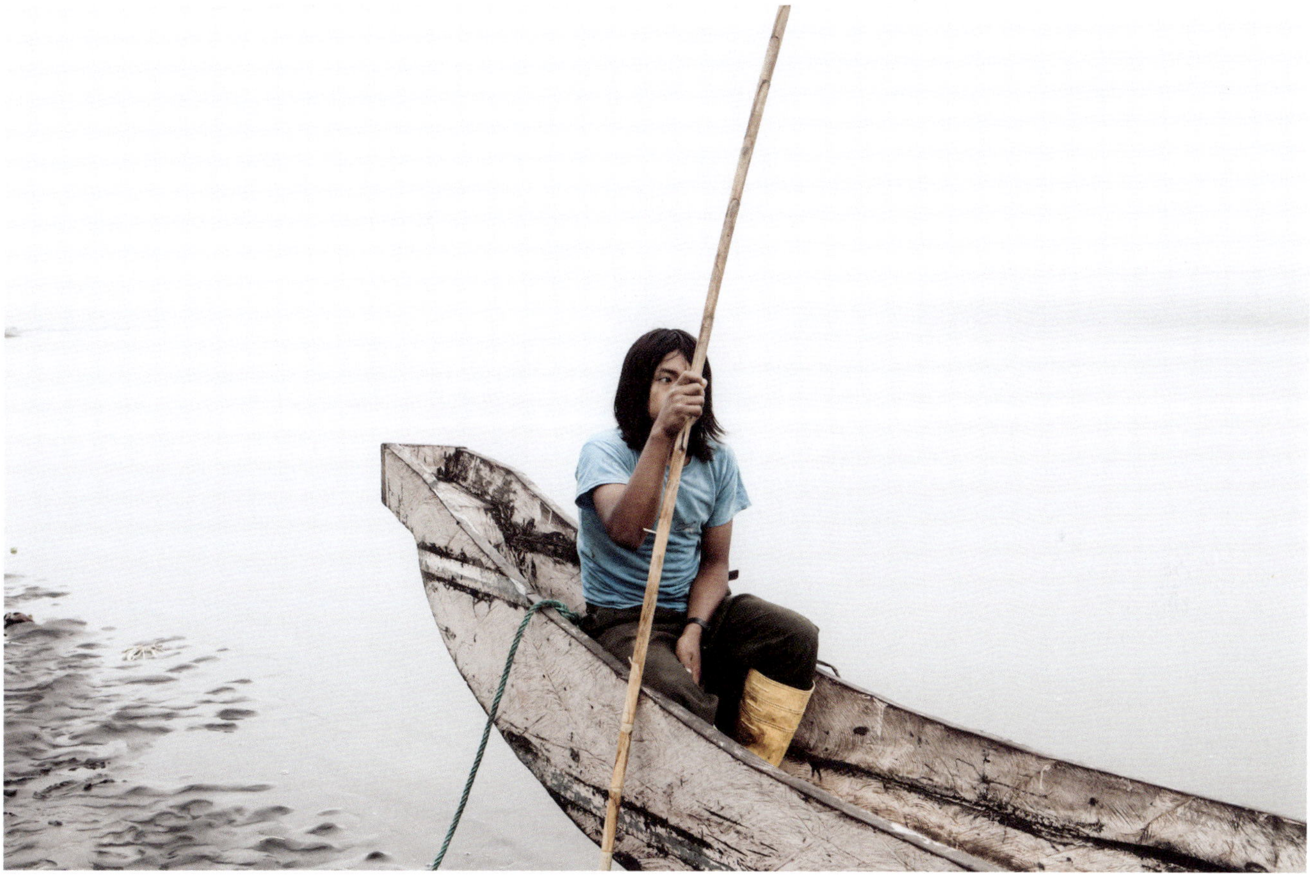

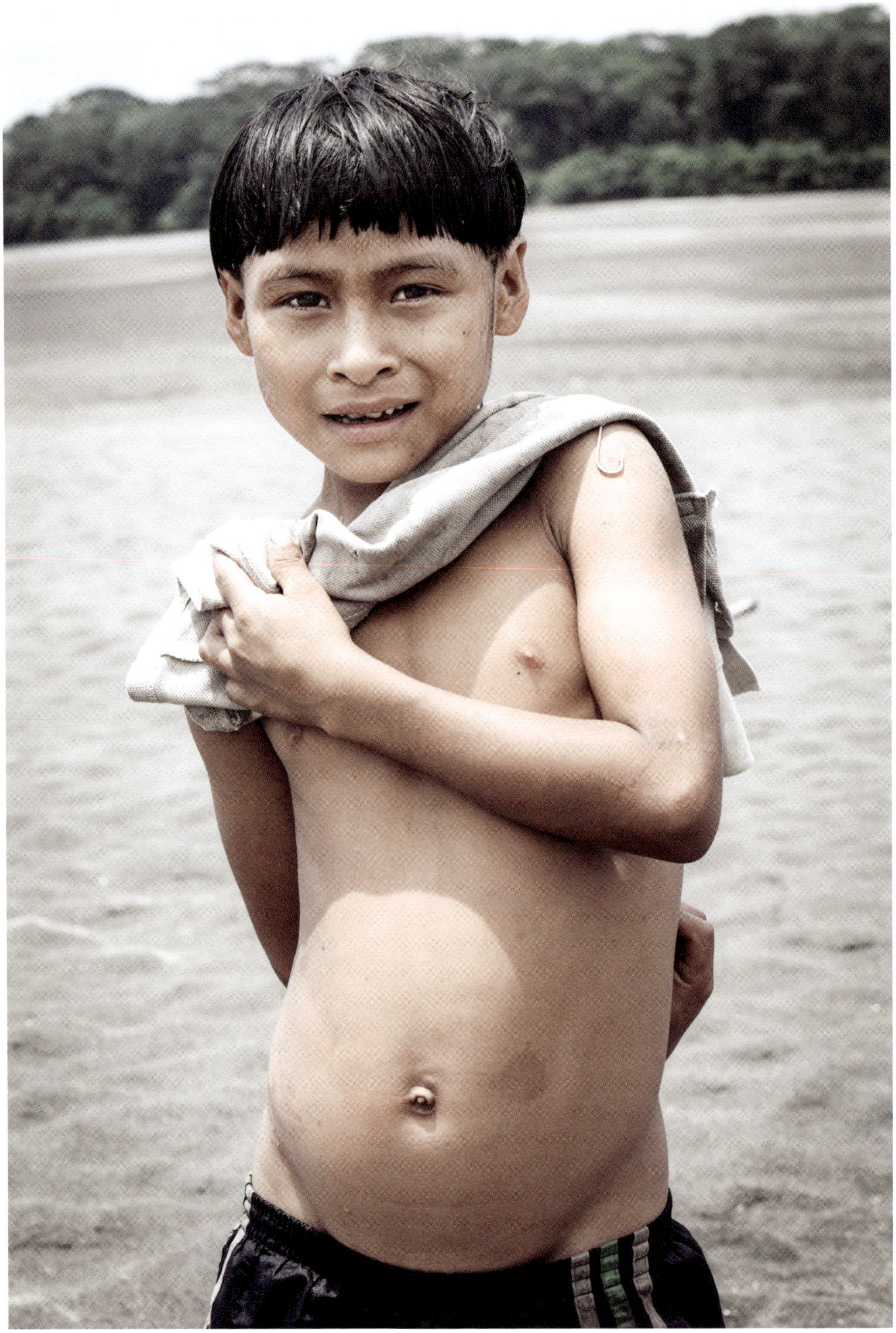

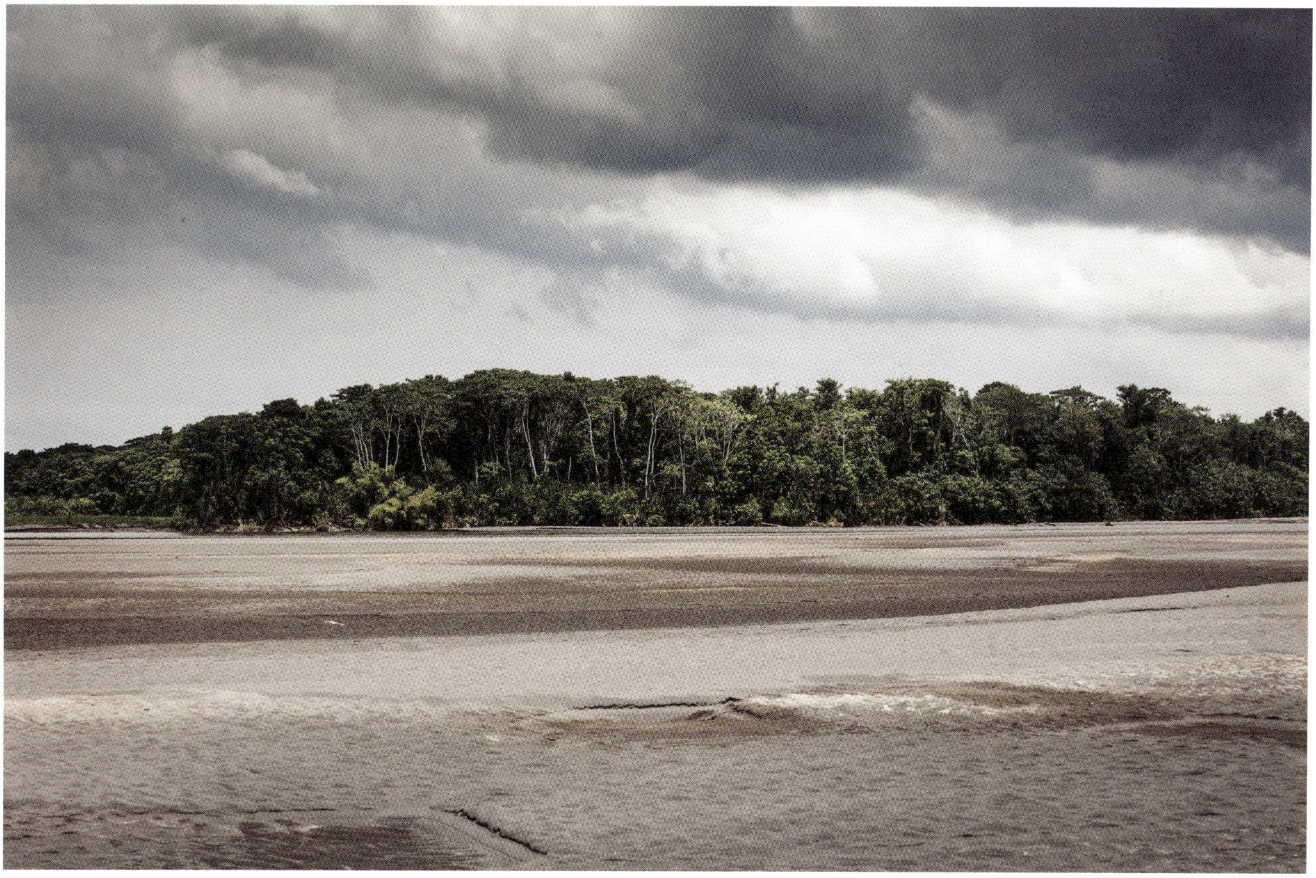

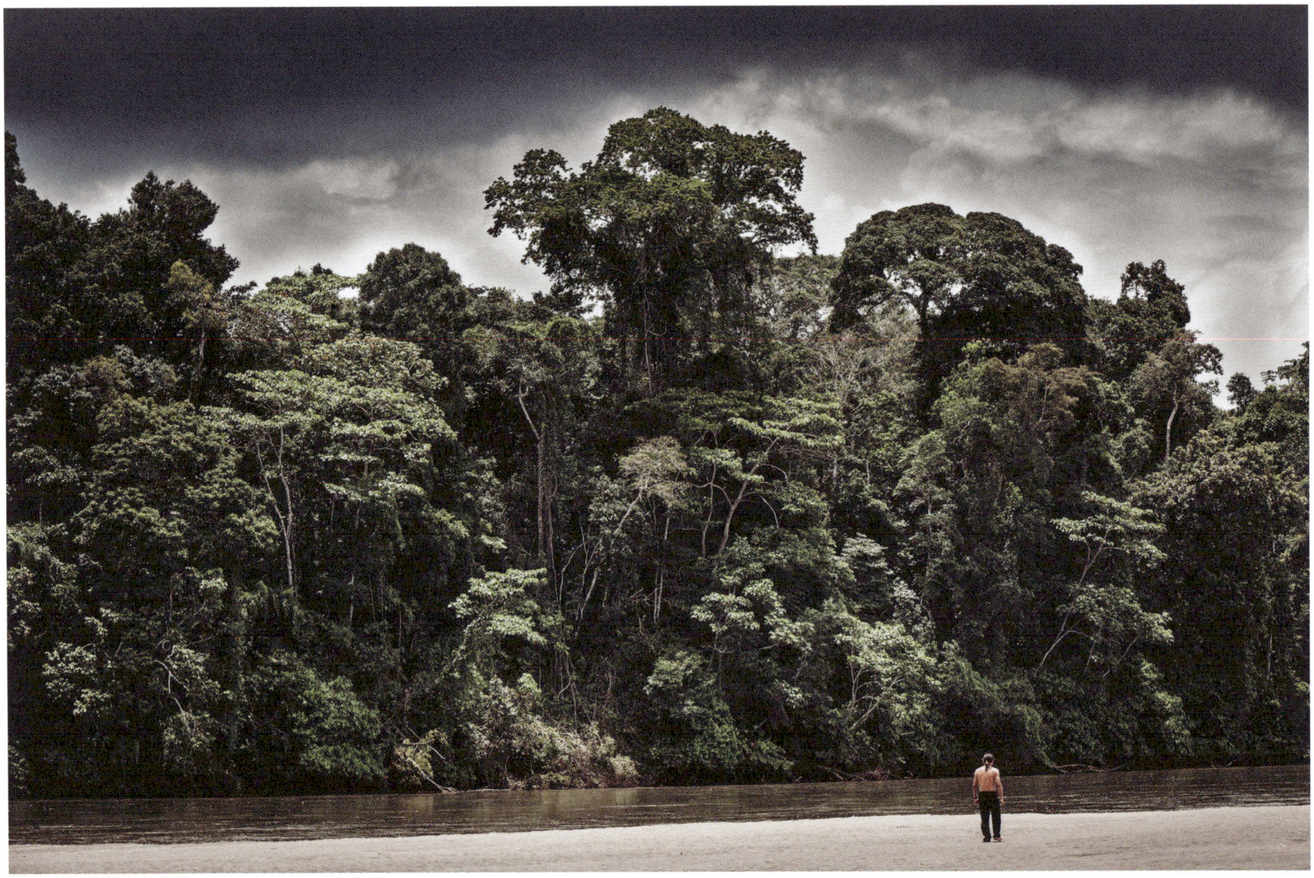

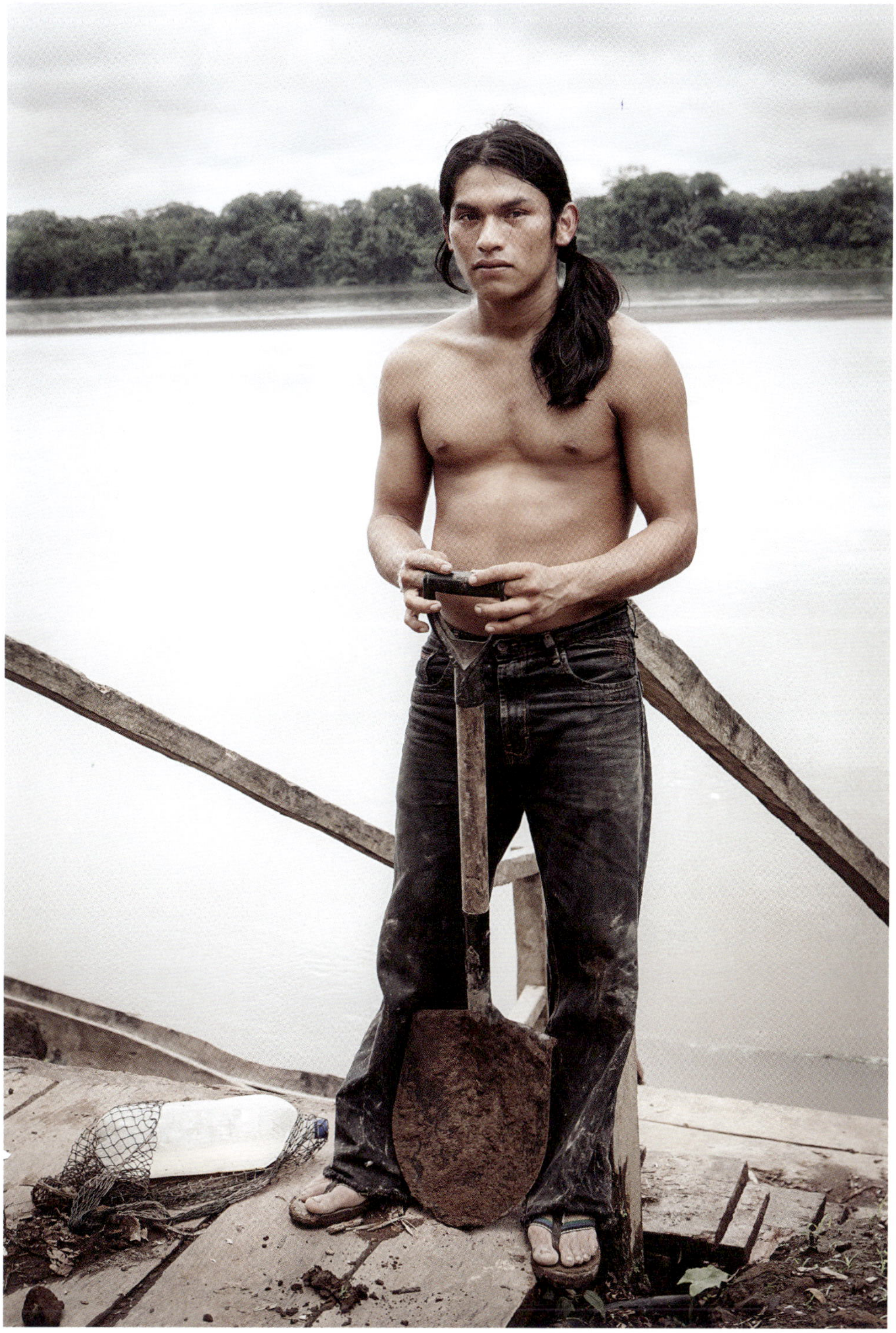

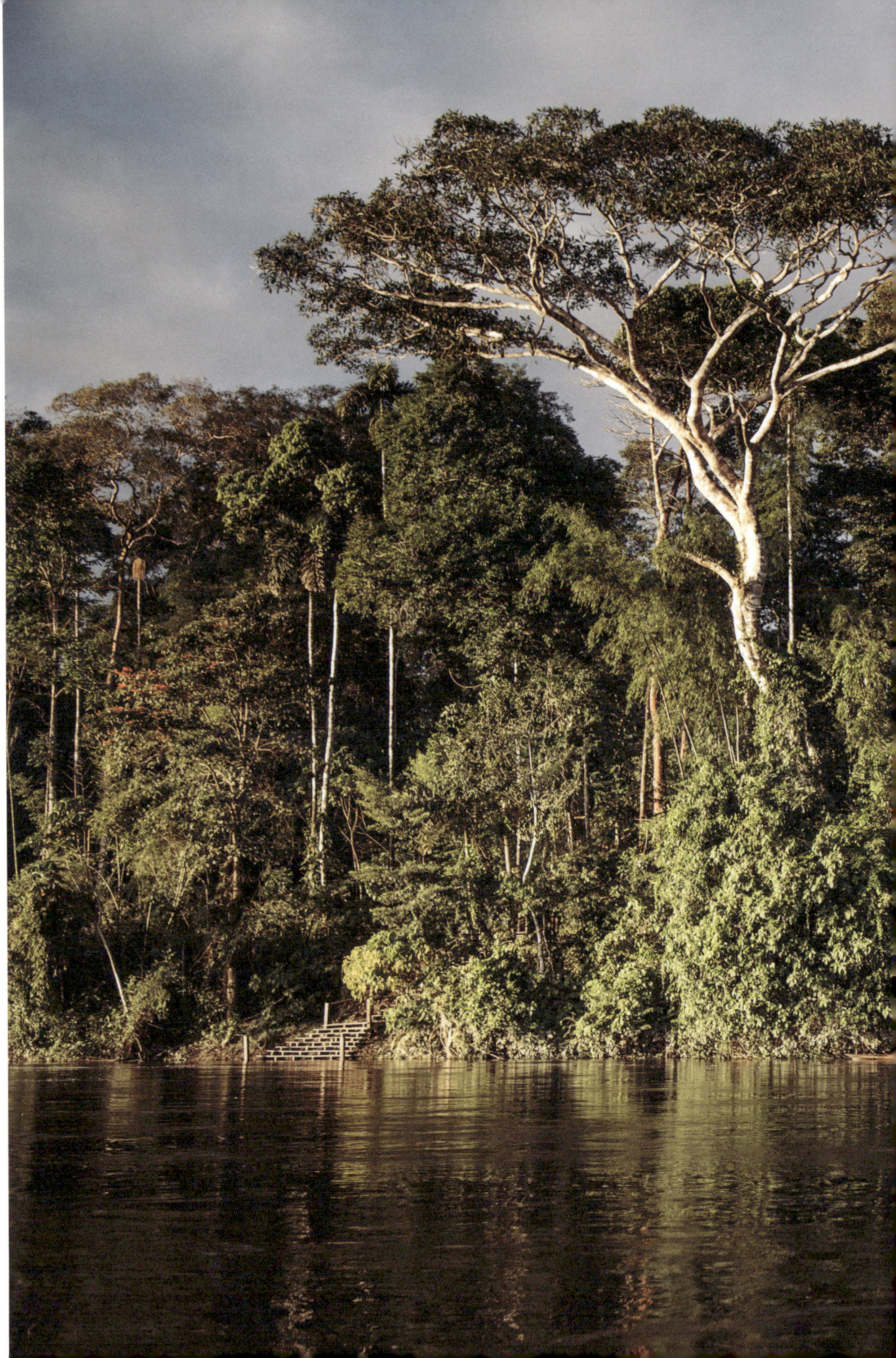

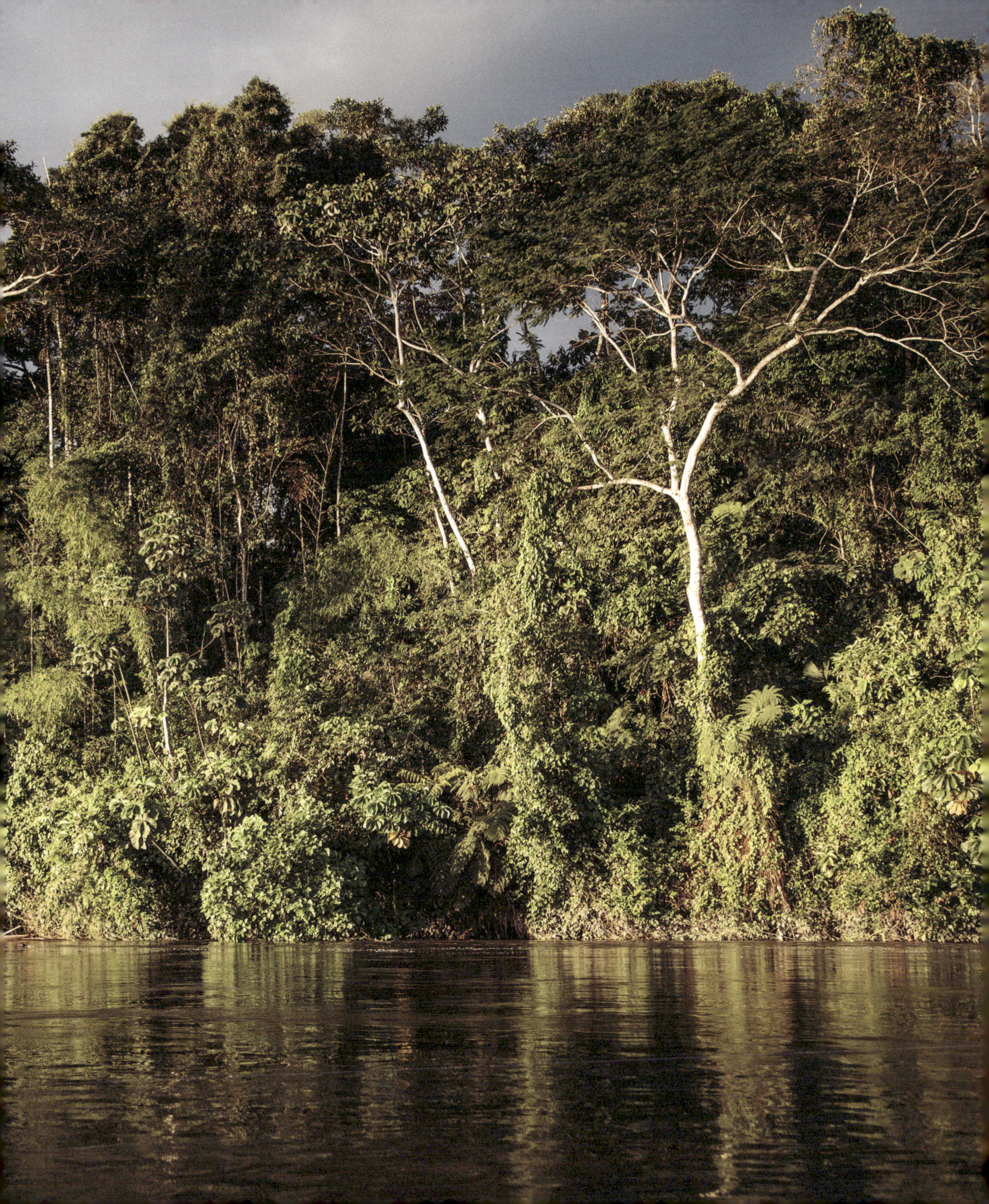

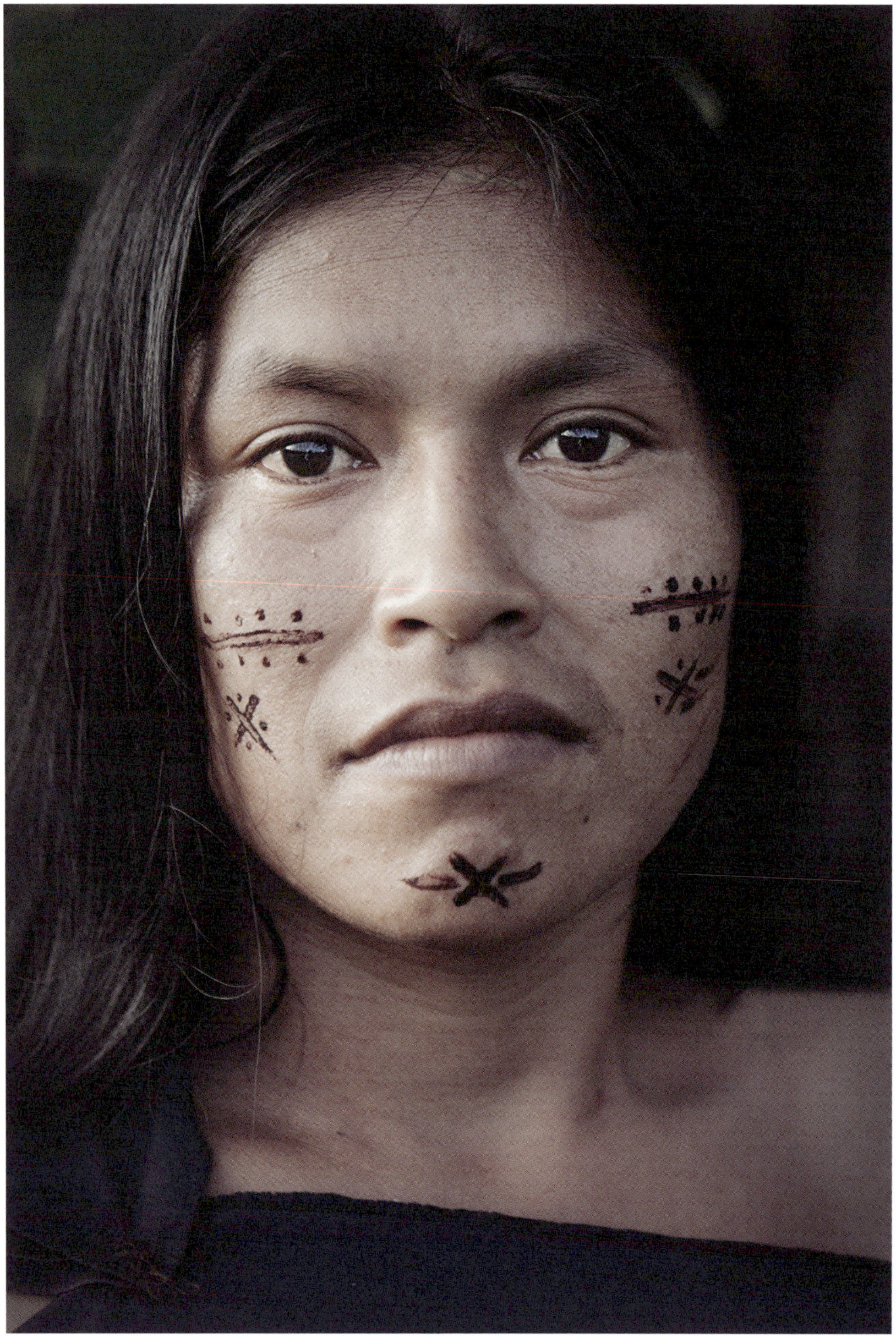

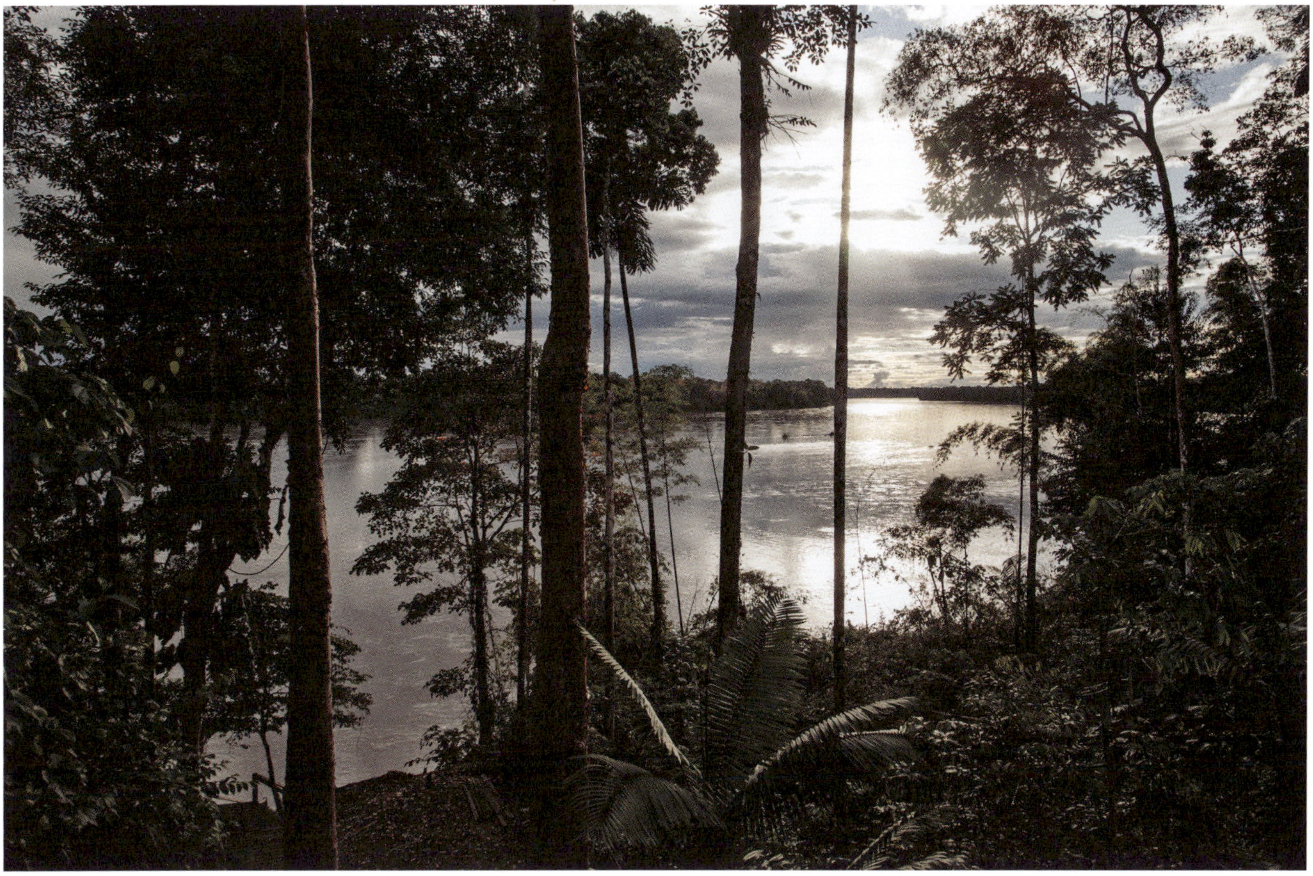

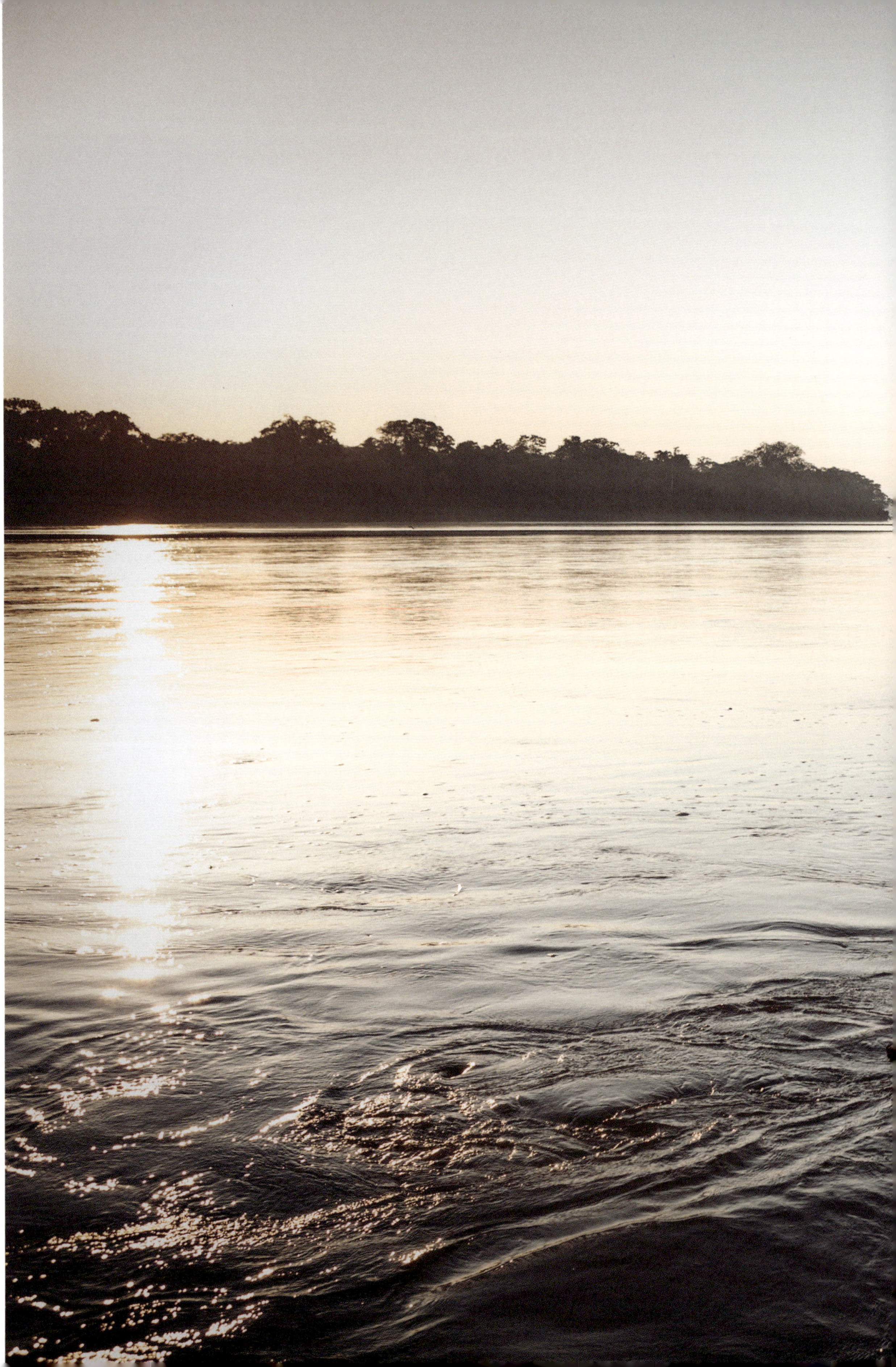

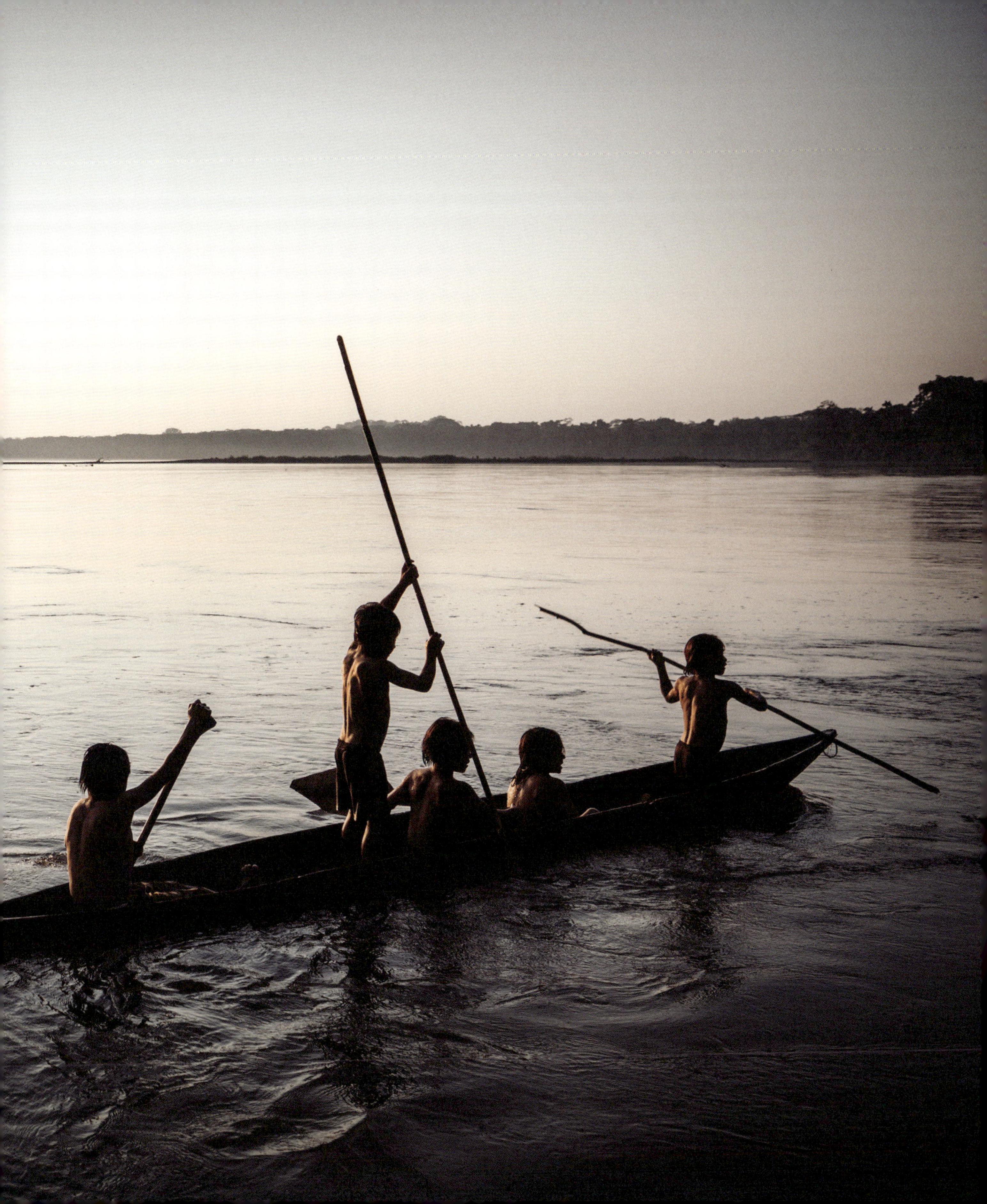

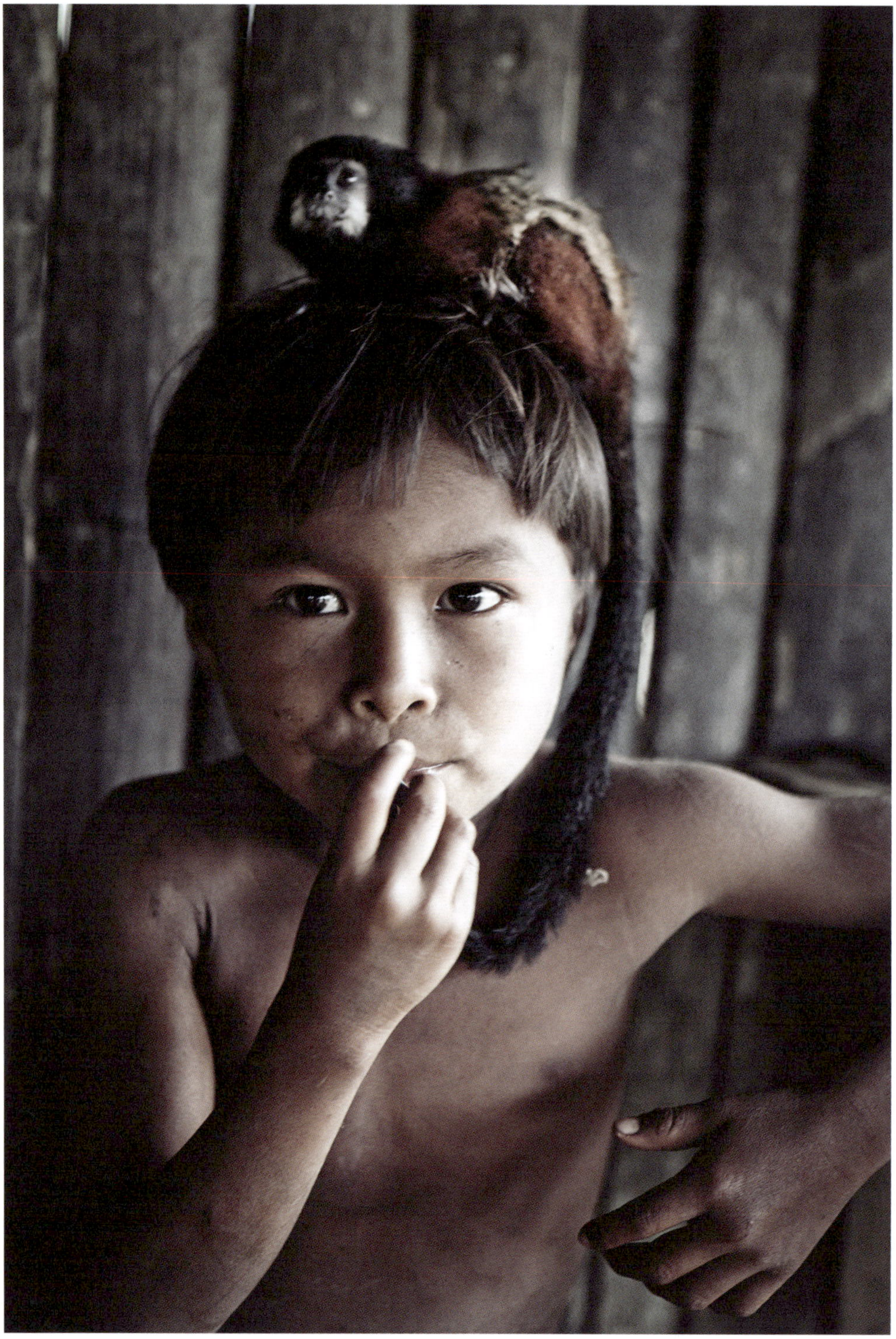

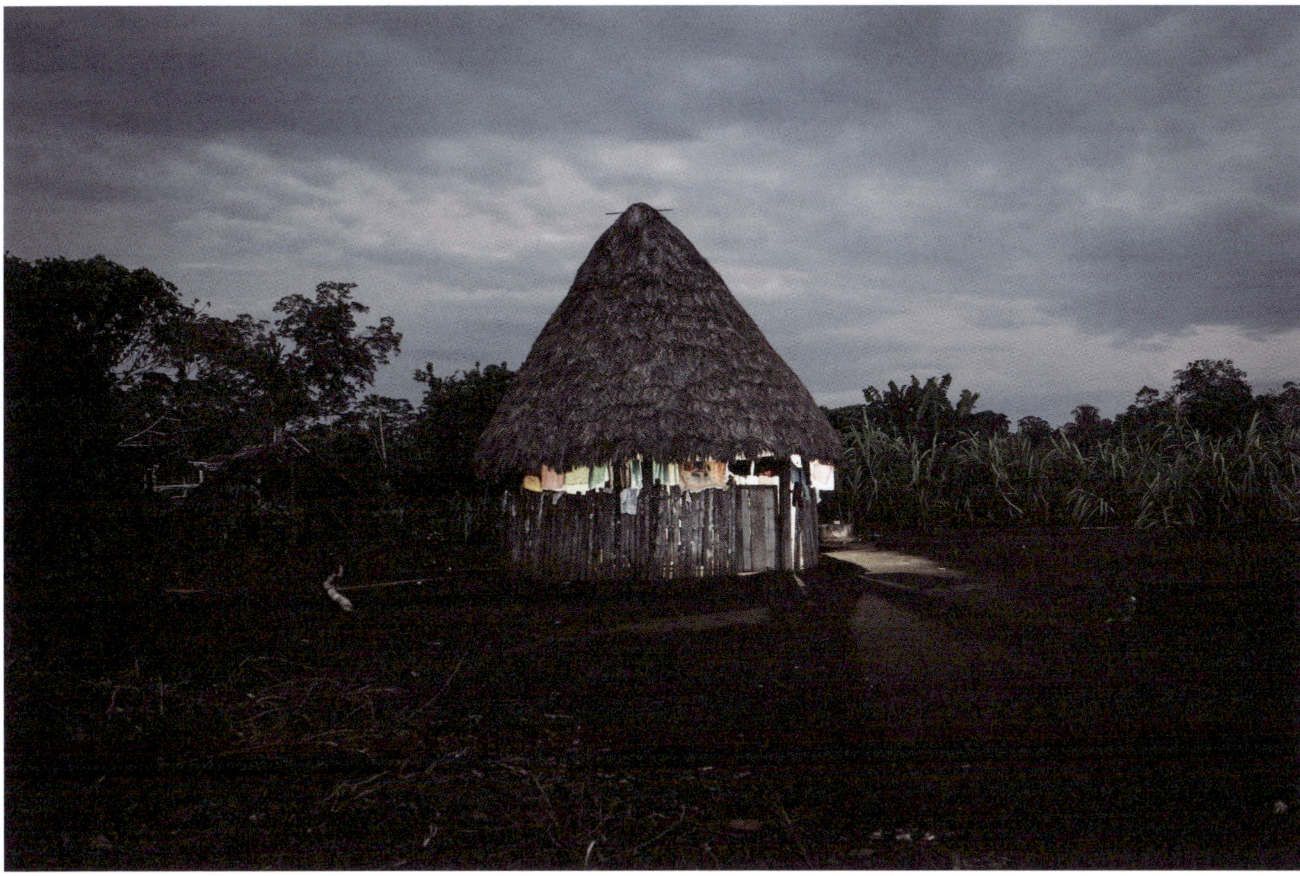

ASIA

MONGOLIA

43° 34′ 6.676″ N 104° 24′ 50.8″ E

MONGOLIA

I visited Mongolian nomads in the Gobi Desert.

Mongolia is located between Russia in the north and the People's Republic of China in the south. Its first origins as a national territory lie in the unification of several tribes of nomadic horsemen over 2000 years ago. Genghis Khan founded the Mongol Empire in 1206 after a campaign of conquest lasting several years, which spread over large parts of Asia and as far as Europe in the course of his reign. The area of today's Mongolia, although only a fraction of this empire, is almost four and a half times the size of Germany. However, with around three million inhabitants, Mongolia is the most sparsely populated country in the world, and more than 40 percent of its small population live in the capital Ulaanbaatar.

In the countryside, on the steppe and in the Gobi Desert, people live largely nomadically in the ger, the traditional Mongolian tent. The family plays a central role in their everyday lives. Their clan, the lineage of which many of them can trace back over many generations, fills them with great pride.

Originally, Altai shamanism and Tengrism were the predominant beliefs among the inhabitants of the Central Asian steppe. After several waves of conquest and missionary work, Buddhism is now the majority religion of the Mongolians. However, shamanism and Buddhism have become strongly intermingled, so that one can speak of a syncretic belief system. Ancestor worship is still a central component of Mongolian culture and religion.

The Gobi Desert covers a large part of Mongolia and the northern border region of the People's Republic of China. It is not a continuous desert in the true sense of the word, but is made up of different desert, semi-desert and steppe landscapes. With the Sahara, the deserts of Australia and the Arabian Peninsula, the Gobi is among the largest non-polar deserts in the world. It is not the driest, but has the greatest temperature differences. In spring and fall, there are violent storms that carry away earth crumbs and sand. Climate change is intensifying the existing

MONGOLEI

Mein Besuch galt mongolischen Nomadinnen und Nomaden in der Wüste Gobi.

Die Mongolei befindet sich zwischen Russland im Norden und der Volksrepublik China im Süden. Ihre ersten Ursprünge als Staatsgebiet liegen in der Vereinigung mehrerer Stämme von Reiternomaden vor über 2000 Jahren. Dschingis Khan gründete 1206 nach einem mehrere Jahre dauernden Eroberungsfeldzug das Mongolische Reich, das sich im Laufe seiner Herrschaft über große Teile Asiens bis nach Europa ausbreitete. Die Fläche der heutigen Mongolei ist, obwohl nur noch ein Bruchteil dieses Imperiums, knapp viereinhalbmal so groß wie die Deutschlands. Mit rund drei Millionen Einwohnern ist die Mongolei jedoch der am dünnsten besiedelte Staat weltweit, und von der geringen Bevölkerung leben mehr als 40 Prozent in der Hauptstadt Ulaanbaatar.

Auf dem Land, in der Steppe und der Wüste Gobi siedeln die Menschen weitgehend nomadisch im Ger, dem traditionellen Zelt der Mongol:innen. Die Familie spielt für ihren Lebensalltag eine zentrale Rolle. Ihre Clans, deren Linie sie über zahlreiche Generationen zurückverfolgen können, erfüllen viele von ihnen mit großem Stolz.

Ursprünglich waren unter den Bewohnern der zentralasiatischen Steppe der Altai-Schamanismus und der Tengrismus die vorherrschenden Glaubensrichtungen. Nach mehreren Eroberungs- und Missionierungsschüben stellt heute der Buddhismus die Mehrheitsreligion der Mongol:innen dar. Schamanismus und Buddhismus haben sich allerdings stark vermischt, sodass man von einem synkretistischen Glaubenssystem sprechen kann. Zentraler Bestandteil der mongolischen Kultur und Religion ist nach wie vor der Ahnenkult.

Die Wüste Gobi erstreckt sich über einen Großteil der Mongolei und die nördliche Grenzregion der Volksrepublik China. Sie ist im eigentlichen Sinne keine durchgehende Wüste, sondern setzt sich aus unterschiedlichen Wüsten-, Halbwüsten- und Steppenlandschaften zusammen. Mit der Sahara, den Wüsten Australiens und der

weather extremes. It leads to a decline in biodiversity, threatens the existence of nomads and their livestock and increases the spread of deserts on all continents. According to the United Nations, 41 percent of the earth is dryland, 20 percent of which is completely barren. More than 1.3 billion people in over 100 countries around the world are affected by the increasing destruction of fertile soil. Deserts, which already cover around a third of the Earth's surface, are spreading rapidly: 50,000 to 70,000 square kilometers of soil are lost to them every year. According to the IPCC, this is an agricultural area the size of Switzerland.

arabischen Halbinsel gehört die Gobi zu den größten nichtpolaren Wüsten. Sie ist nicht die trockenste, weist aber die größten Temperaturunterschiede auf. Im Frühjahr und Herbst gibt es heftige Stürme, die Erdkrumen ab- und den Sand weitertragen. Der Klimawandel verstärkt die bestehenden Wetterextreme. Er führt so zu einem Rückgang der biologischen Vielfalt, bedroht die Existenz der Nomad:innen und ihres Viehs und verstärkt die Ausbreitung der Wüsten auf allen Kontinenten. 41 Prozent der Erde sind nach Angaben der Vereinten Nationen Trockenzonen, 20 Prozent davon sind völlig unfruchtbar. Von der zunehmenden Zerstörung fruchtbaren Bodens sind mehr als 1,3 Milliarden Menschen in über 100 Ländern der Erde betroffen. Die Wüsten, die schon heute rund ein Drittel der Oberfläche der Erde bedecken, breiten sich schnell aus: 50.000 bis 70.000 Quadratkilometer Boden gehen jedes Jahr an sie verloren. Das ist laut dem IPCC eine landwirtschaftliche Fläche von der Größe der Schweiz.

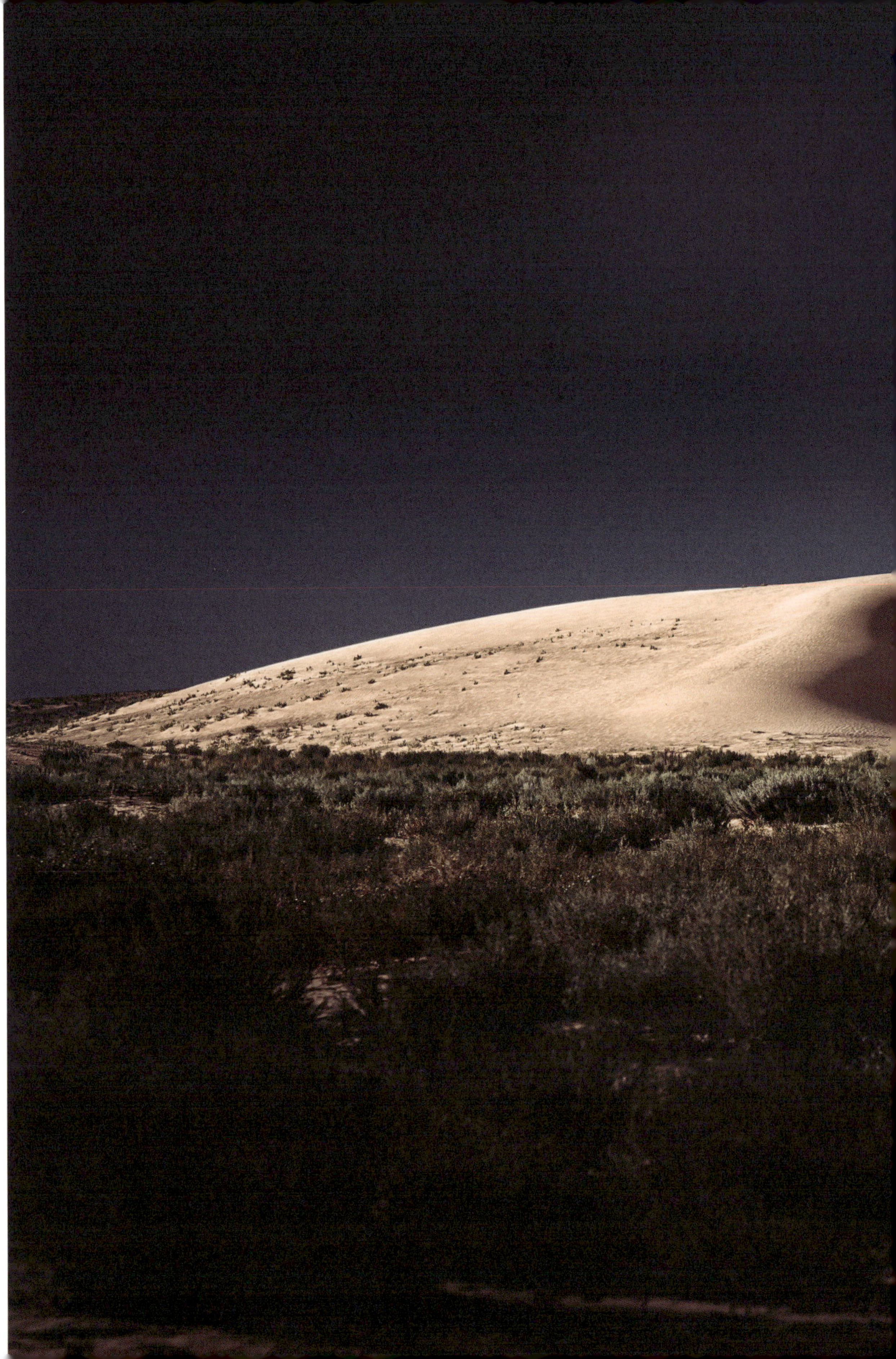

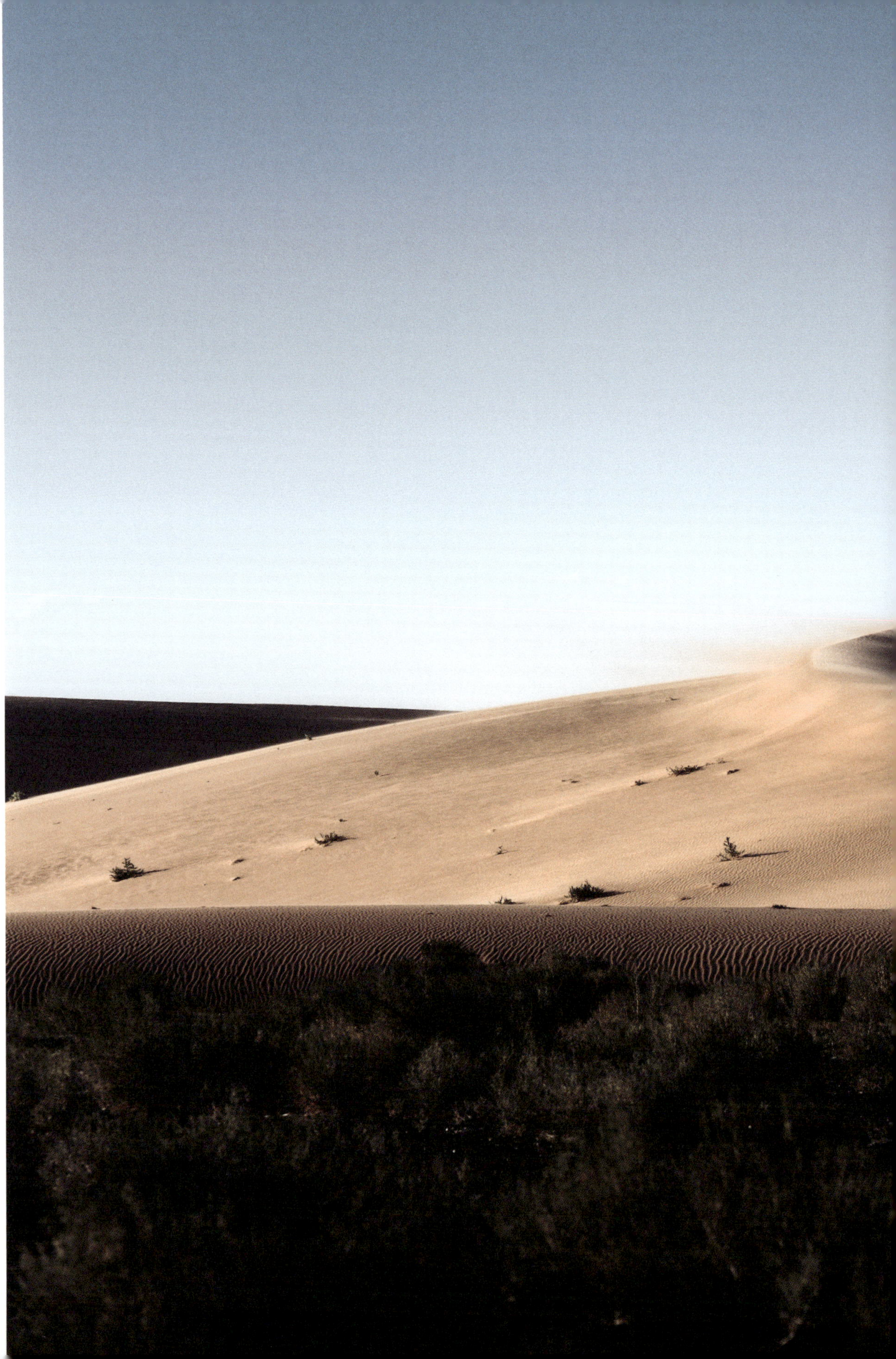

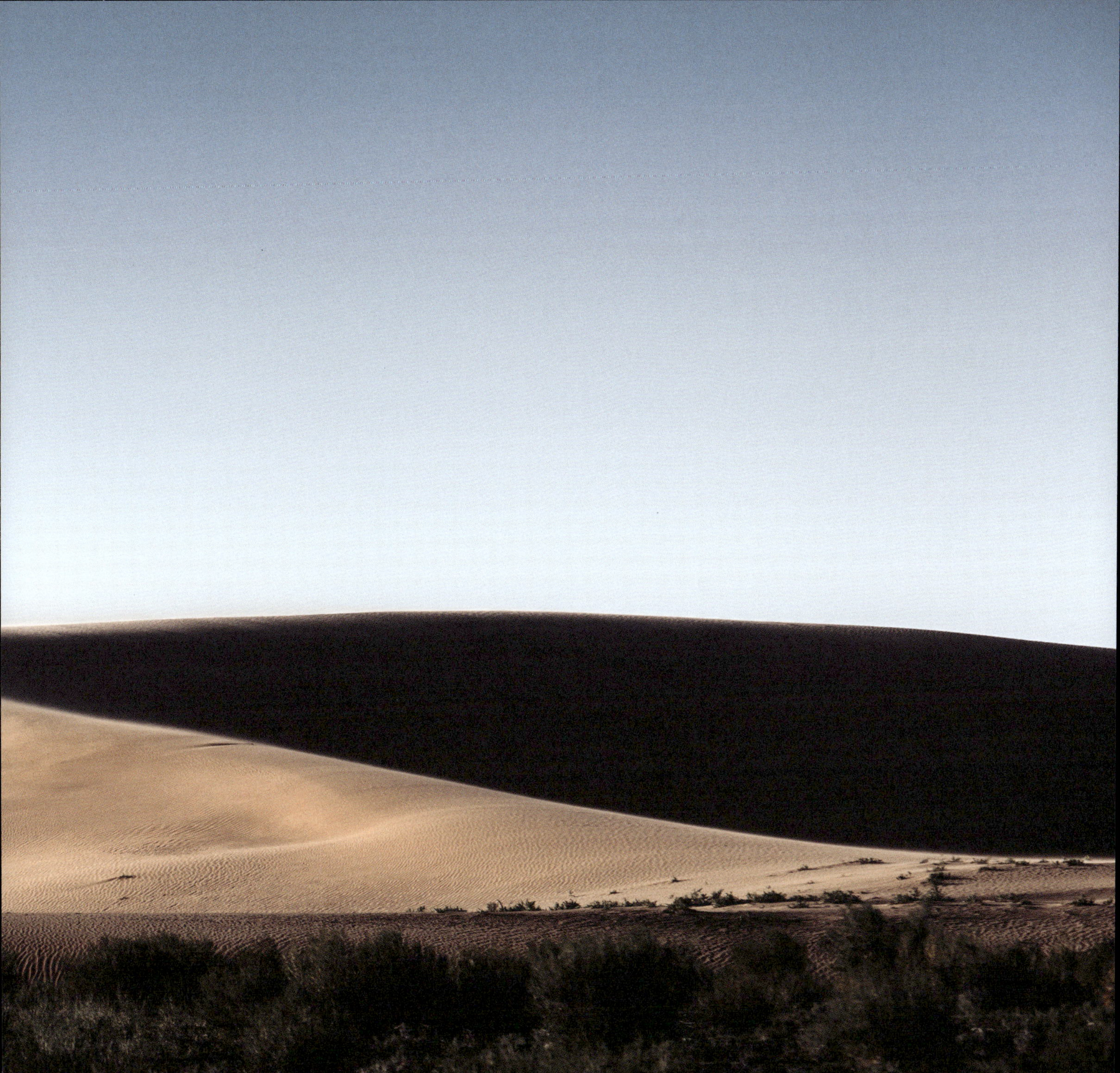

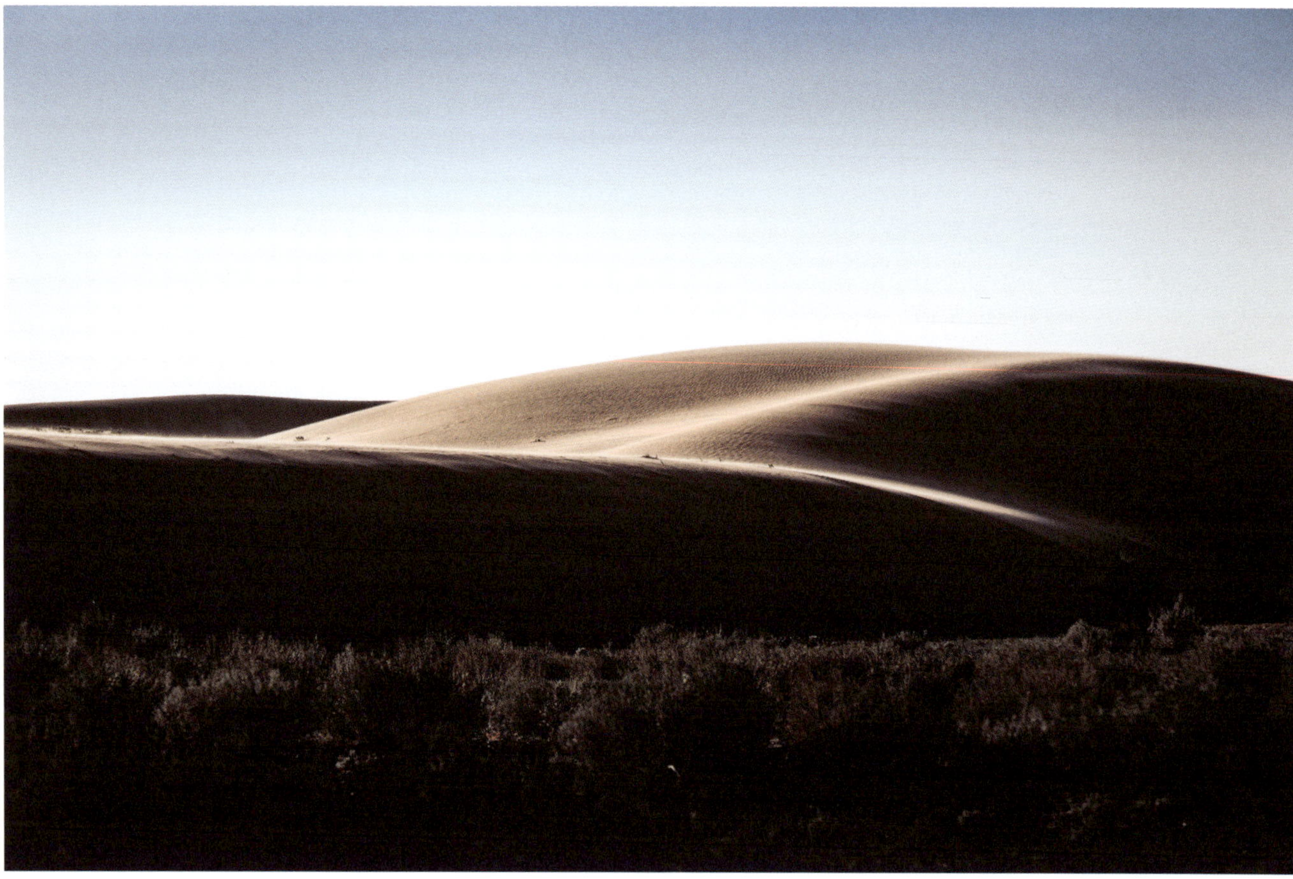

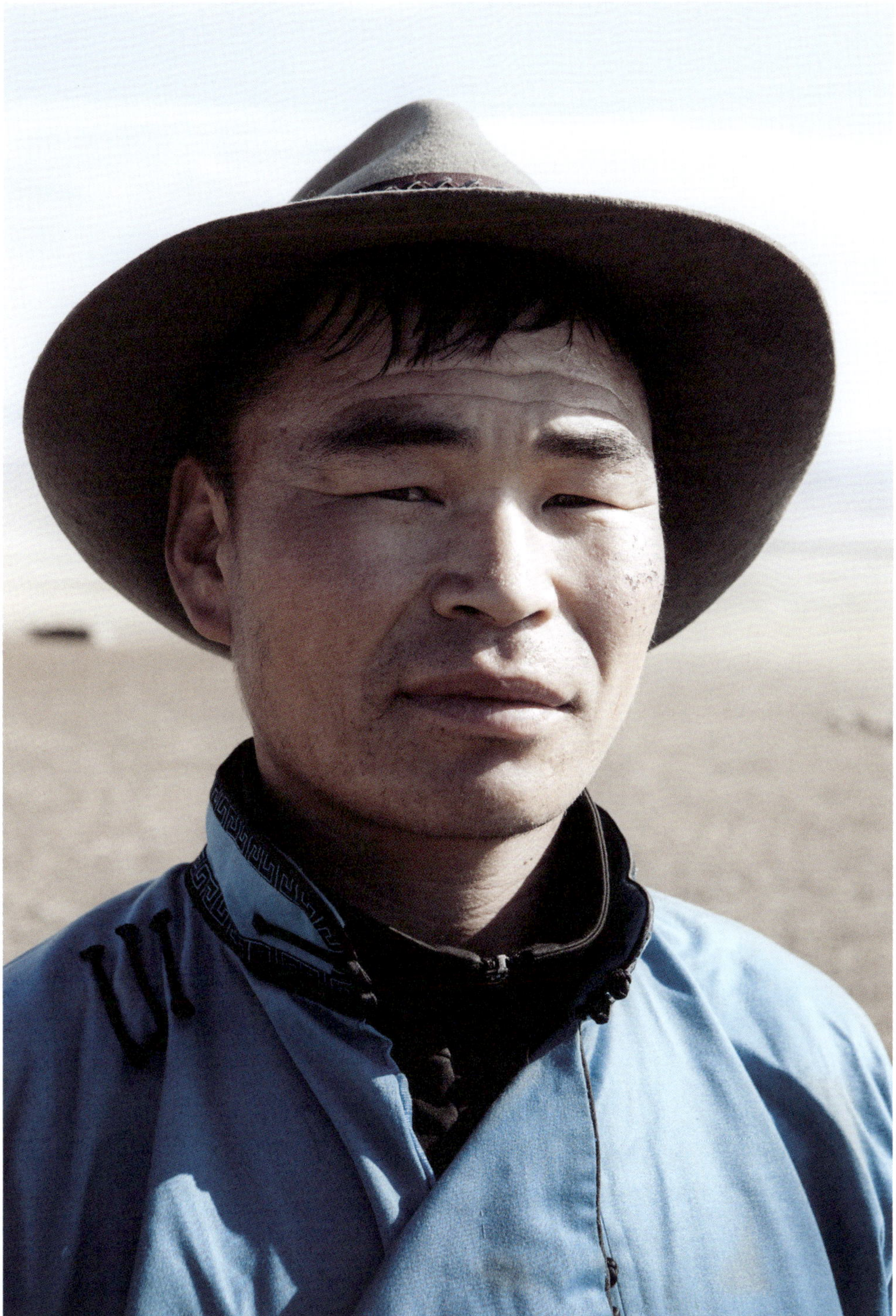

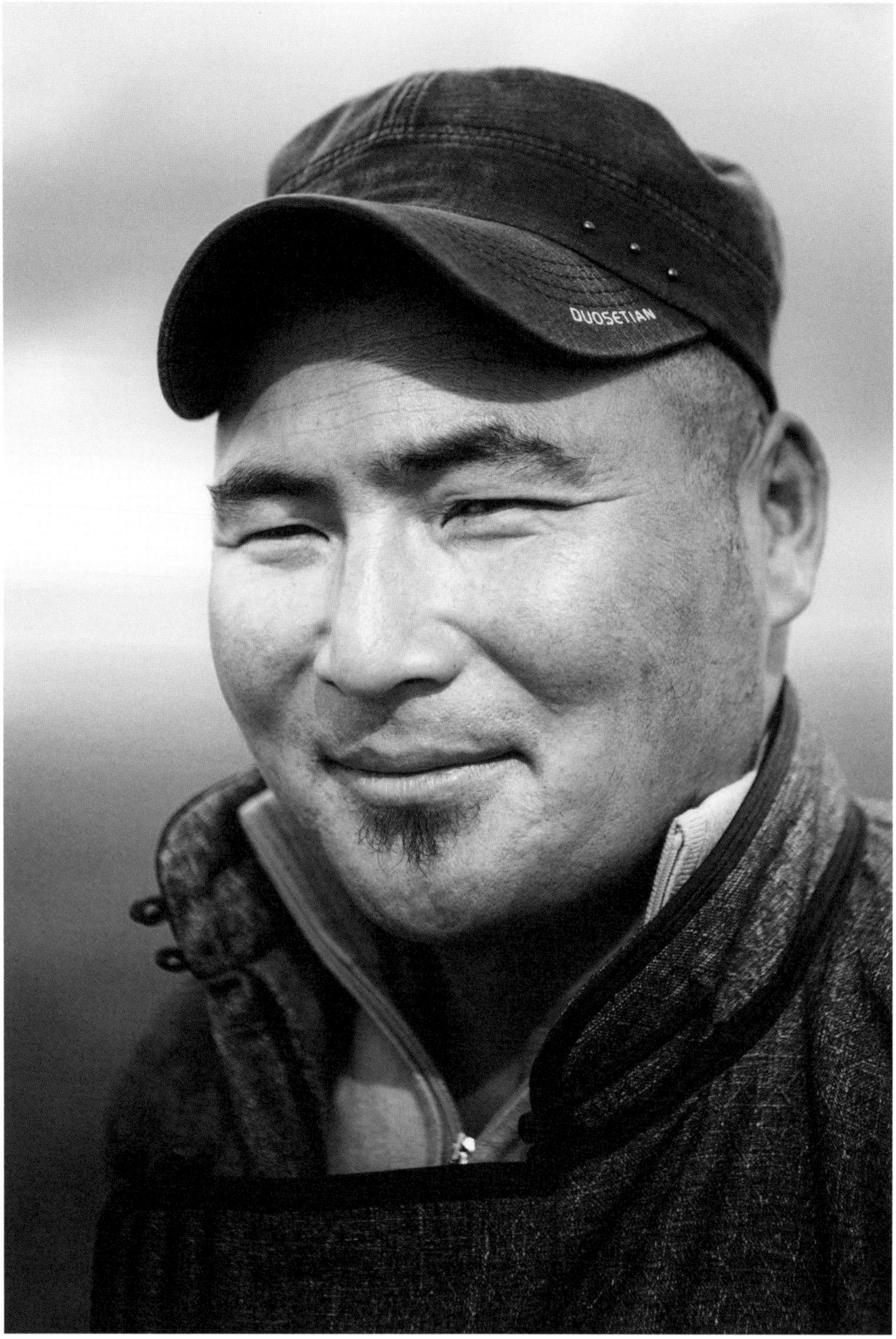
DUOSETIAN

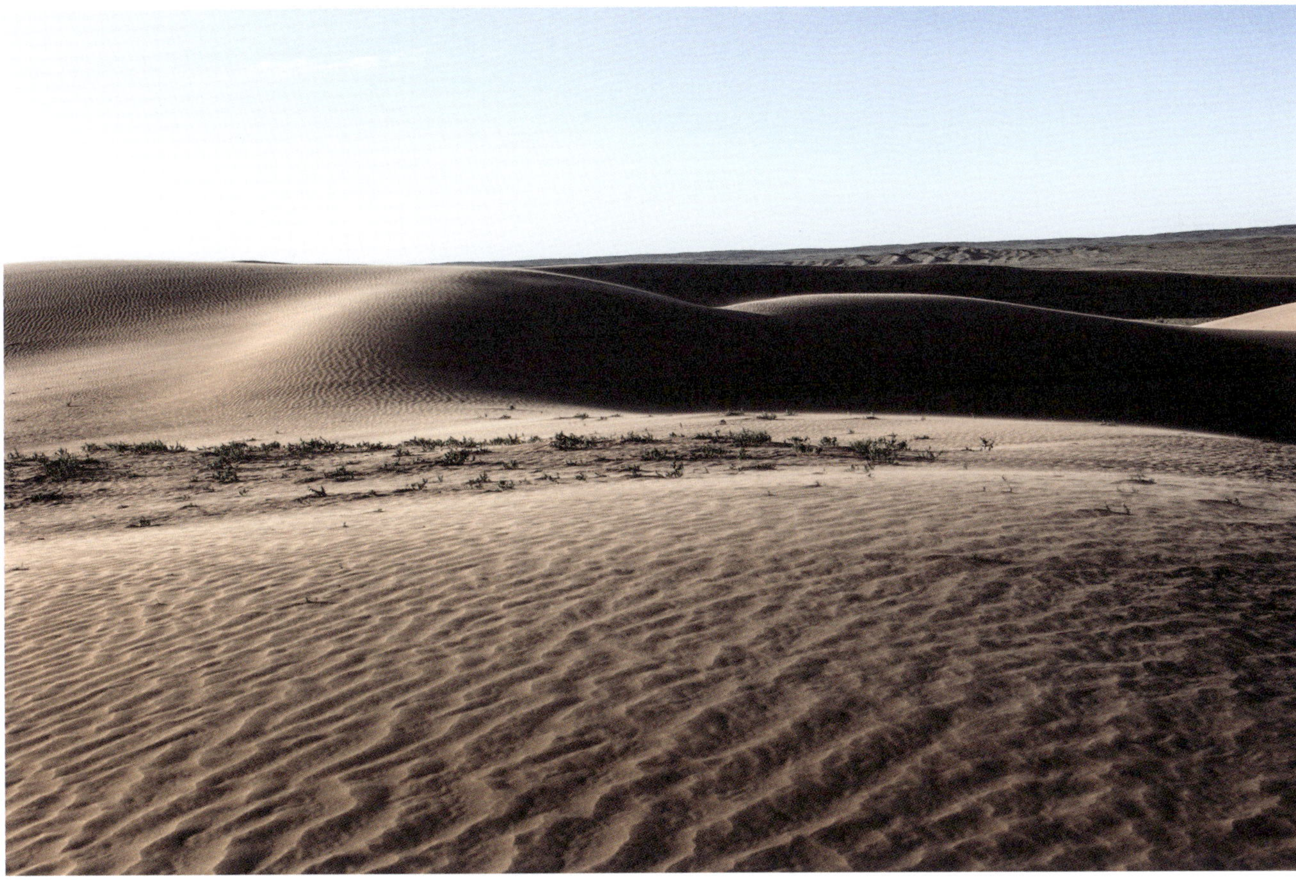

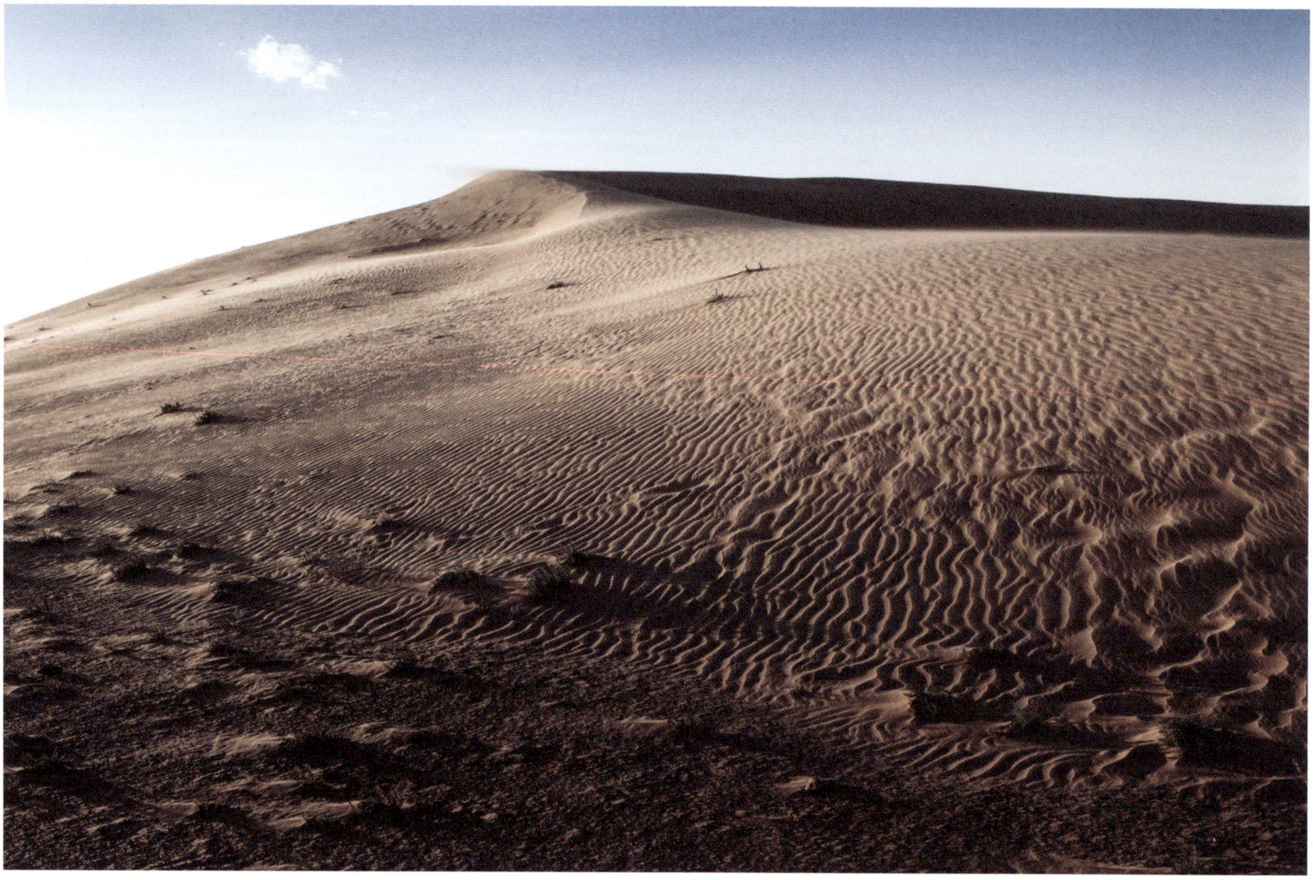

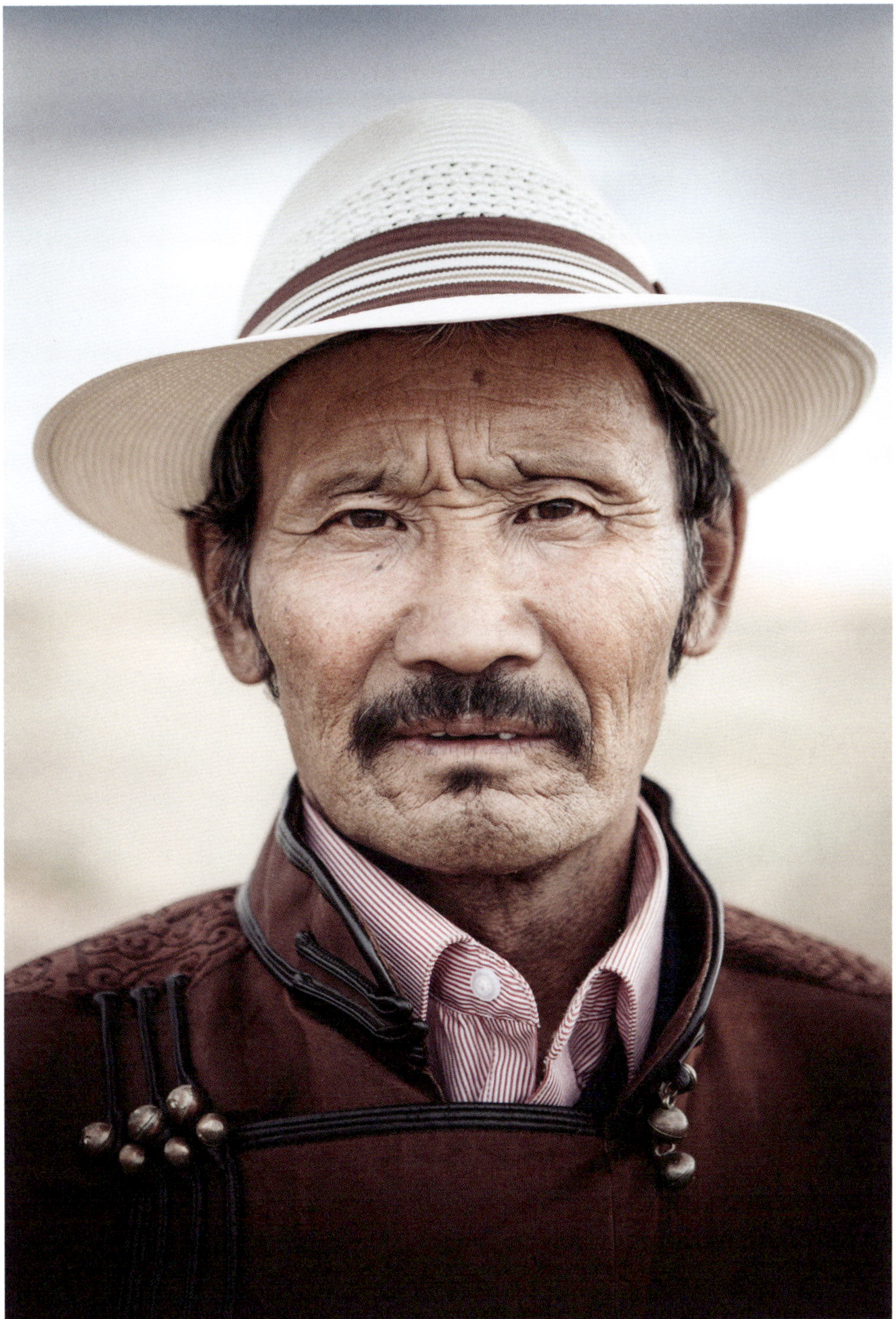

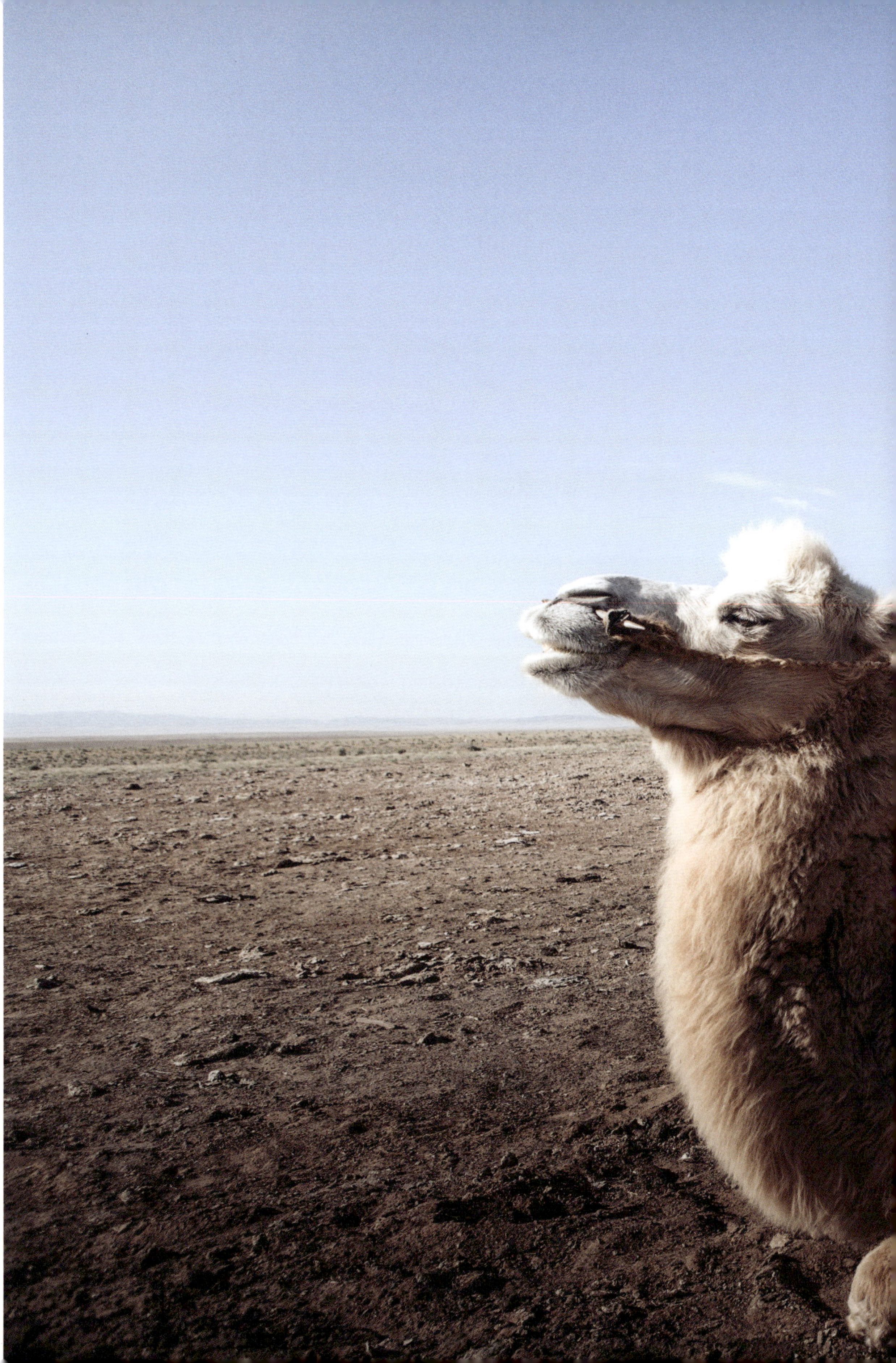

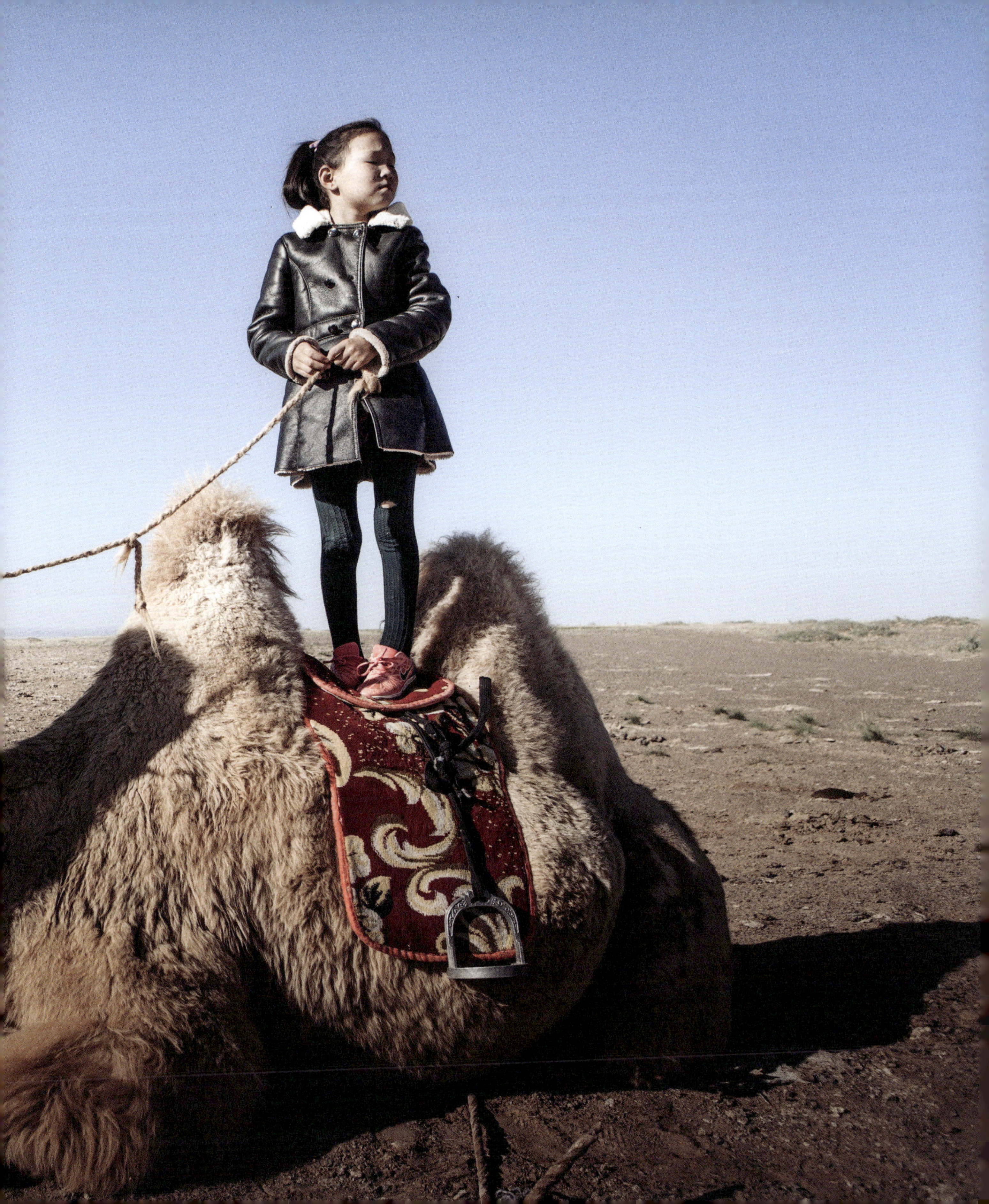

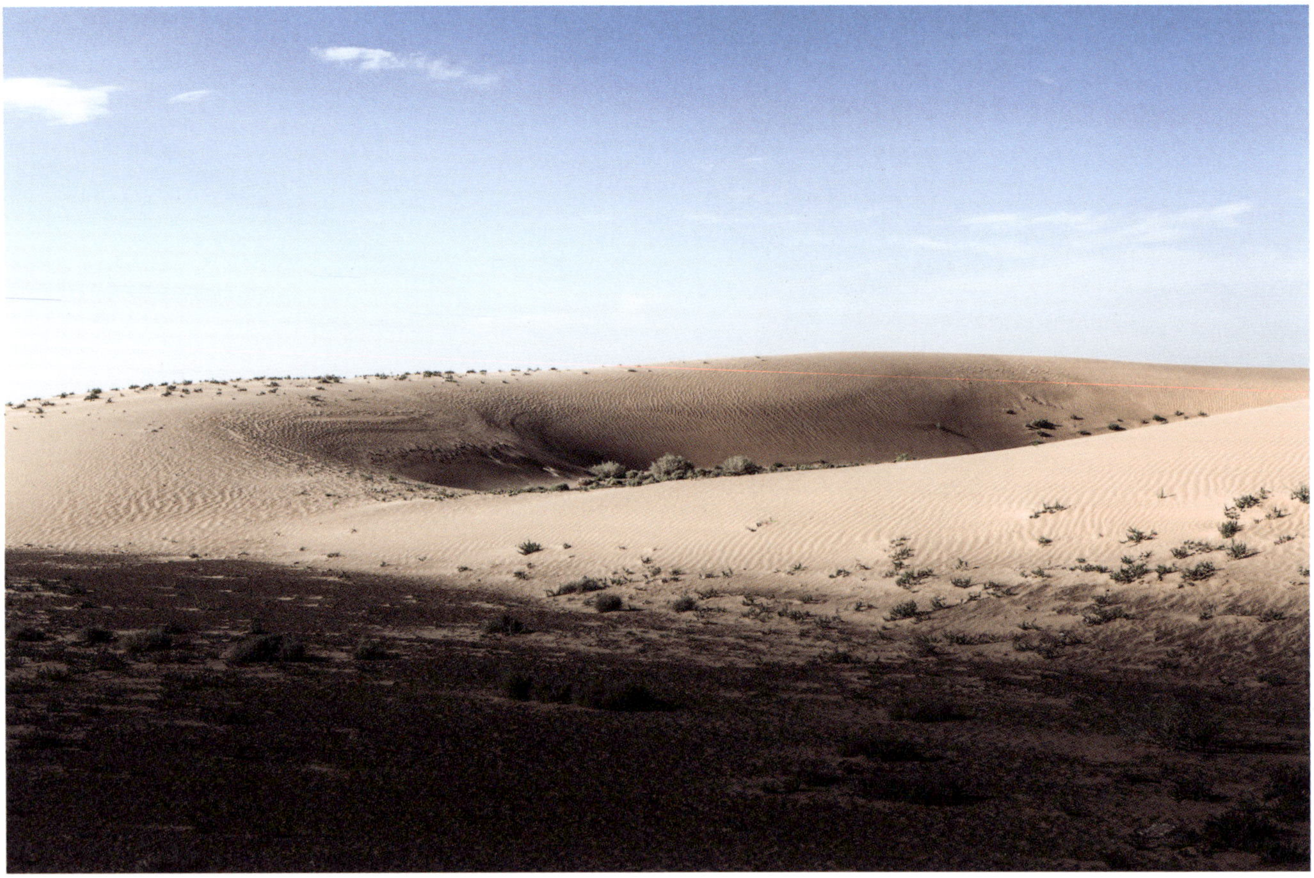

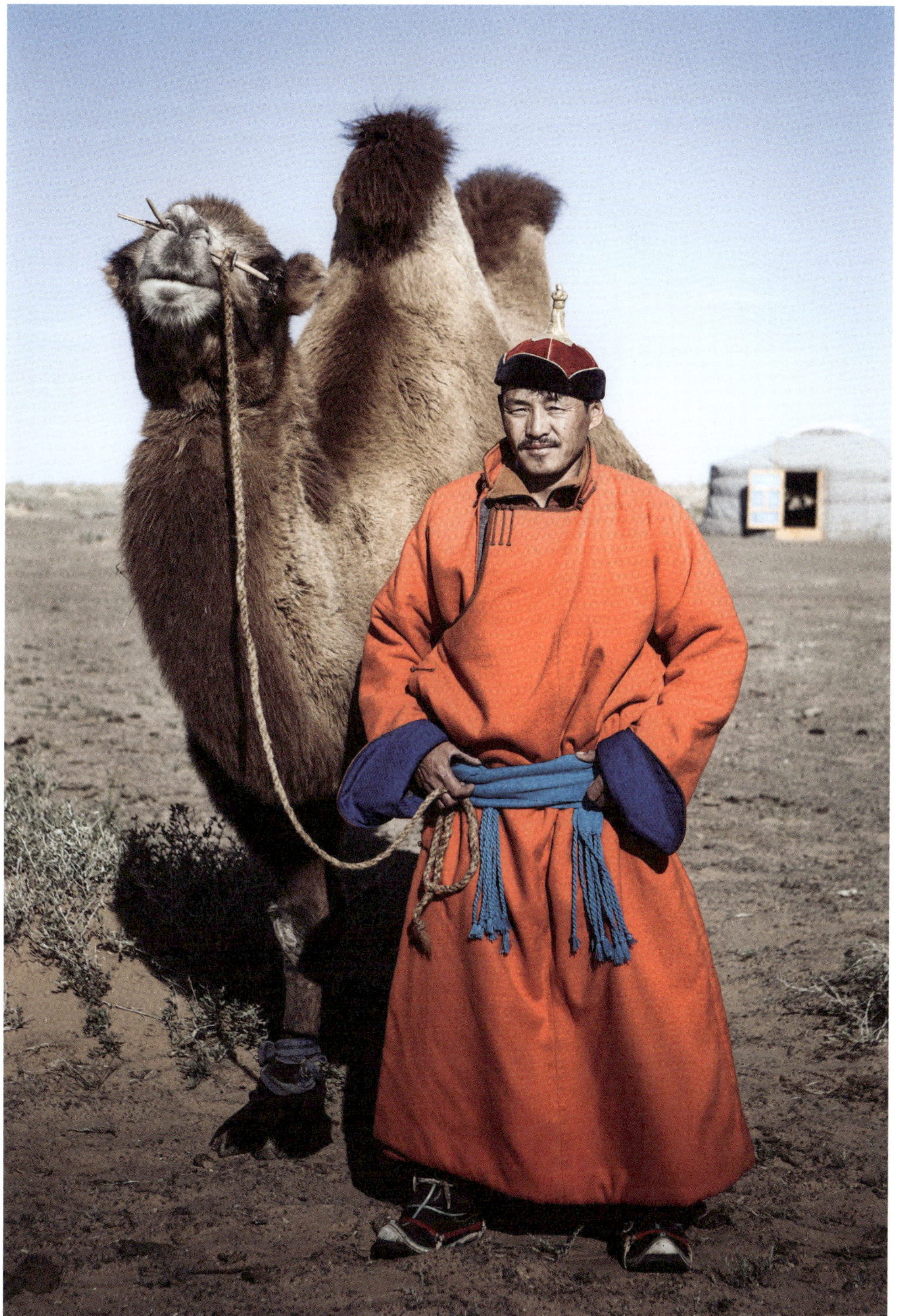

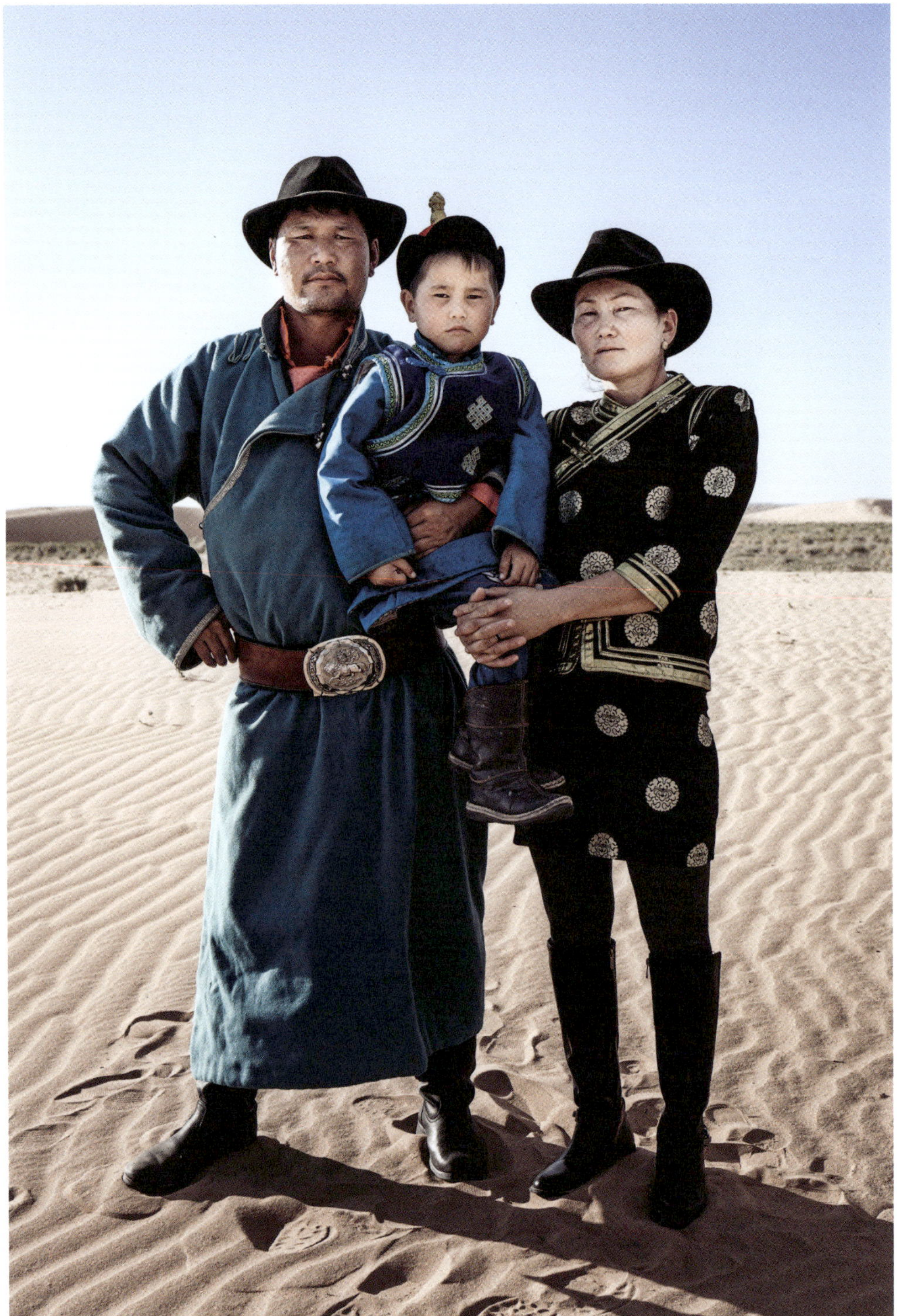

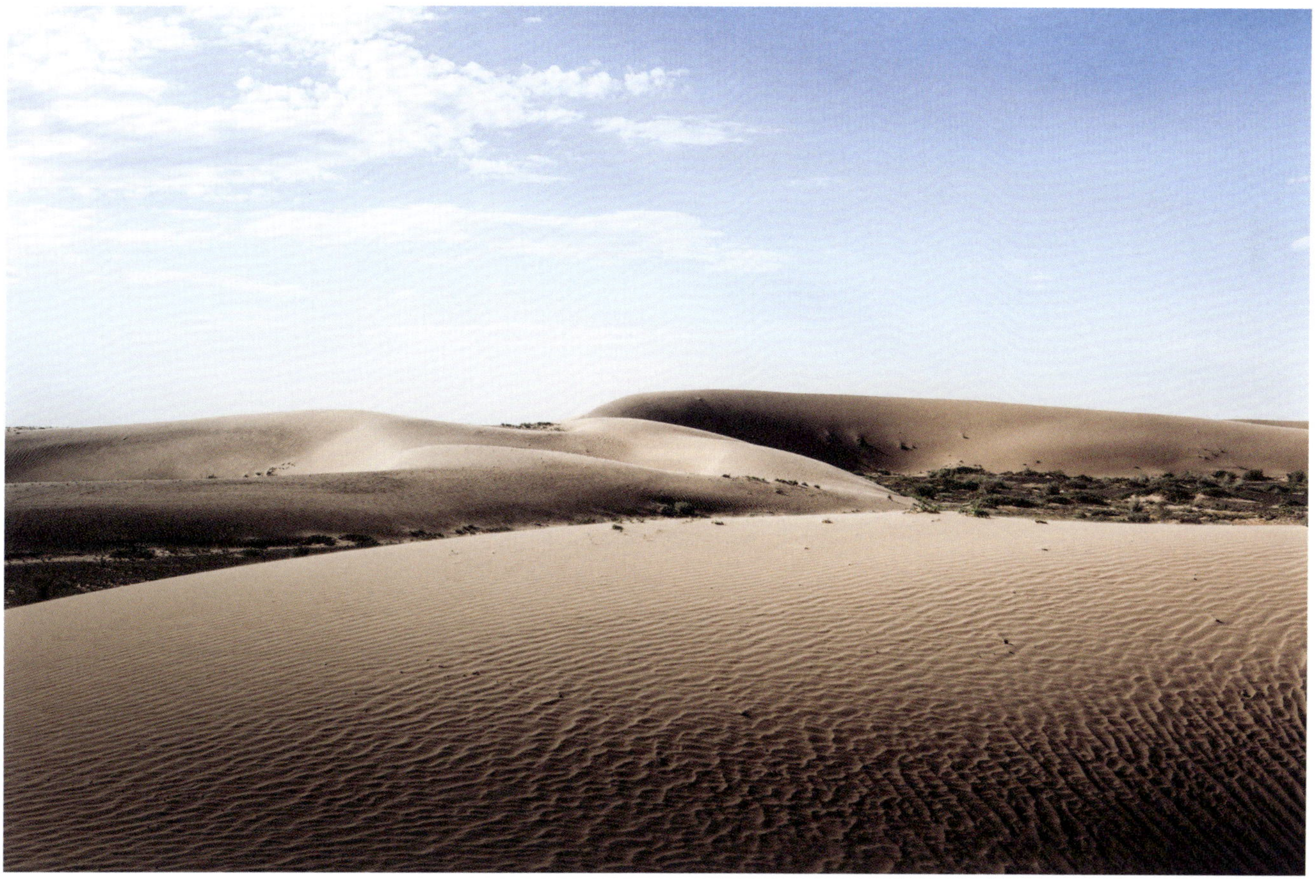

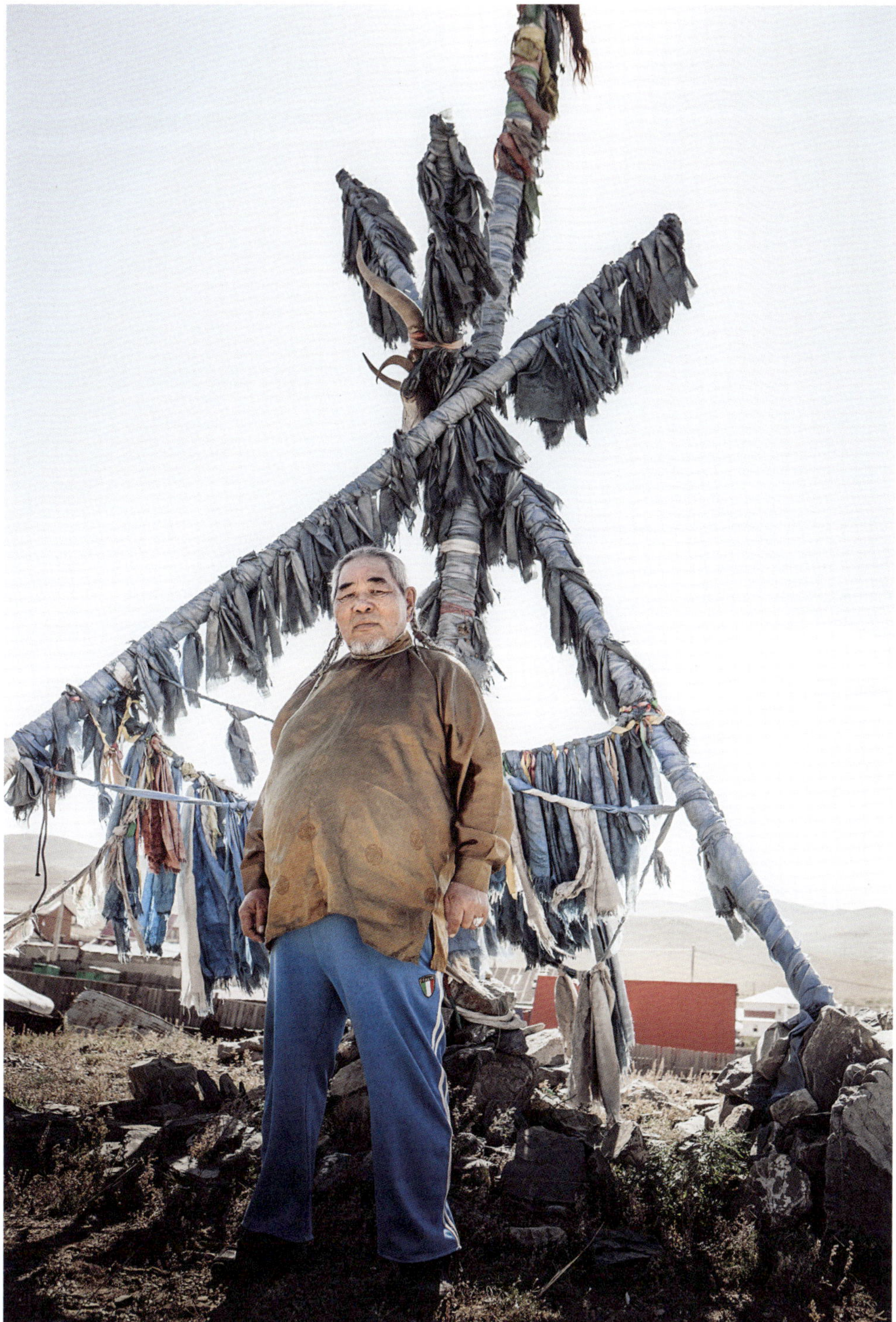

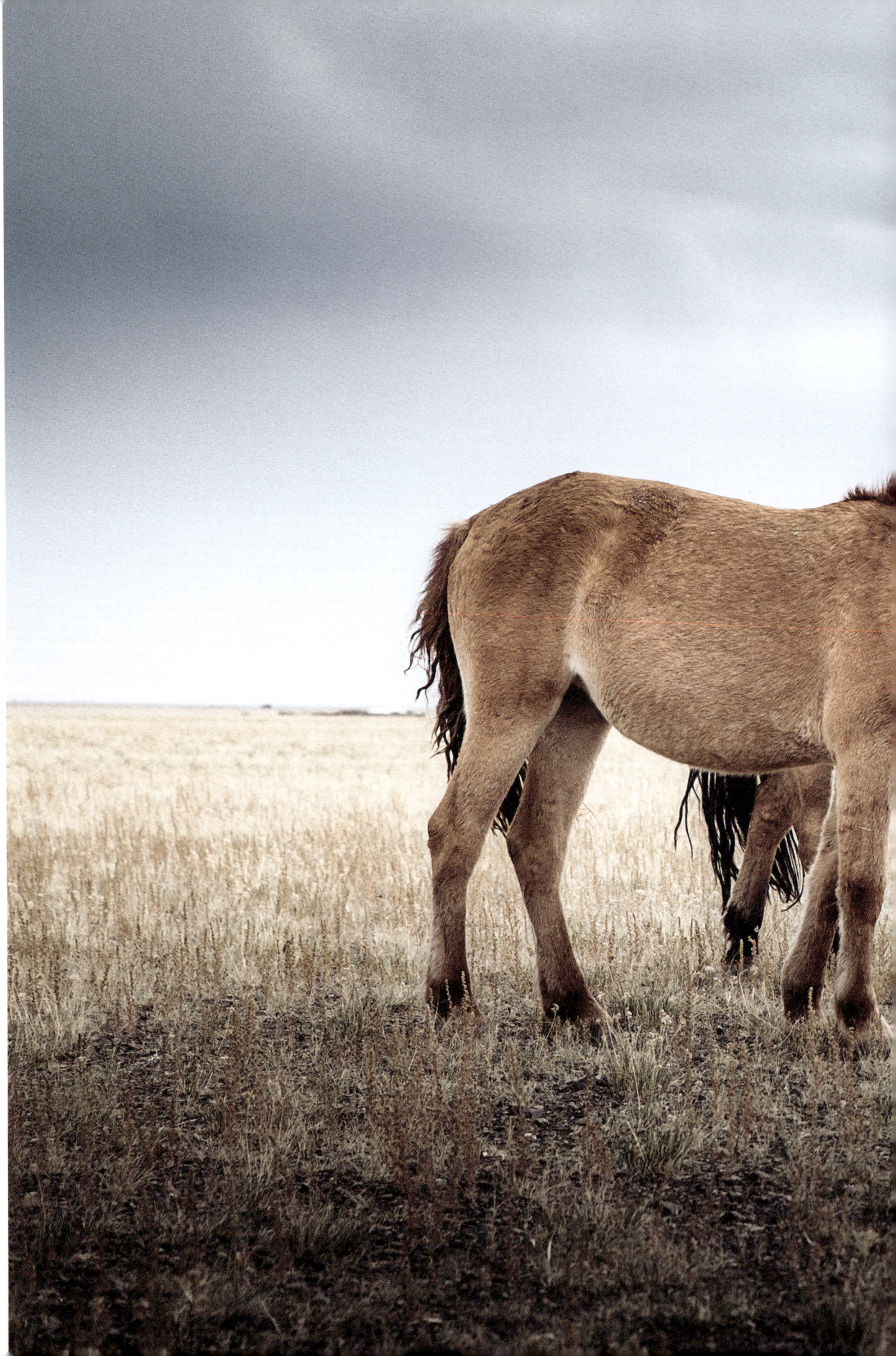

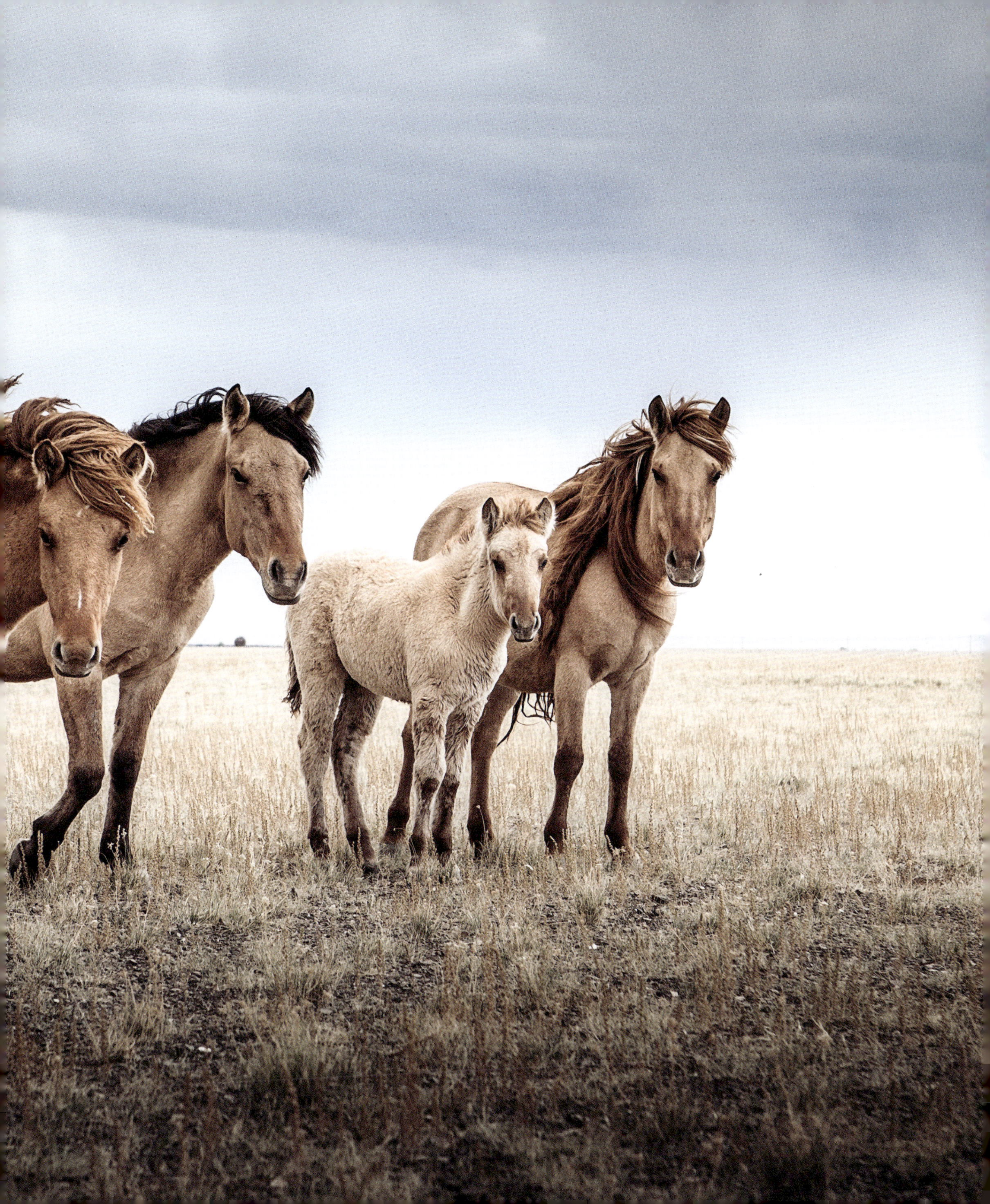

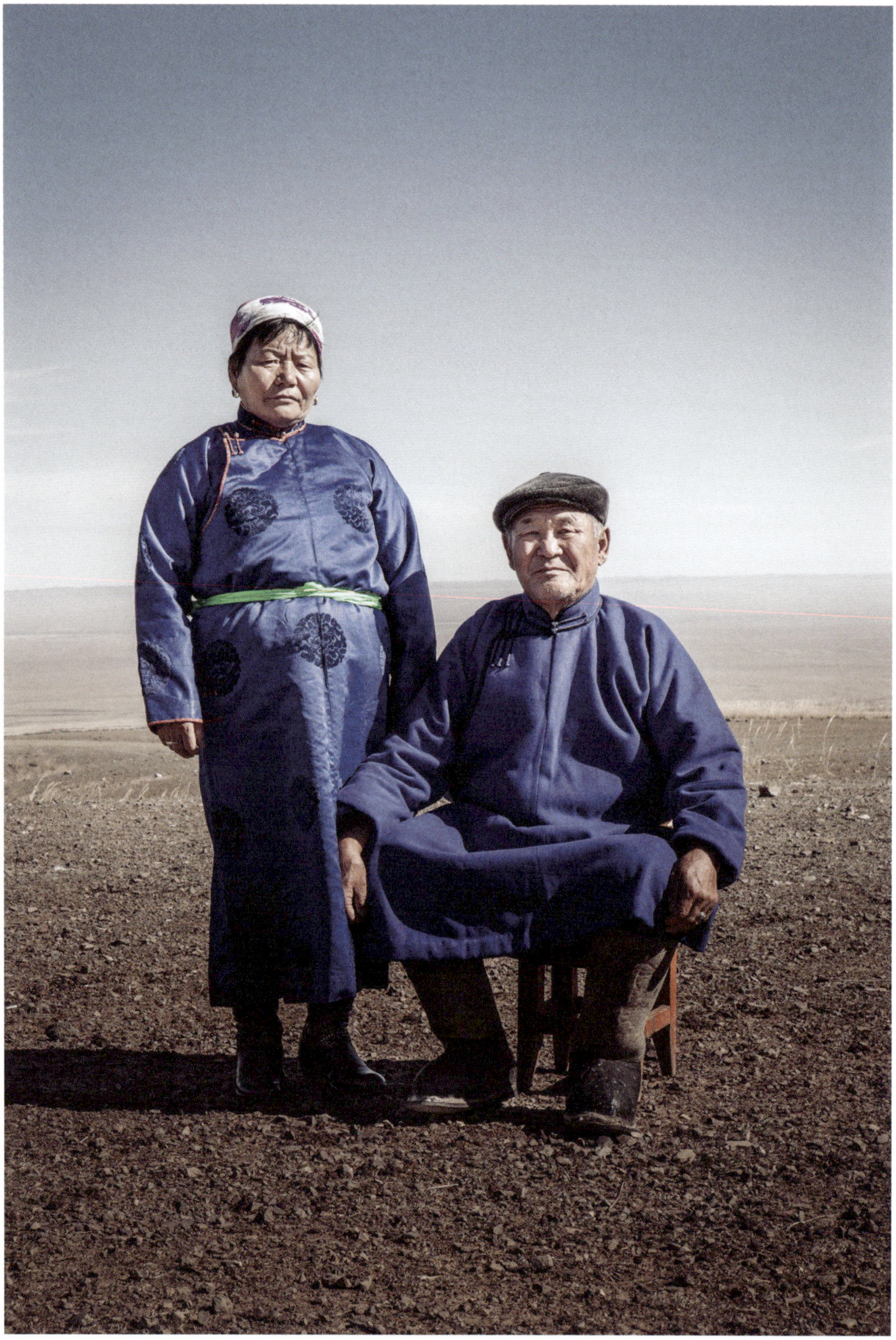

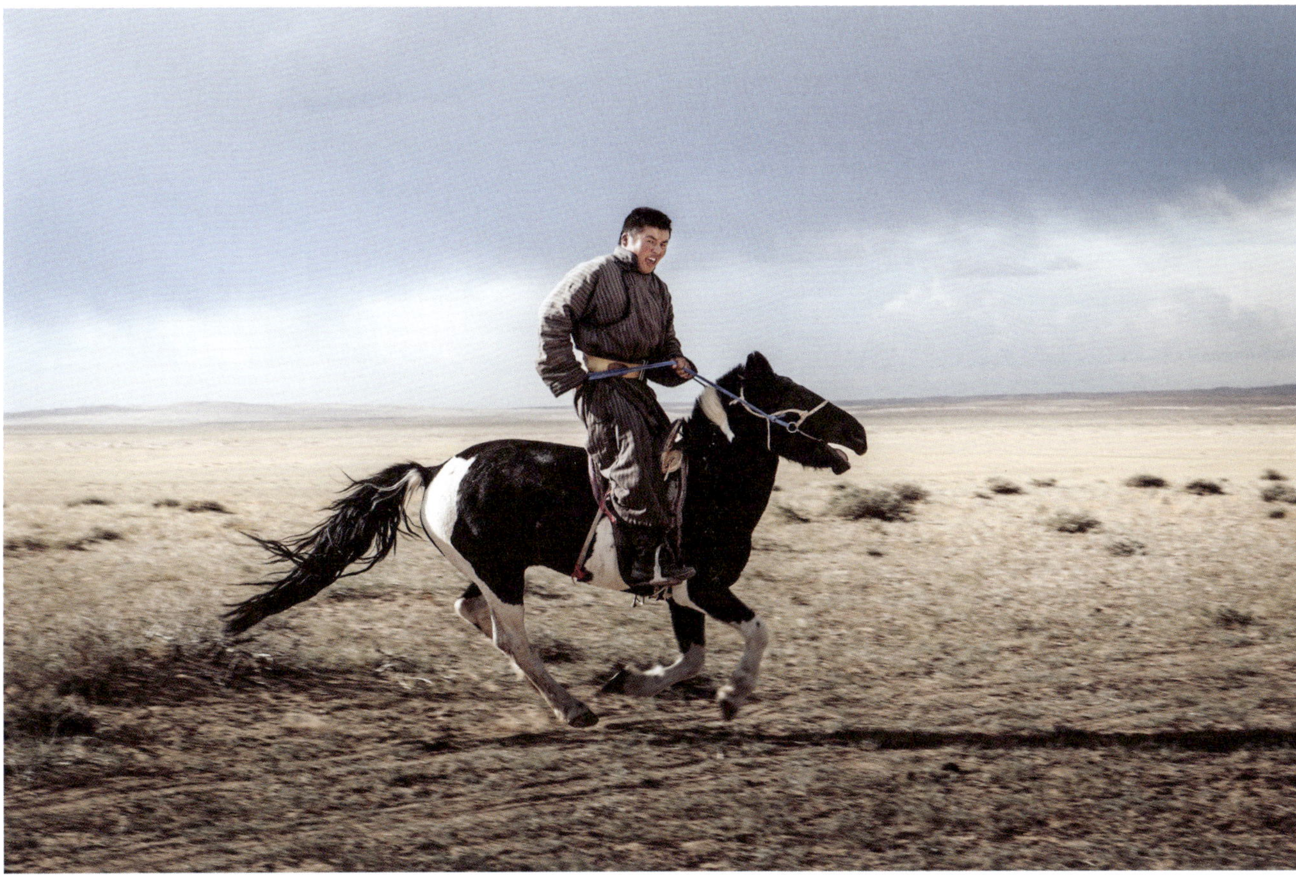

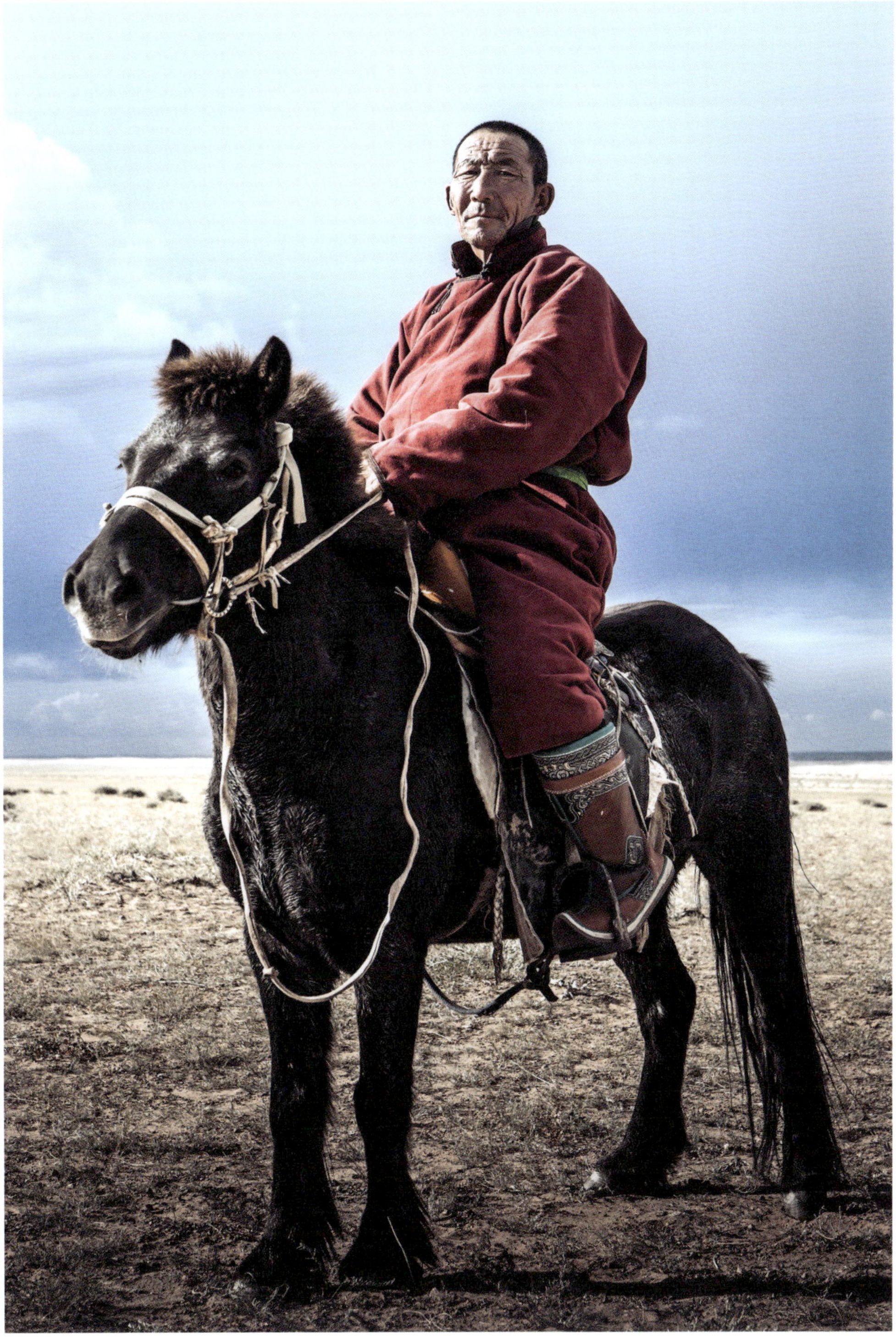

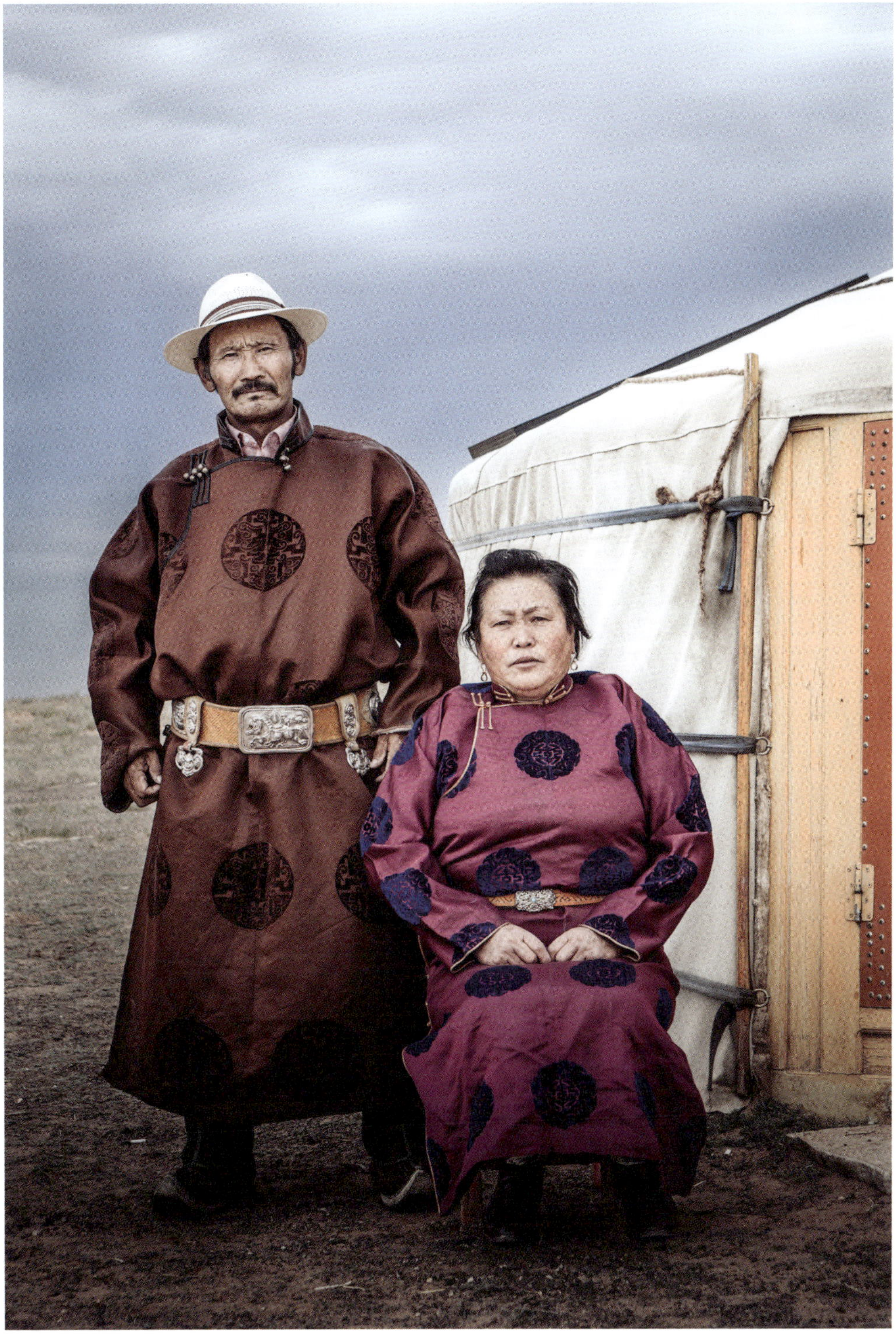

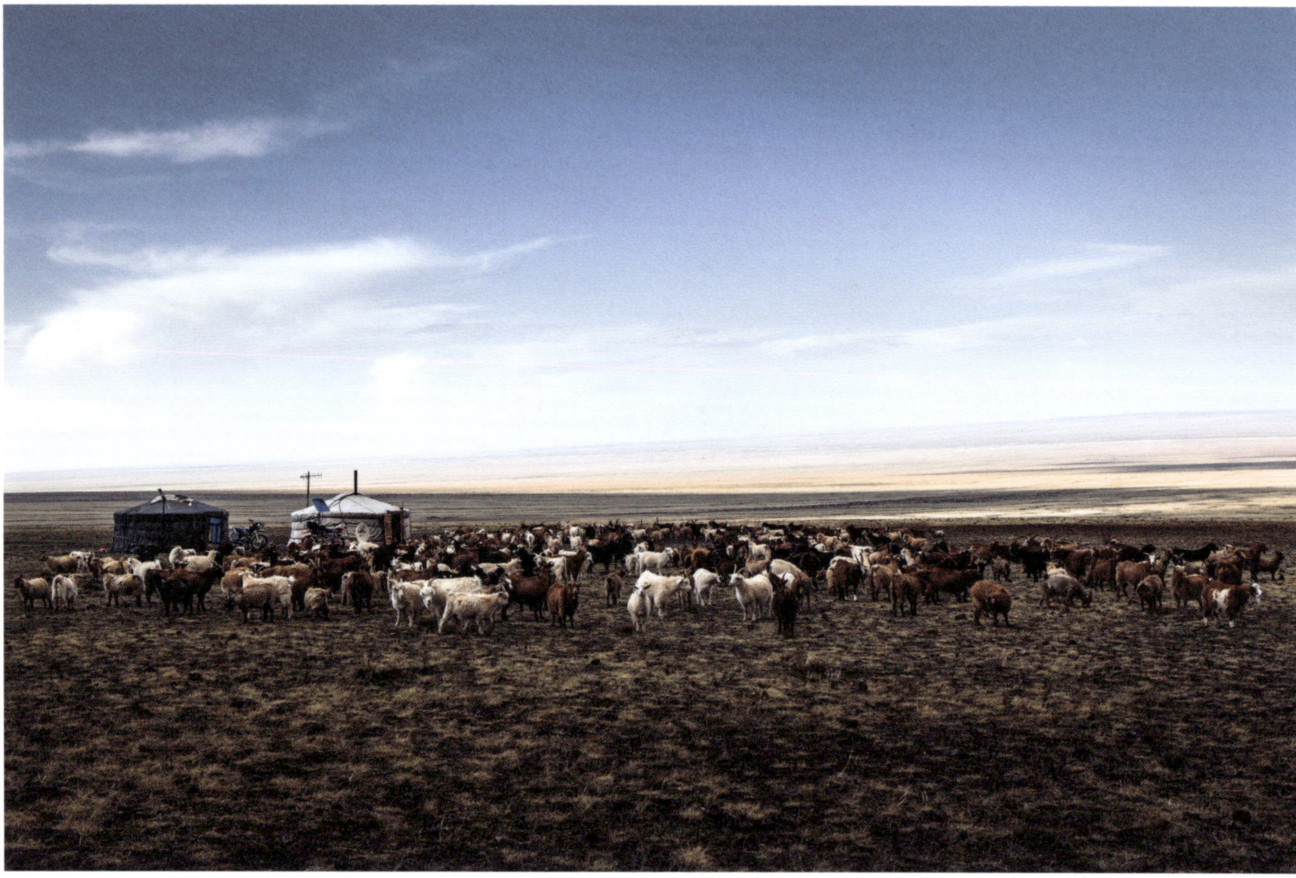

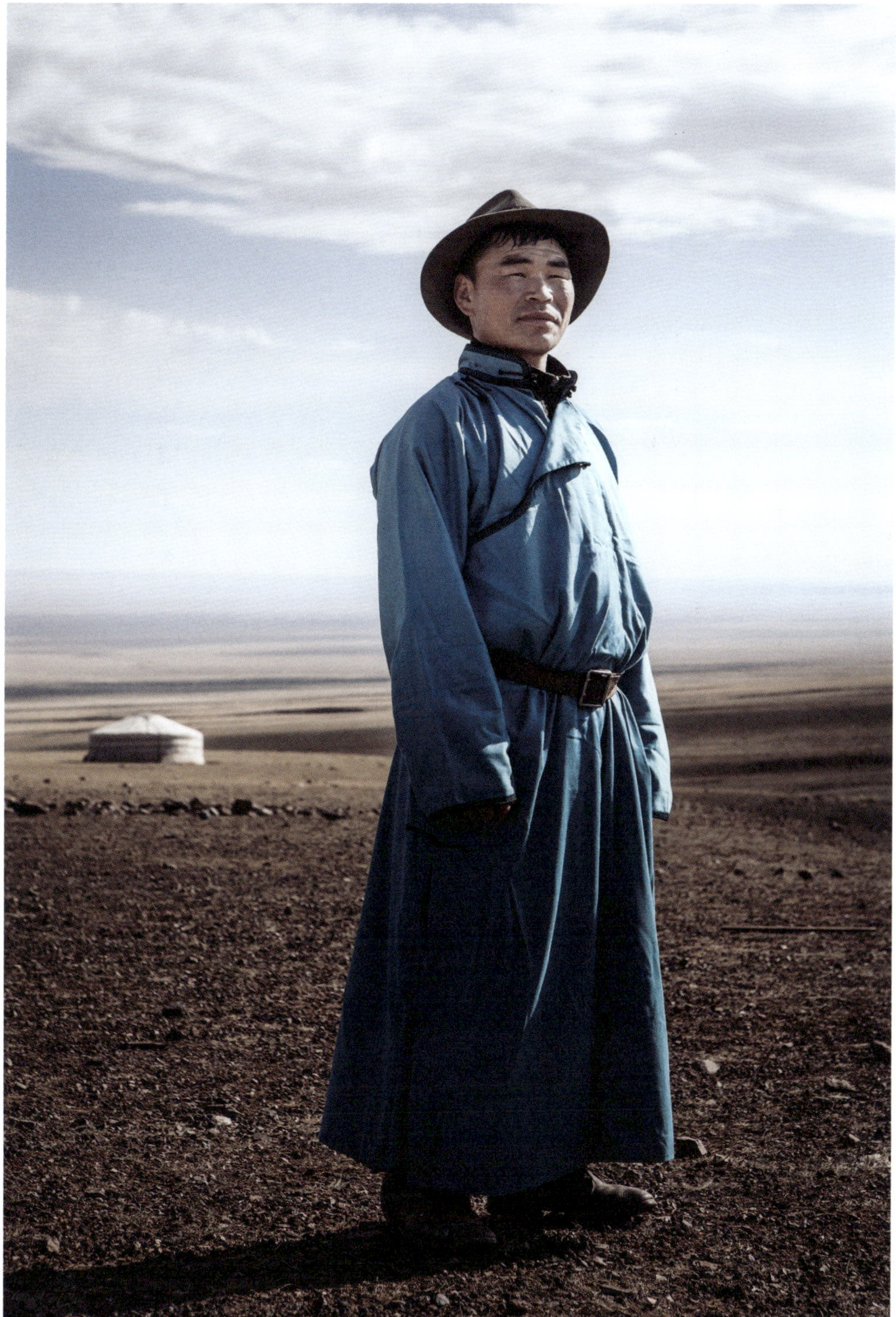

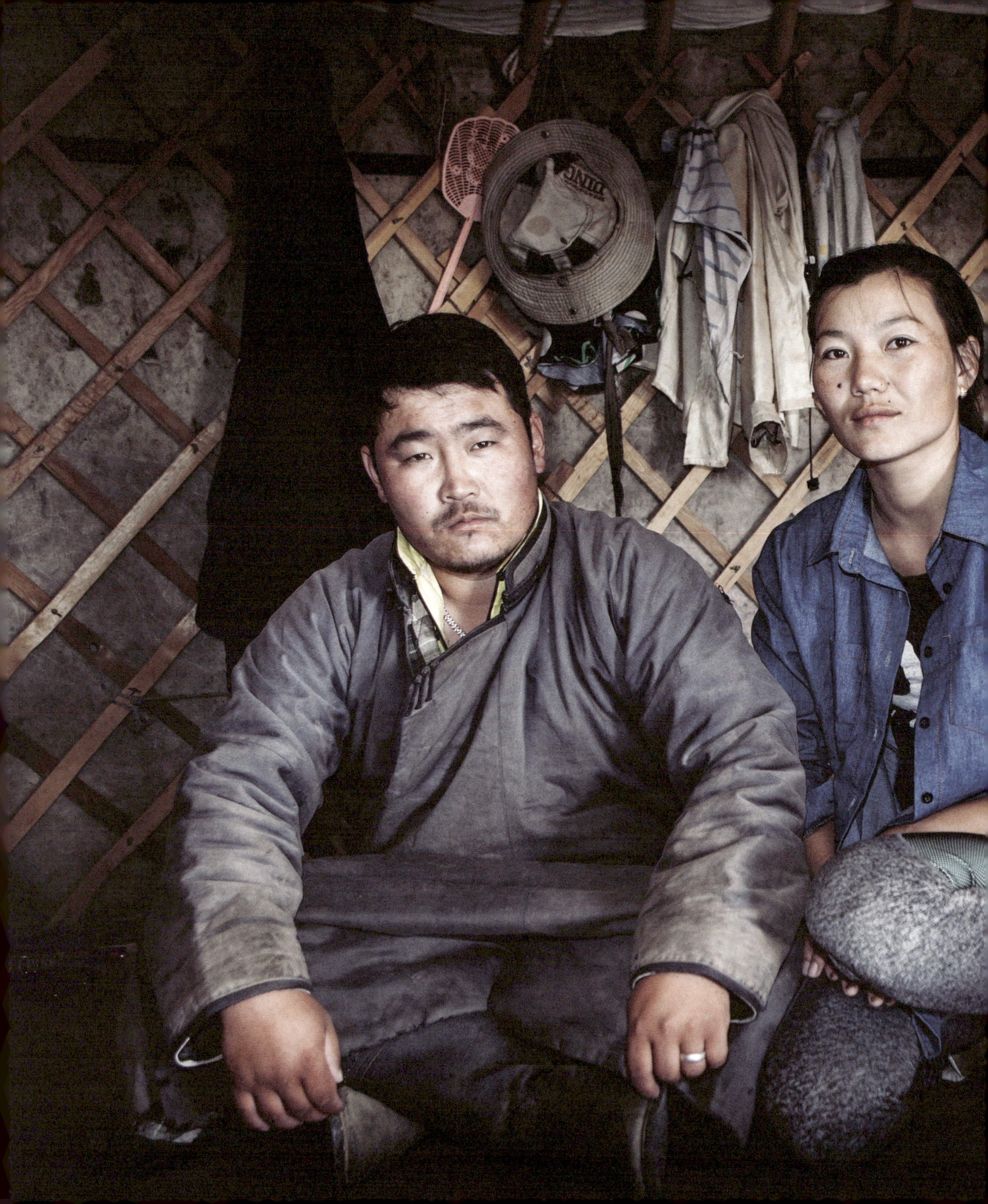

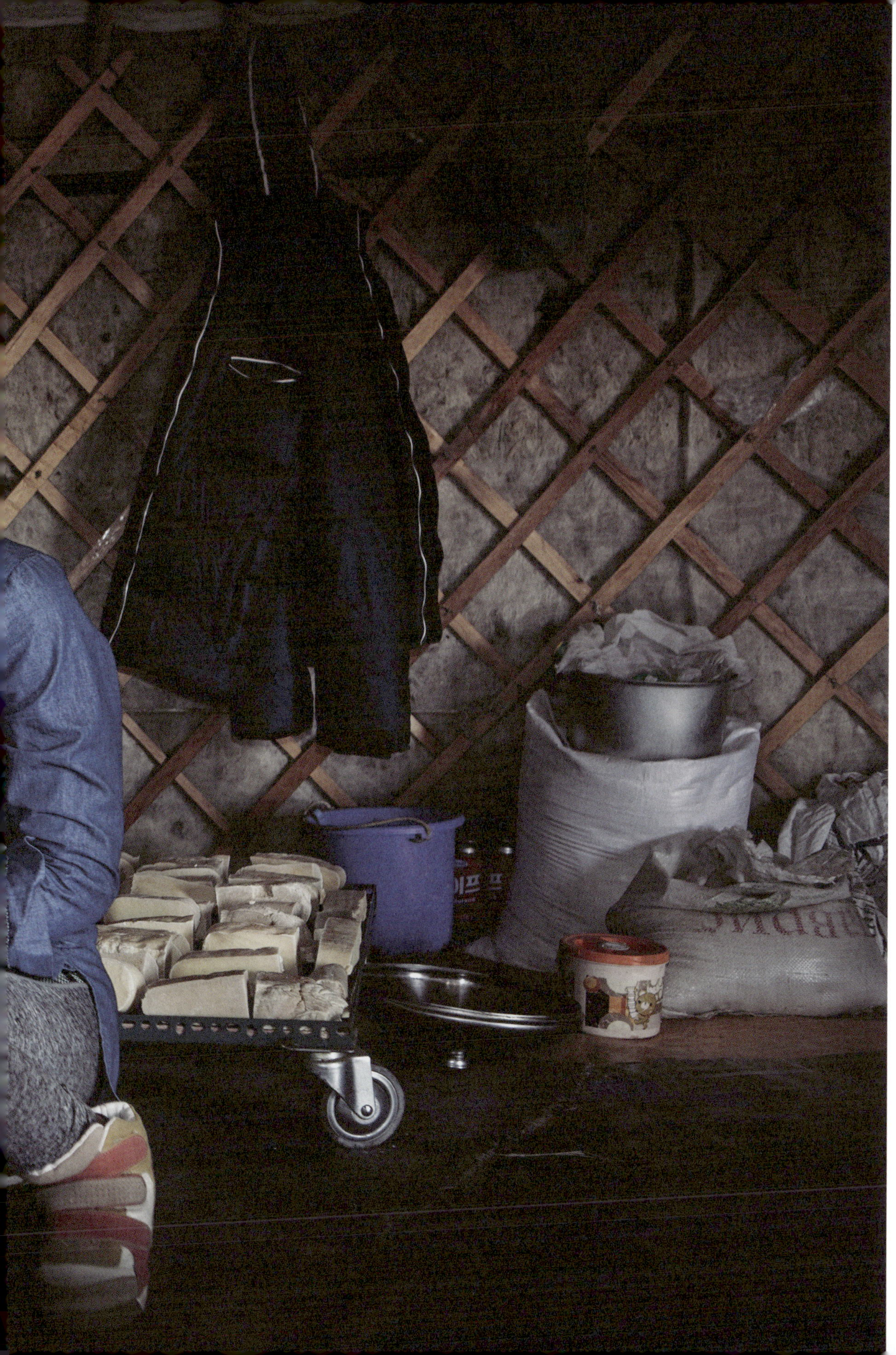

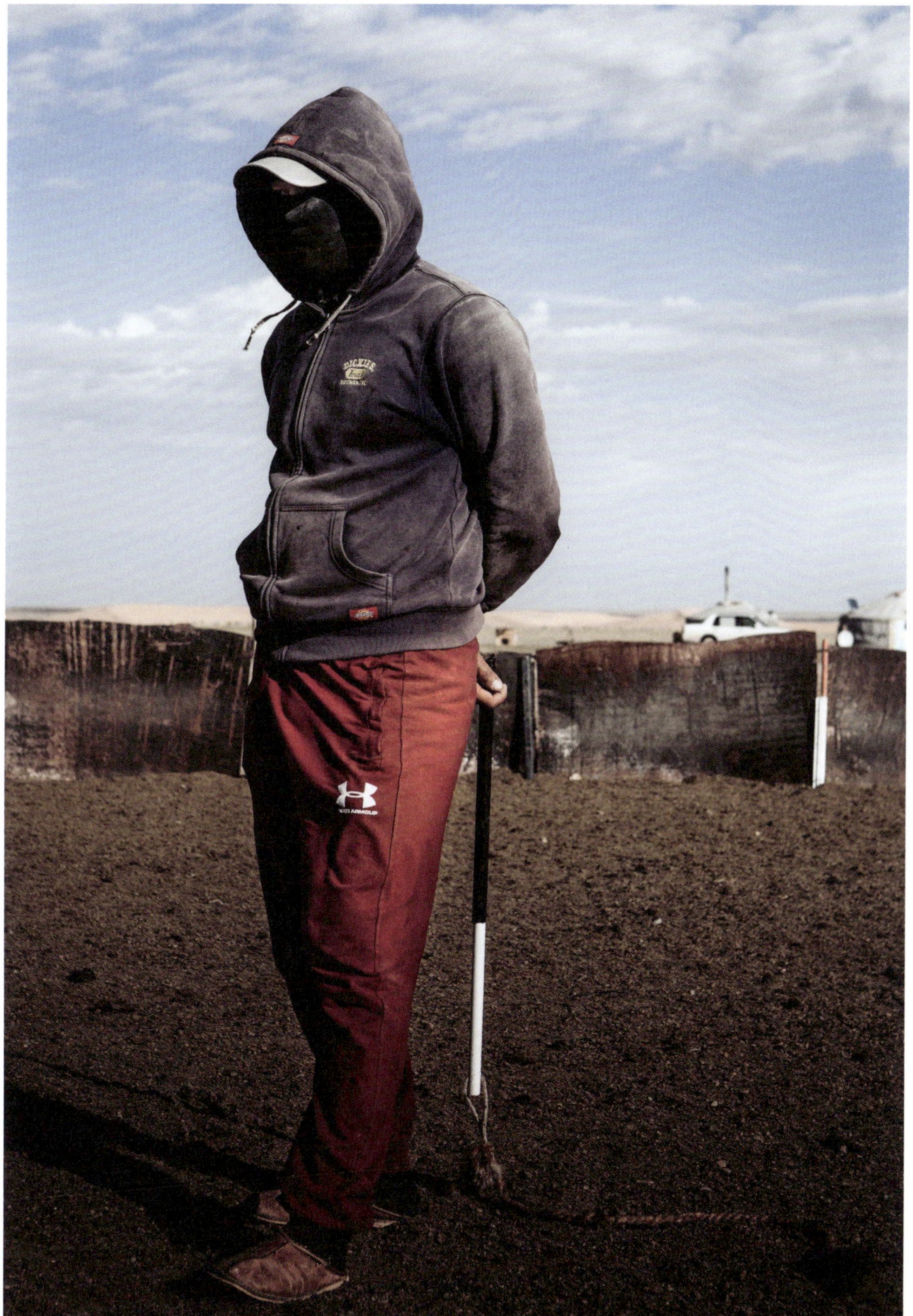

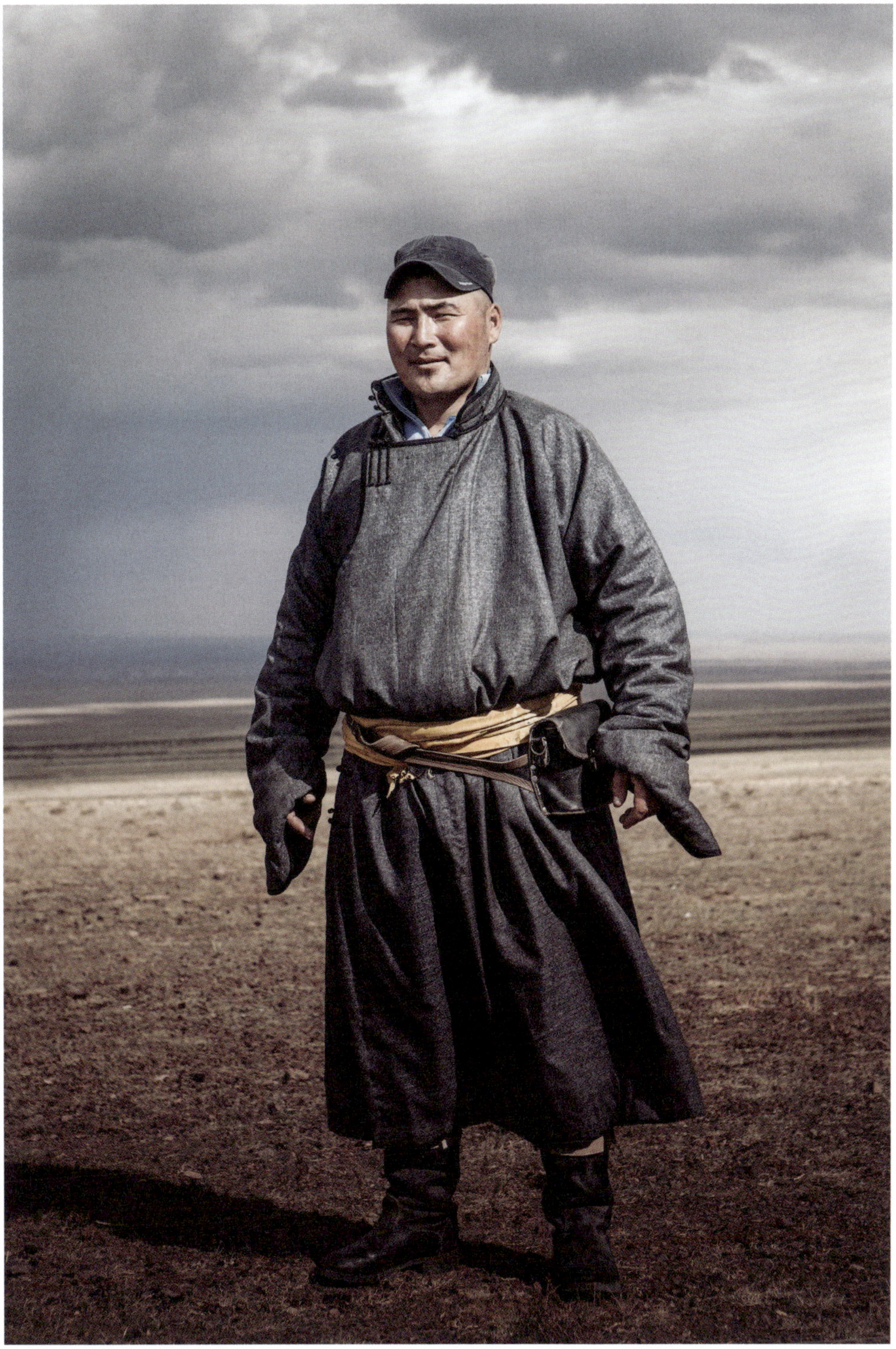

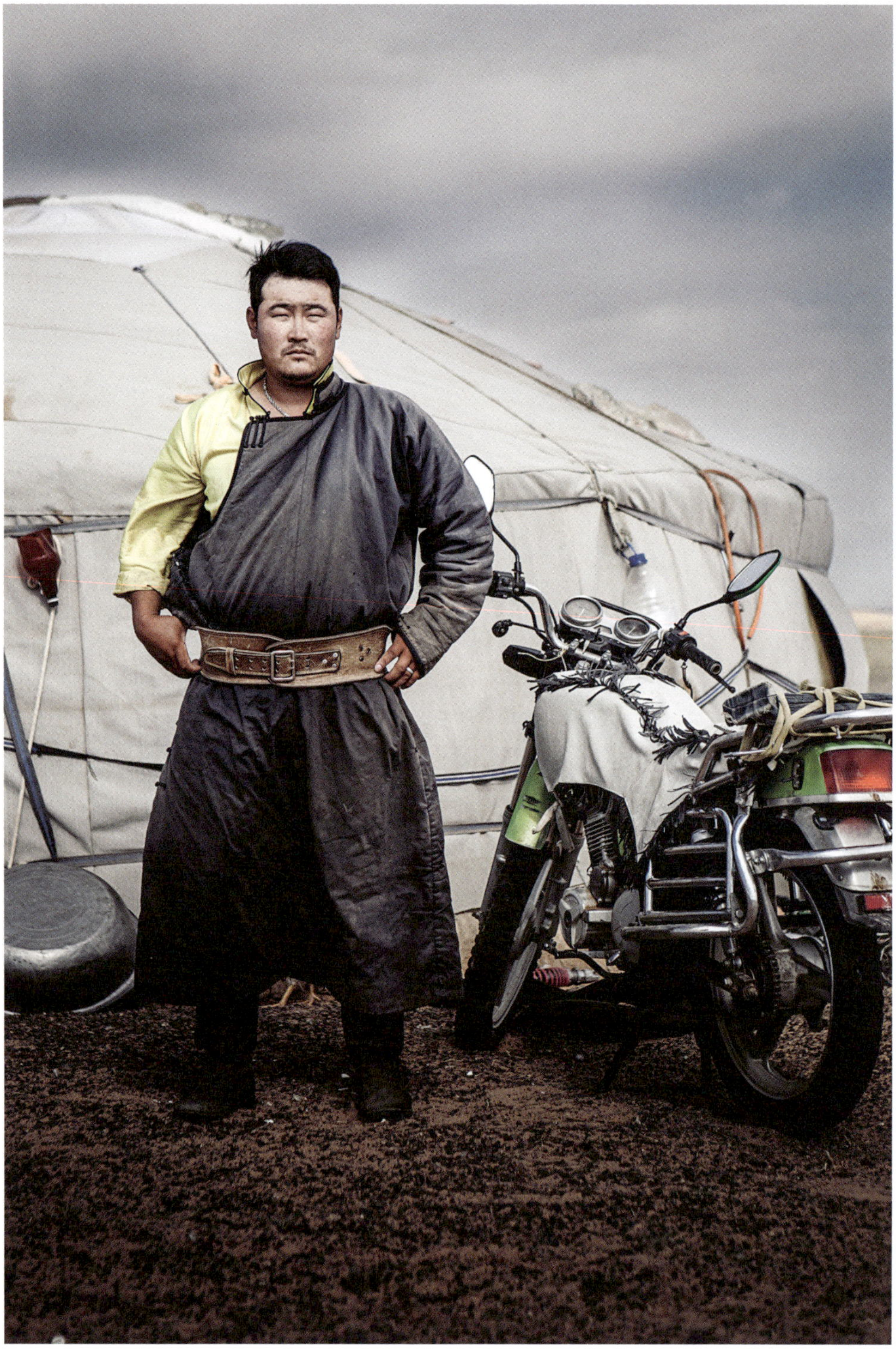

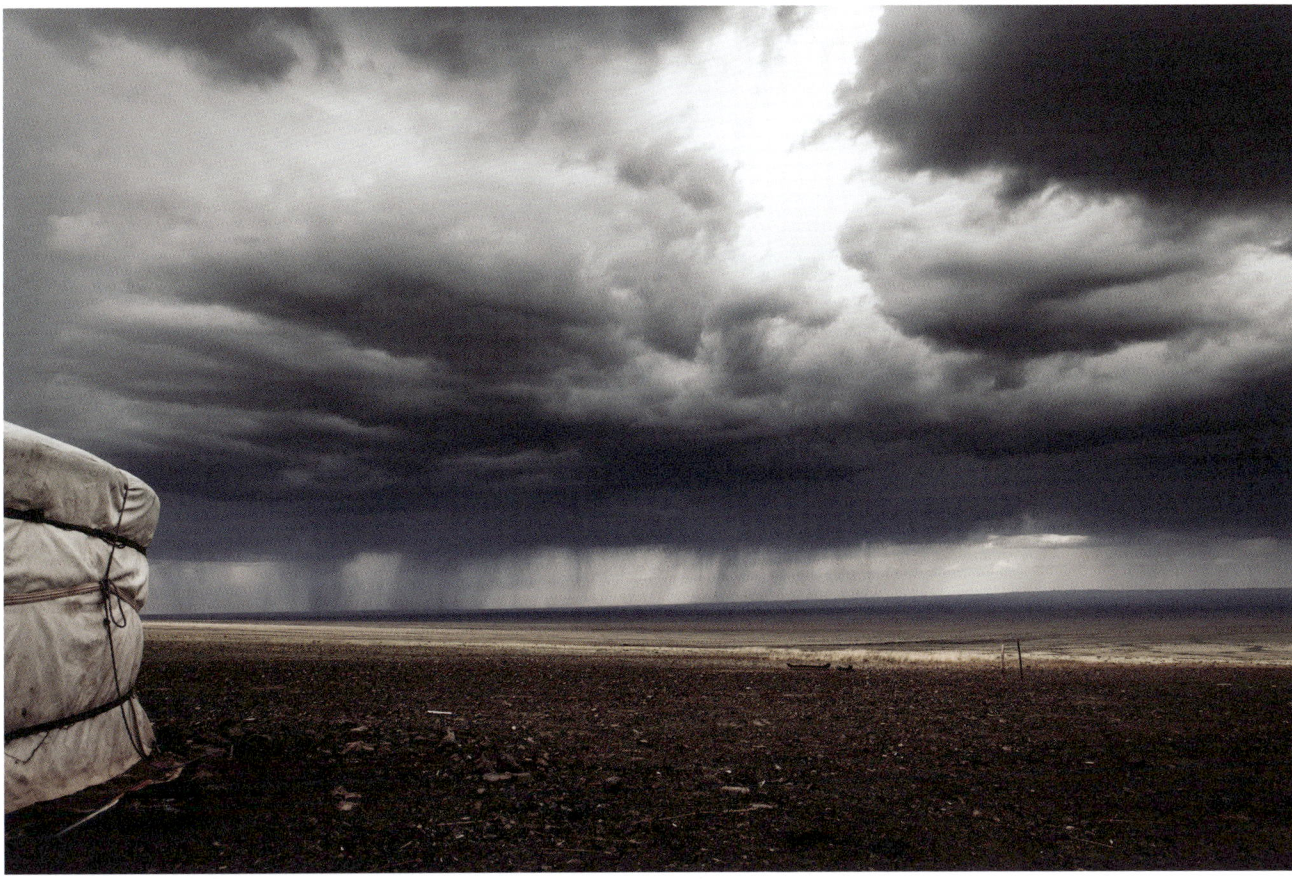

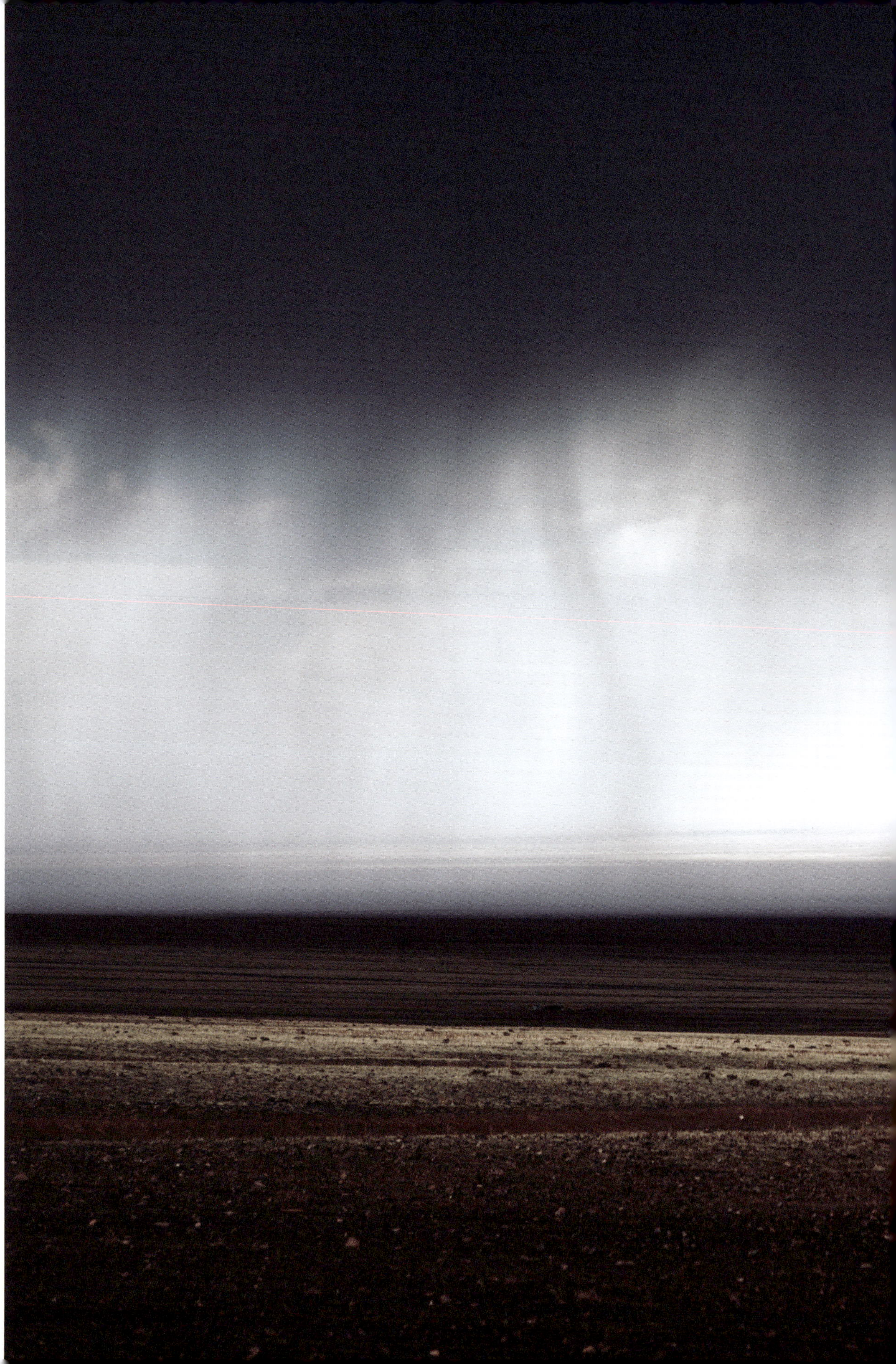

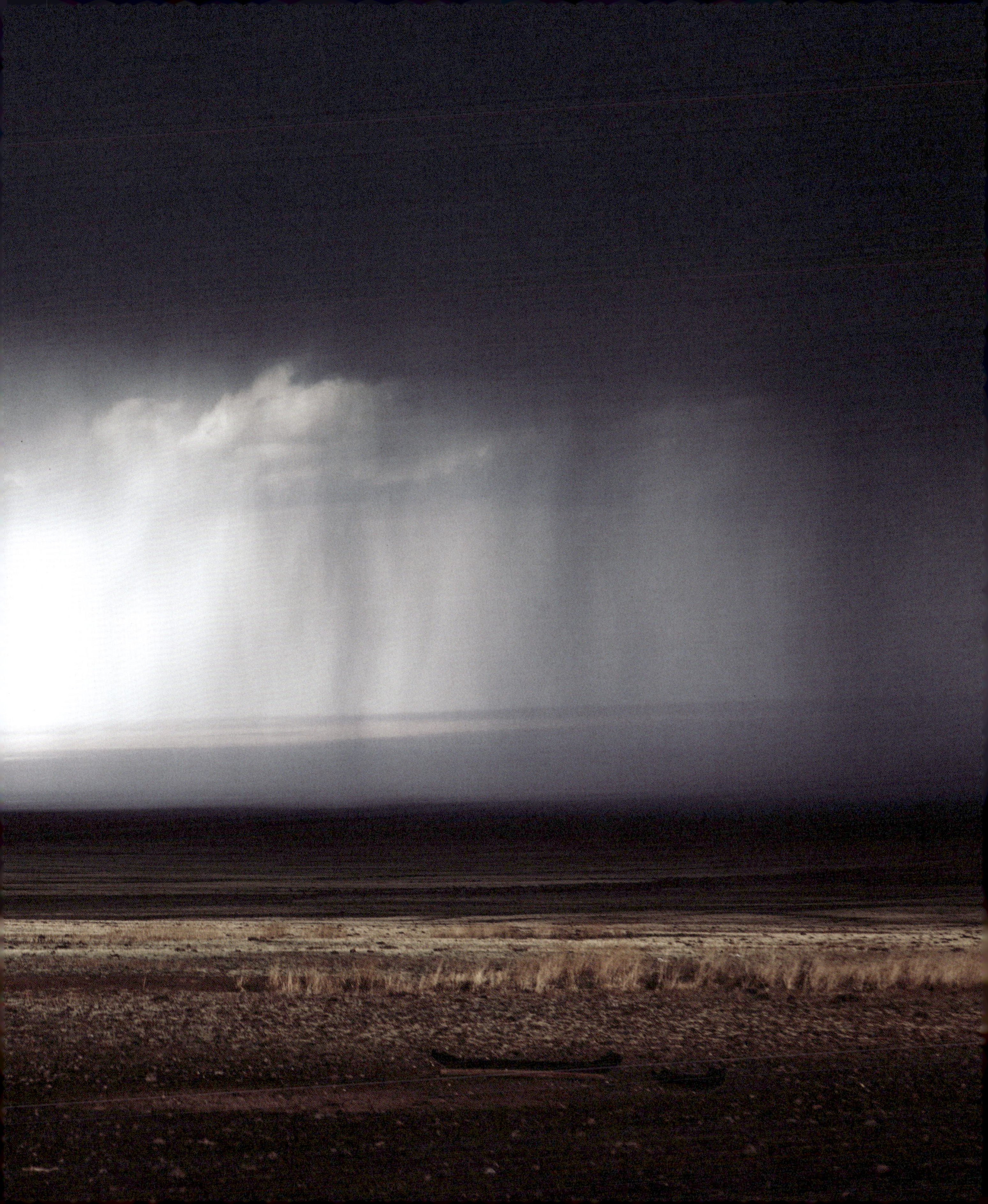

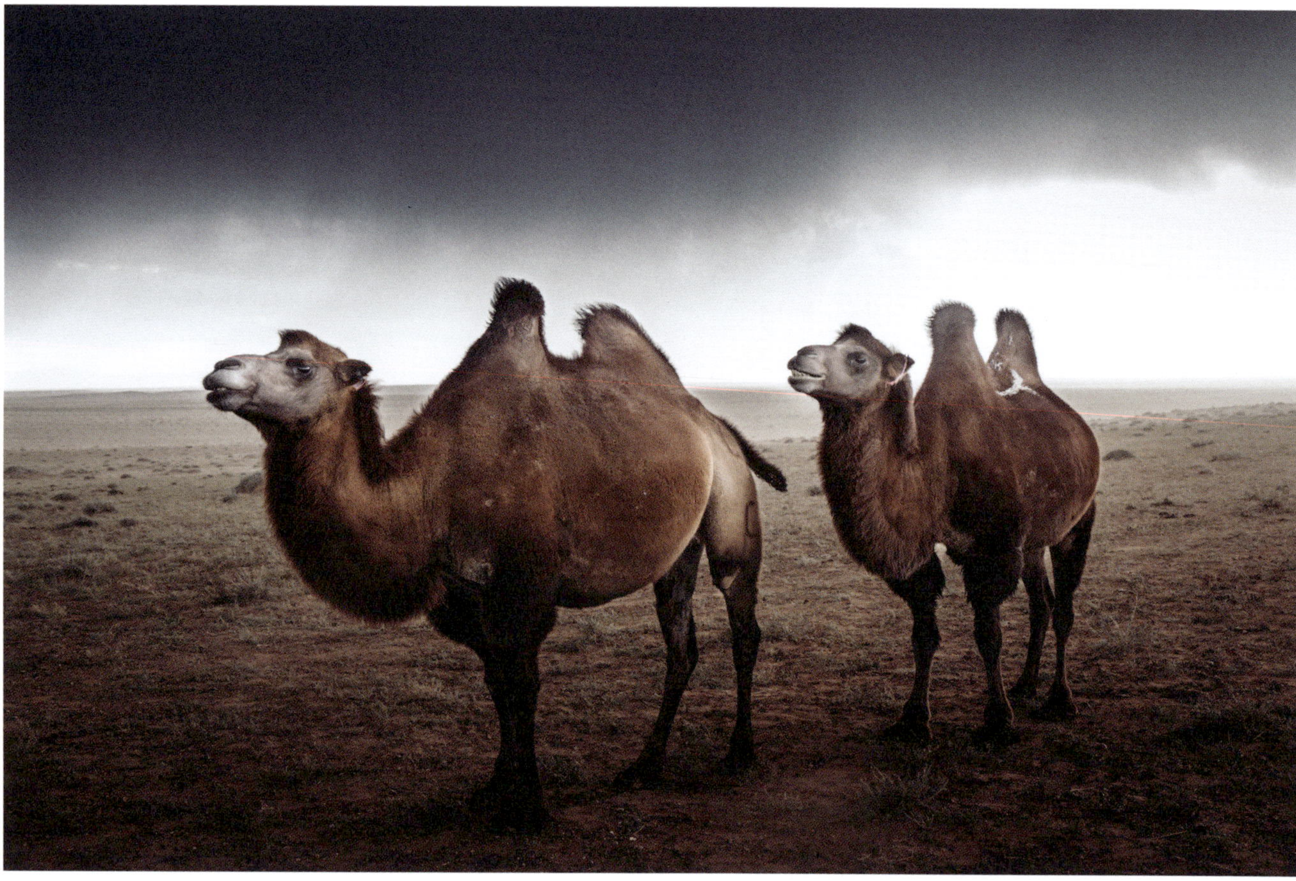

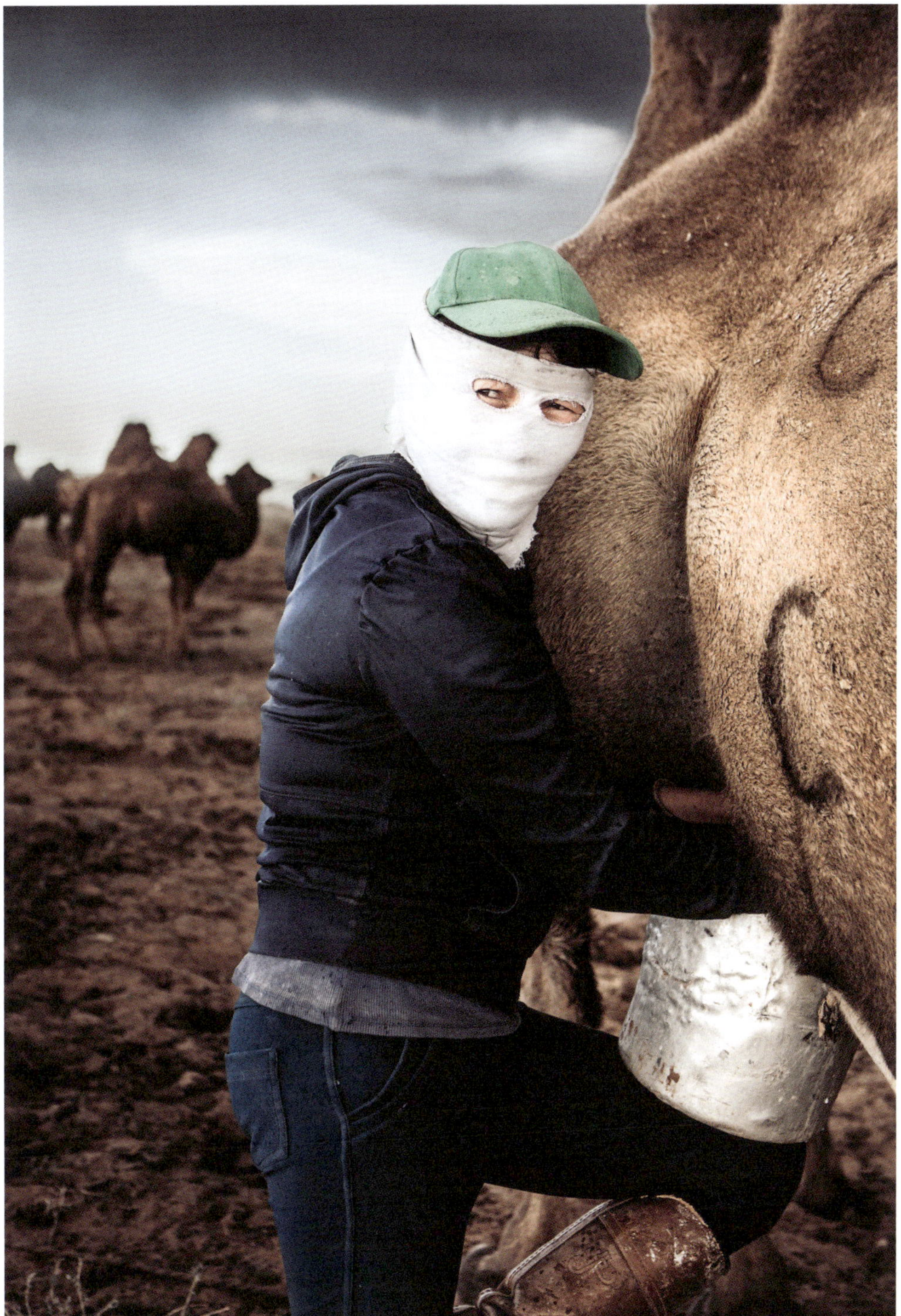

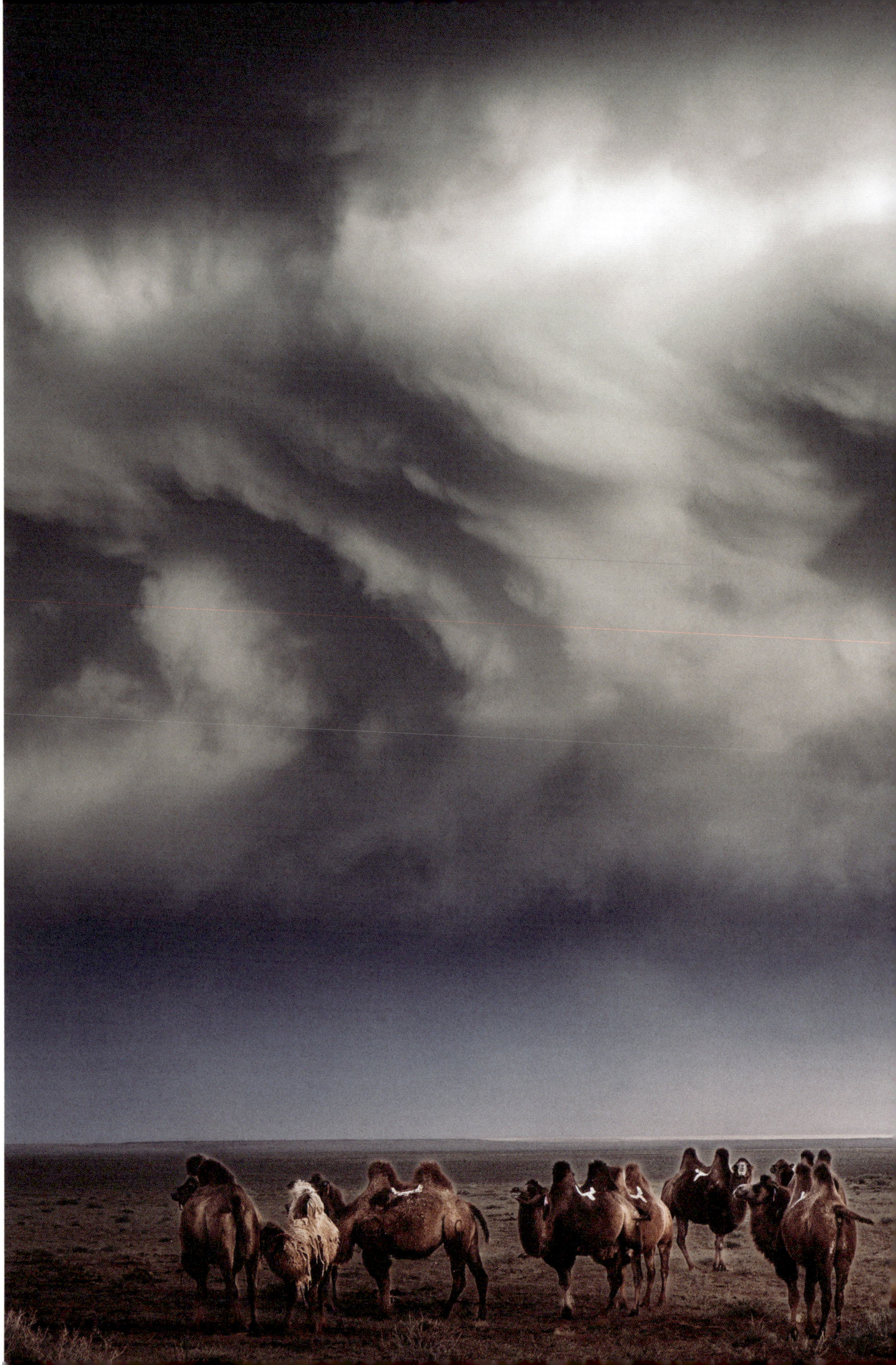

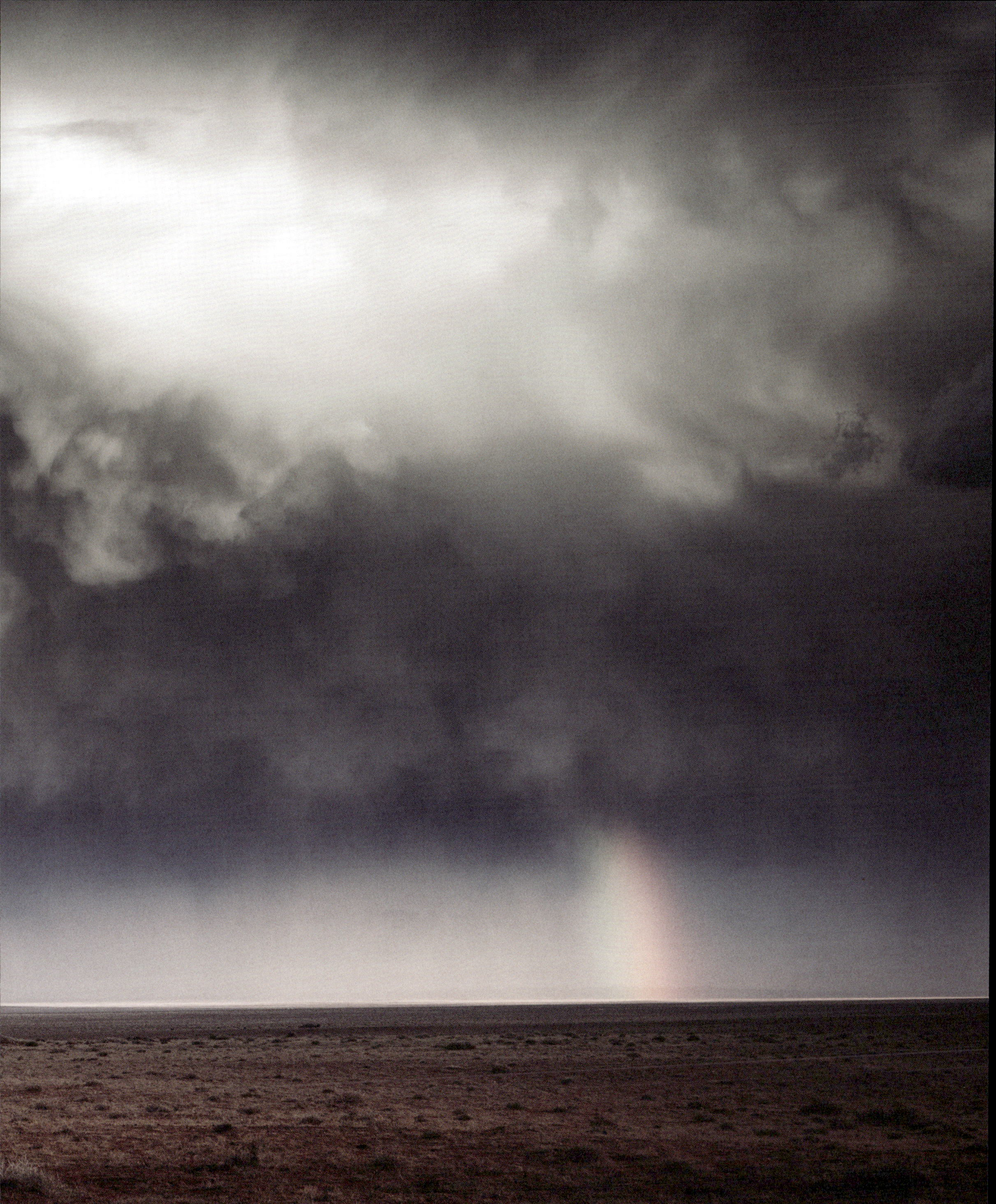

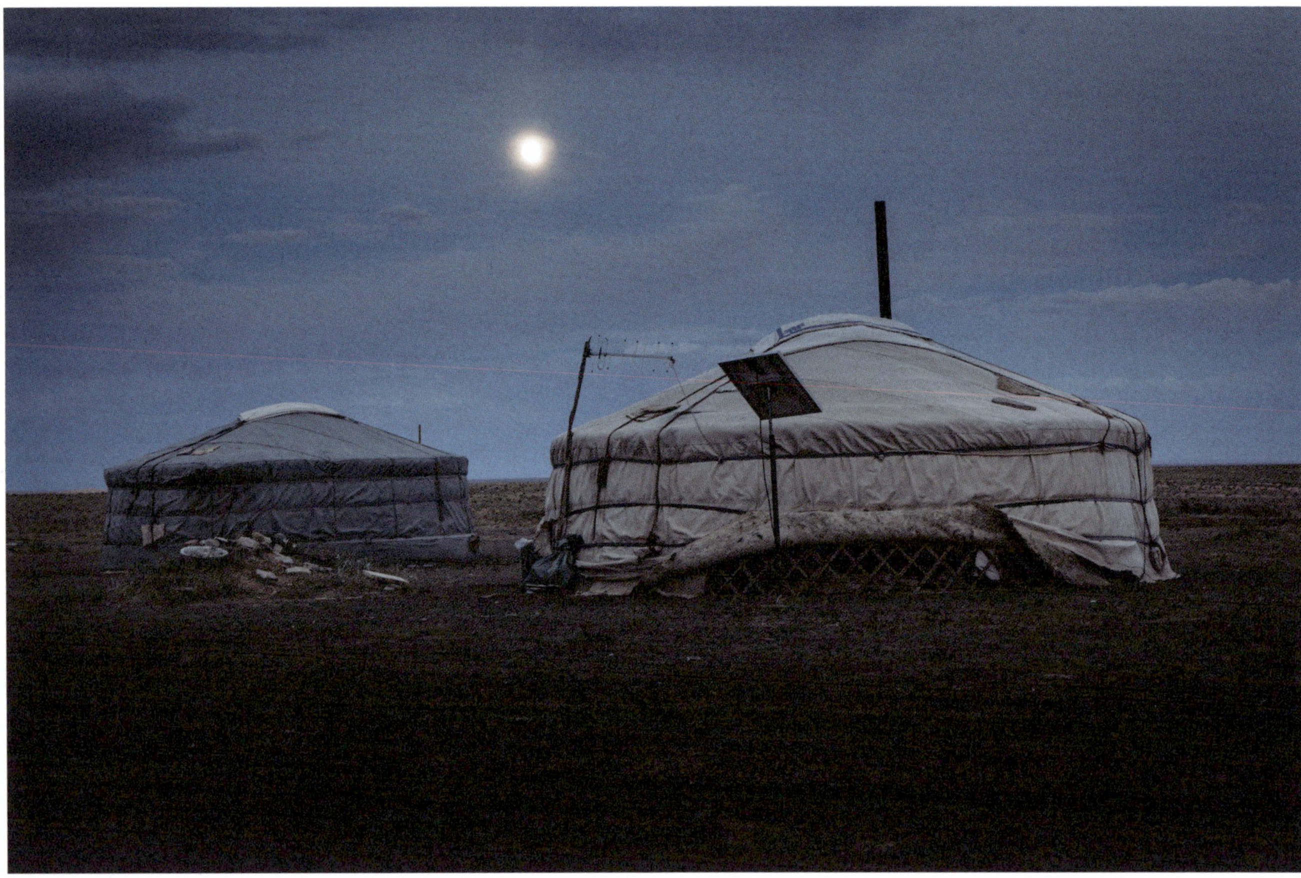

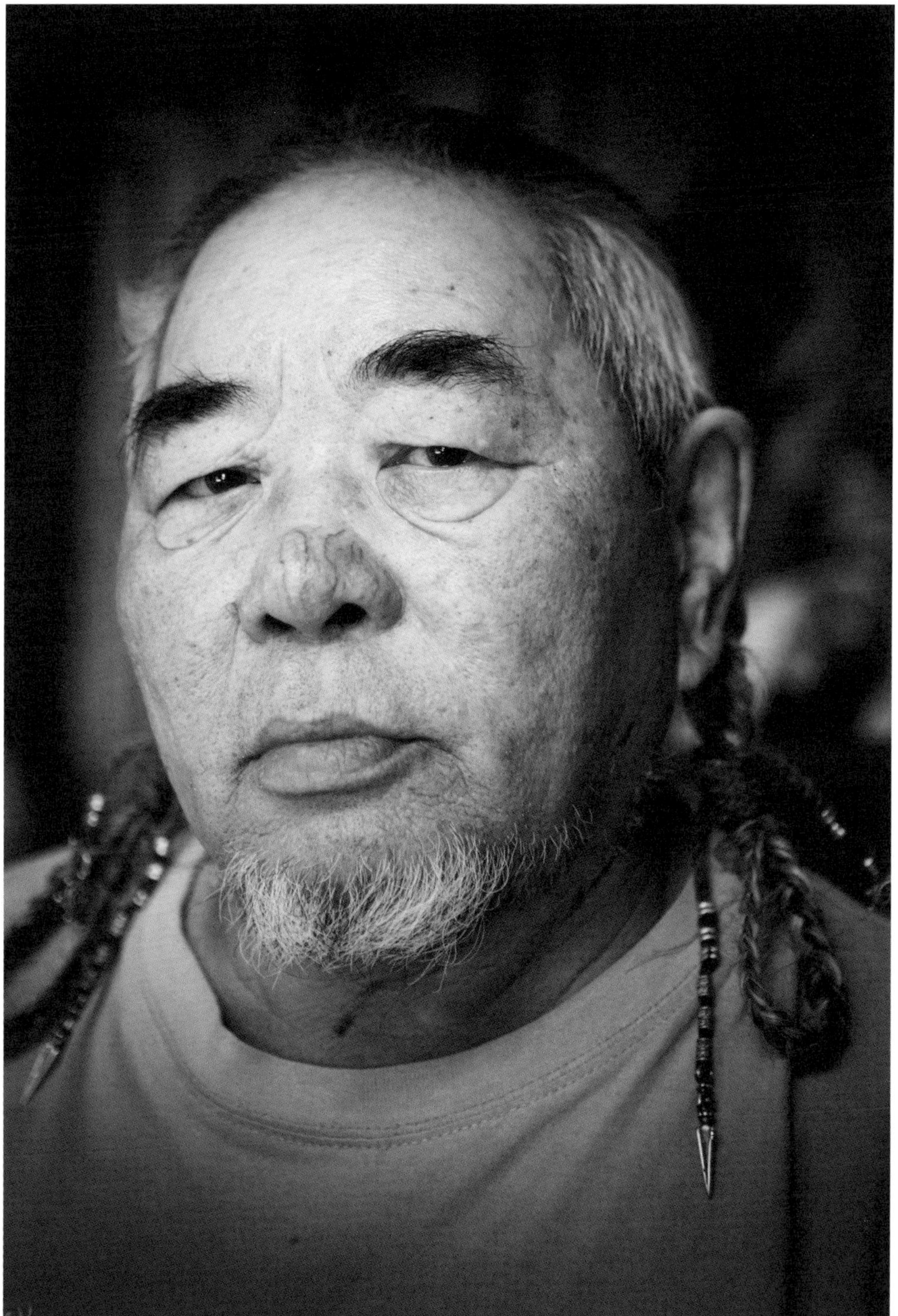

OCEANIA

KIRIBATI

1° 28′ 45.448″ N 173° 1′ 49.048″ E

KIRIBATI

I visited the I-Kiribati in the village of Tabonibara on the main island of Tarawa.

Kiribati, an island nation in the South Pacific, is made up of 32 coral atolls, comparted in three island groups, and one island. Of these, 21 islands are inhabited by around 120,000 people in total. The equator and the 180th meridian run through Kiribati, making it the only country in the world to lie in the northern, southern, western and eastern hemisphere at the same time. Kiribati was colonized around 2000–3000 years ago by Micronesians, who were subsequently mixed with Polynesian influences through further smaller immigrations. This resulted in a culturally relatively homogeneous population that still makes up over 96 percent of Kiribati's inhabitants today. Their members call themselves "I-Kiribati" in both singular and plural. Part of the national territory was first set foot on by Europeans in 1606 and named the Gilbert Islands around 1820. The islands became a British protectorate in 1892 and a British colony in 1916.

With the first European settlers, the Christian missionization of the islanders began in the mid-19th century. However, there is still a close connection to the gods and ancestors, especially in the remote villages.

The country has hardly any natural resources. Originally existing phosphate deposits were completely exploited during the period of British colonization and were already exhausted at the time of Kiribati's independence in 1979. Today, the allocation of fishing rights, the catching and processing of fish and the production of copra from coconuts are the most important sources of income.

Kiribati is particularly threatened by climate change. Most of the country's territory is less than two meters above sea level, which is rising particularly quickly in the South Pacific. According to current studies, the island state is likely to be uninhabitable by 2050 and completely submerged by 2070. In preparation for a possibly necessary resettlement in the future, the state purchased around 24 square

KIRIBATI

Hier besuchte ich die I-Kiribati im Dorf Tabonibara auf der Hauptinsel Tarawa.

Kiribati, ein Inselstaat im Südpazifik, setzt sich aus 32 Korallenatollen, verteilt auf drei Inselgruppen, und einer weiteren Insel zusammen. 21 der Inseln werden von insgesamt 120.000 Menschen bewohnt. Durch Kiribati verlaufen der Äquator und auch der 180. Längengrad; damit liegt es als einziger Staat der Welt zugleich in der nördlichen, südlichen, westlichen und östlichen Hemisphäre. Besiedelt wurde Kiribati vor etwa 2000–3000 Jahren durch Mikronesier:innen, in die sich folgend durch weitere kleinere Einwanderungen polynesische Einflüsse mischten. So entstand eine kulturell relativ homogene Bevölkerung, die heute noch über 96 Prozent der Einwohner Kiribatis ausmacht. Ihre Angehörigen nennen sich „I-Kiribati", sowohl im Singular als auch im Plural. Ein Teil des Staatsgebiets wurde erstmals 1606 von Europäern betreten und um 1820 Gilbert-Inseln genannt. 1892 wurden die Inseln britisches Protektorat, ab 1916 britische Kolonie.

Mit den ersten europäischen Siedlern begann ab Mitte des 19. Jahrhunderts die christliche Missionierung der Inselbewohner. Vor allem in den weit abgelegenen Dörfern existiert jedoch noch heute ein enge Verbindung zu den traditionellen Göttern und Ahnen.

Der Staat verfügt kaum über natürliche Ressourcen. Ursprünglich vorhandene Phosphatdepots wurden während der Zeit der britischen Kolonialisierung vollständig ausgebeutet und waren zum Zeitpunkt der staatlichen Unabhängigkeit Kiribatis 1979 bereits erschöpft. Heute bilden die Vergabe von Fischereirechten, der Fang und die Verarbeitung von Fisch sowie die Produktion von Kopra aus Kokosnüssen die wichtigsten Einnahmequellen.

Kiribati ist in besonderem Maße vom Klimawandel bedroht. Der größte Teil des Staatsgebiets liegt nicht einmal zwei Meter über dem Meeresspiegel, und im Südpazifik erhöht sich dieser besonders schnell. Aktuellen Studien zufolge dürfte

kilometers of land on the island of Vanua Levu, which is around 2,000 kilometers away and belongs to Fiji, back in 2012. The country, which is around 3 hours away by plane, is the only one that has so far agreed to take in climate refugees from Kiribati if necessary.

Rising sea levels are already causing the freshwater bubbles under the islands to become salty, which is leading to the death of the coconut palms. The coral reefs that protect the islands are also under threat. Coral bleaching is closely linked to water temperature. An IPCC report on the consequences of the climate crisis predicts that tropical coral reefs will be almost completely lost with a warming of just 2° Celsius. However, according to the USA's National Oceanic and Atmospheric Administration (NOAA), more than half a billion people worldwide depend on coral reefs for food, income or, like the inhabitants of Kiribati, to protect their coasts.

der Inselstaat bis 2050 unbewohnbar und bis 2070 vollständig versunken sein. In Vorbereitung einer möglicherweise notwendigen Umsiedelung in der Zukunft hat der Staat bereits im Jahr 2012 etwa 24 Quadratkilometer Land auf der rund 2.000 Kilometer entfernten Insel Vanua Levu, die zu Fidschi gehört, gekauft. Das etwa 3 Flugstunden entfernte Land hat sich bisher als einziger Staat bereiterklärt, bei Bedarf Klimaflüchtlinge aus Kiribati aufzunehmen.

Bereits heute beginnen durch den ansteigenden Meeresspiegel die Süßwasserblasen unter den Inseln zu versalzen, was zu einem Absterben der Kokospalmen führt. Auch die den Inseln vorgelagerten und sie schützenden Korallenriffe sind bedroht. Die Korallenbleiche hängt eng mit der Wassertemperatur zusammen. Ein Bericht des IPCC zu den Folgen der Klimakrise rechnet schon bei 2° Celsius Erwärmung mit einem beinahe vollständigen Verlust der tropischen Korallenriffe. Weltweit sind laut einer Analyse der Ozeanbehörde der USA (NOAA) jedoch mehr als eine halbe Milliarde Menschen für Nahrung, Einkommen oder, wie die Einwohner Kiribatis, für den Schutz ihrer Küsten auf Korallenriffe angewiesen.

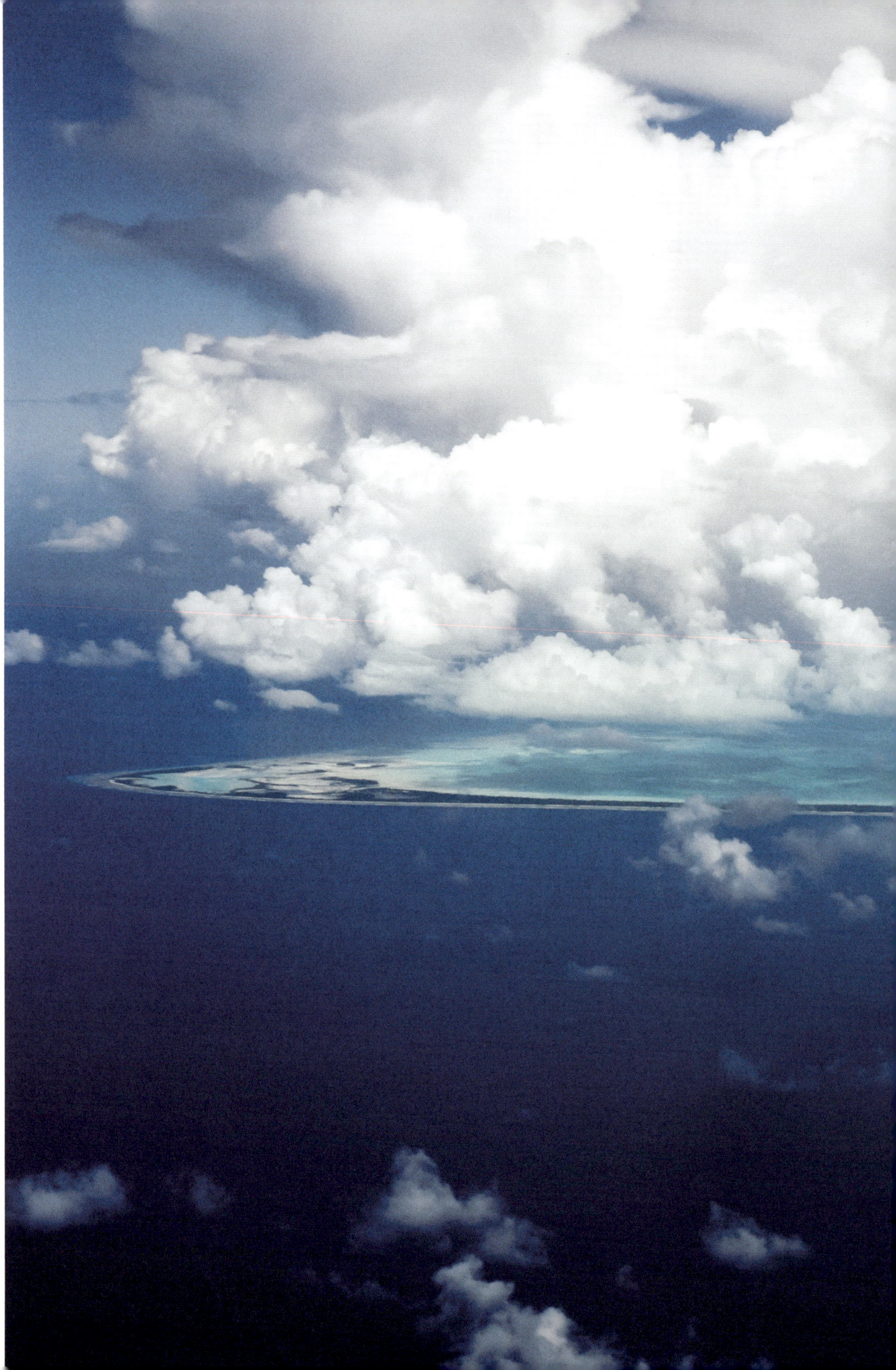

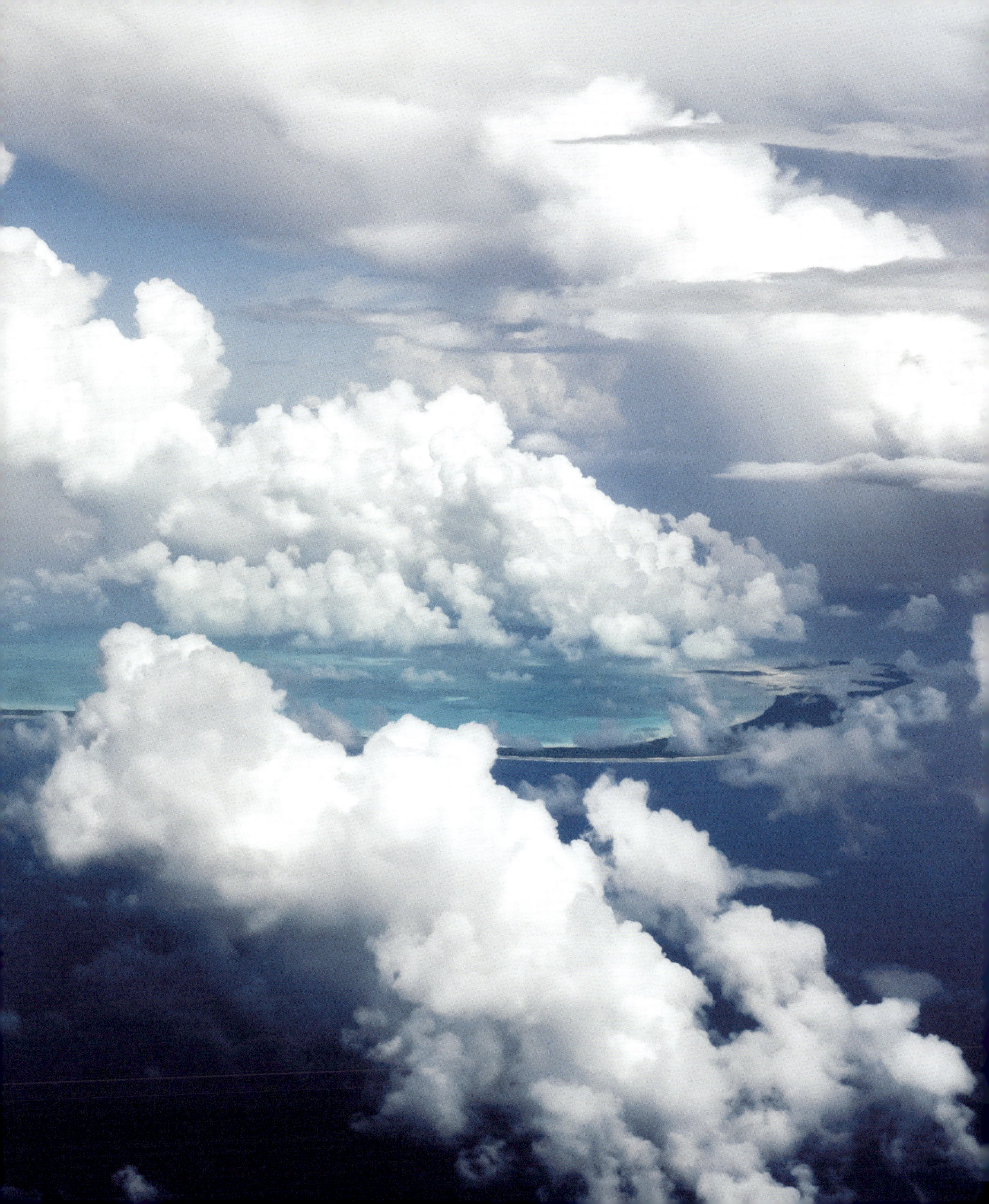

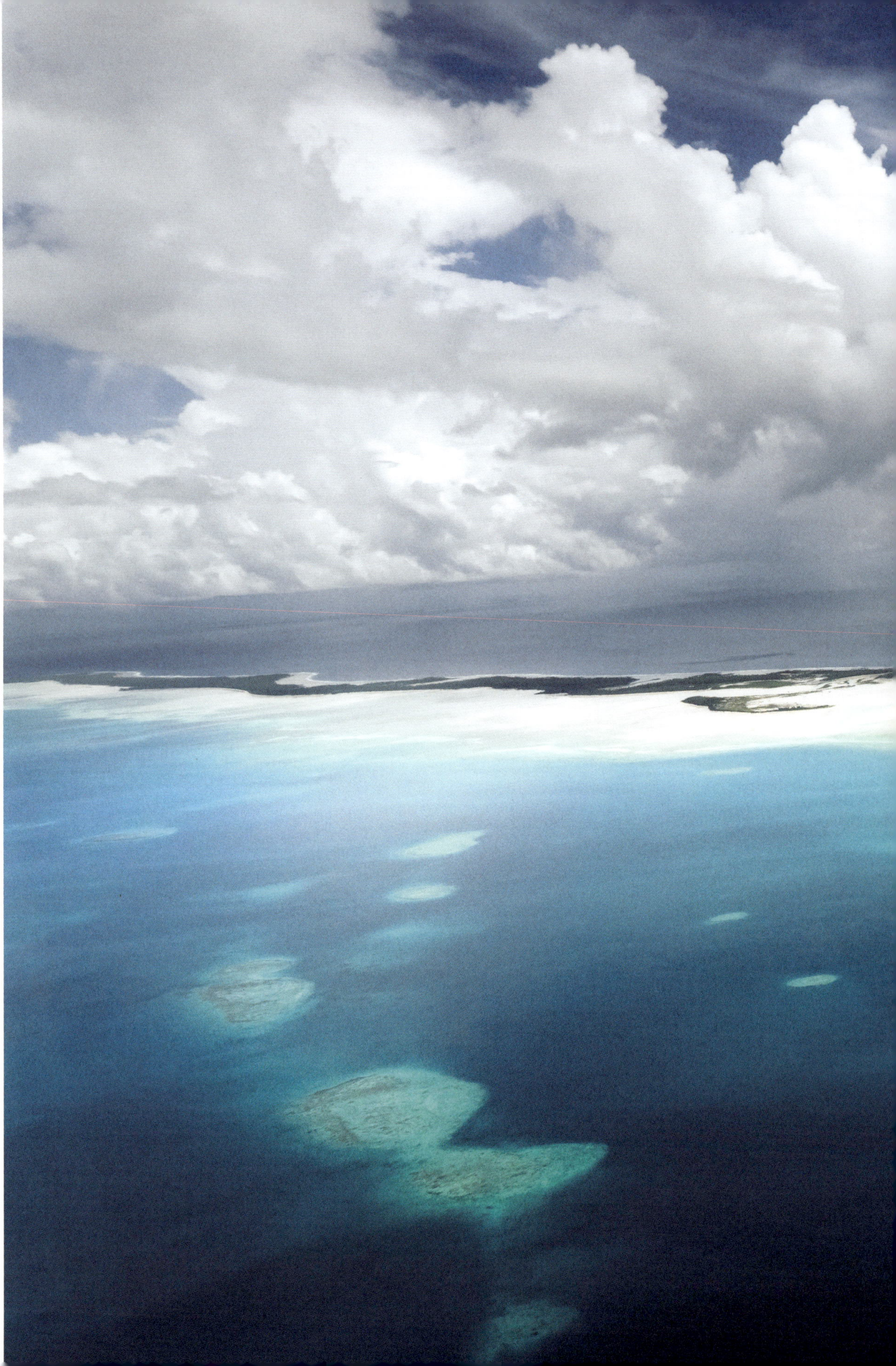

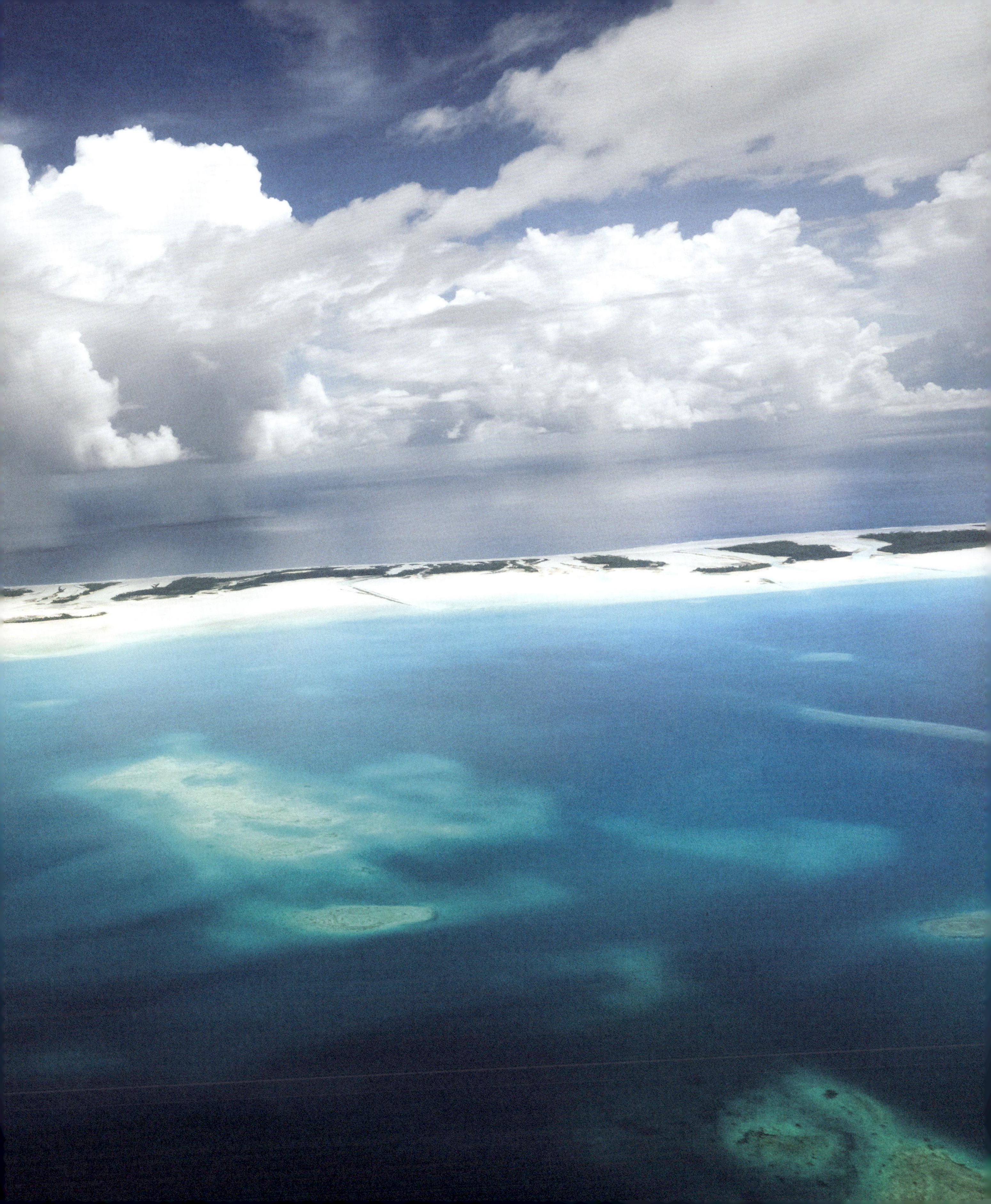

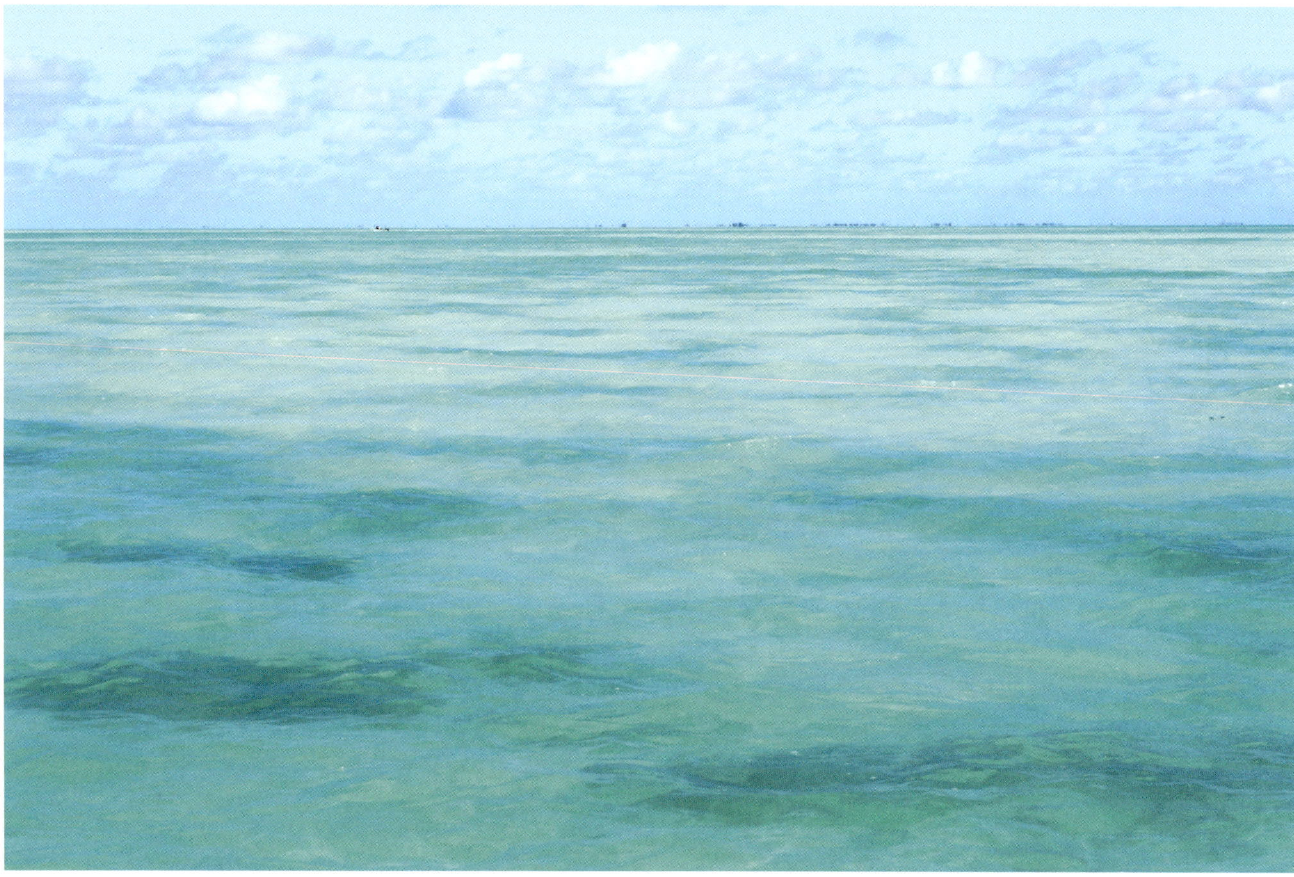

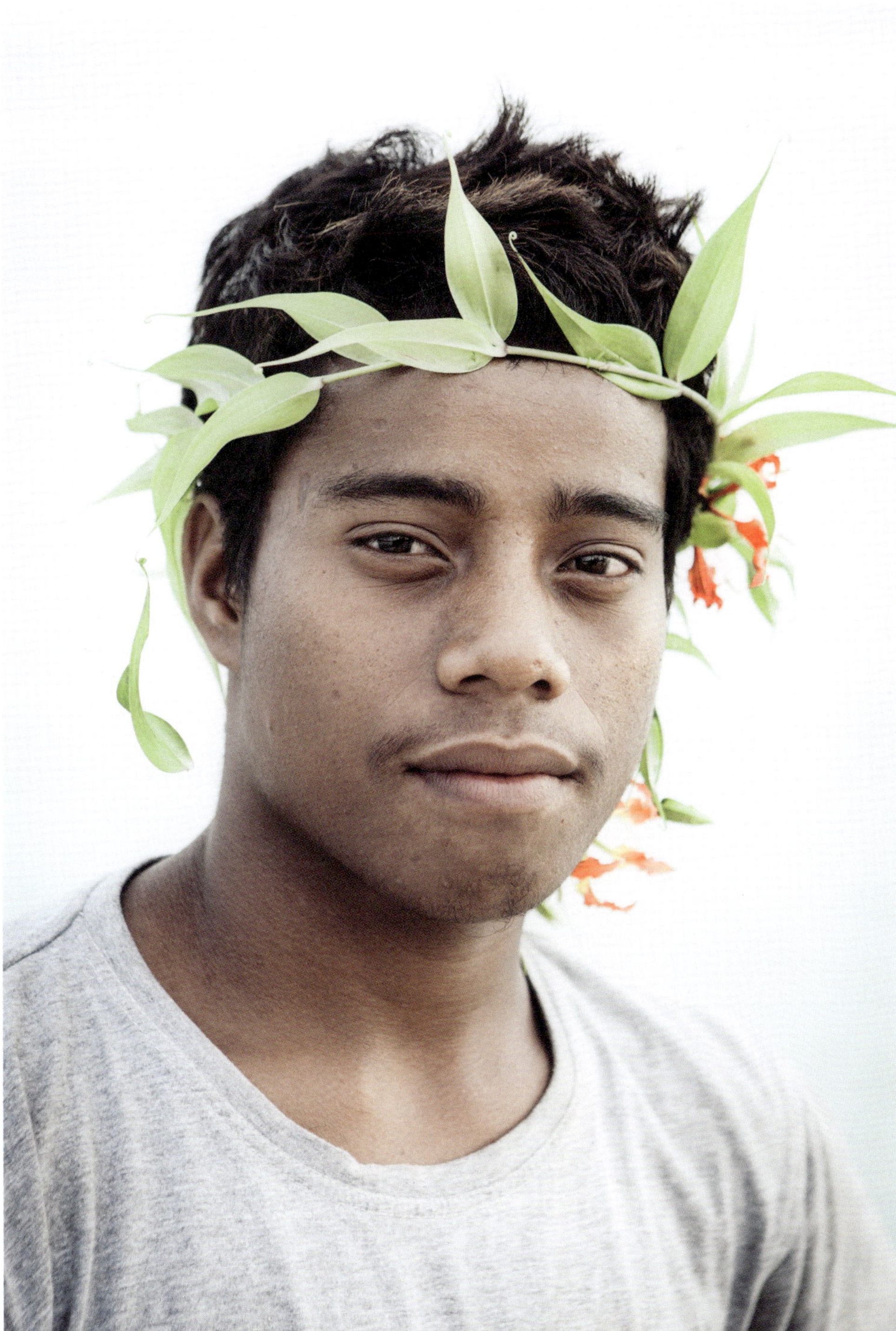

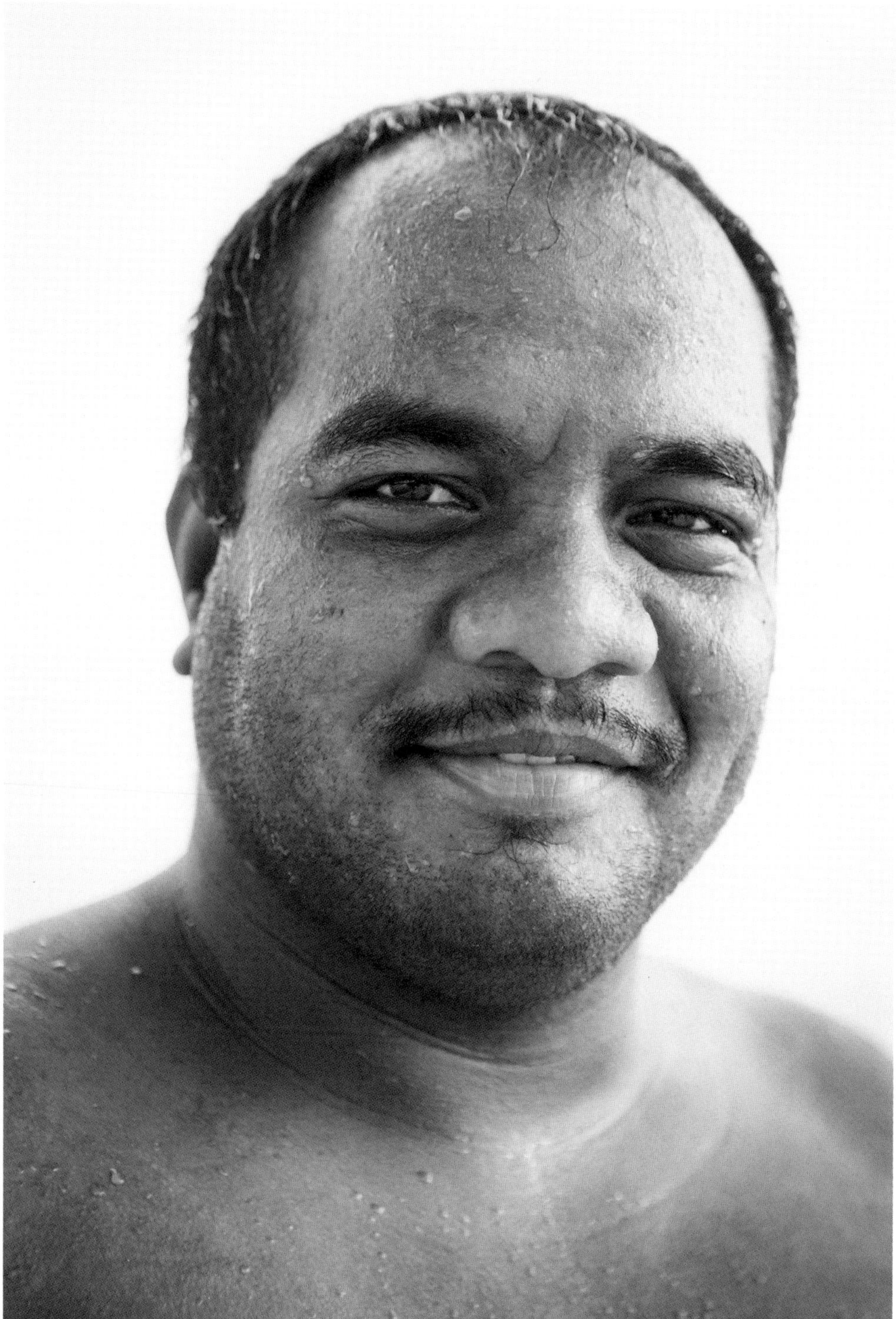

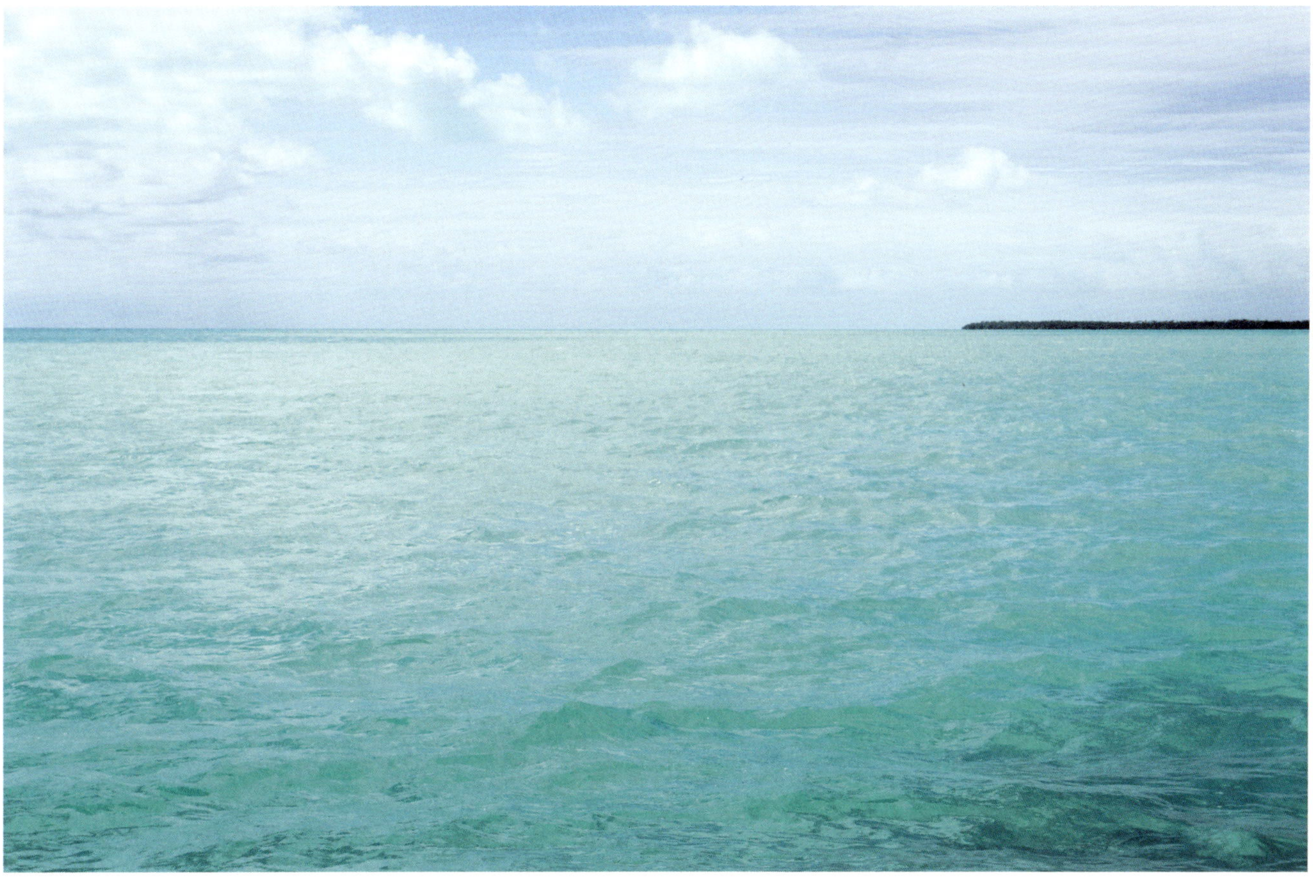

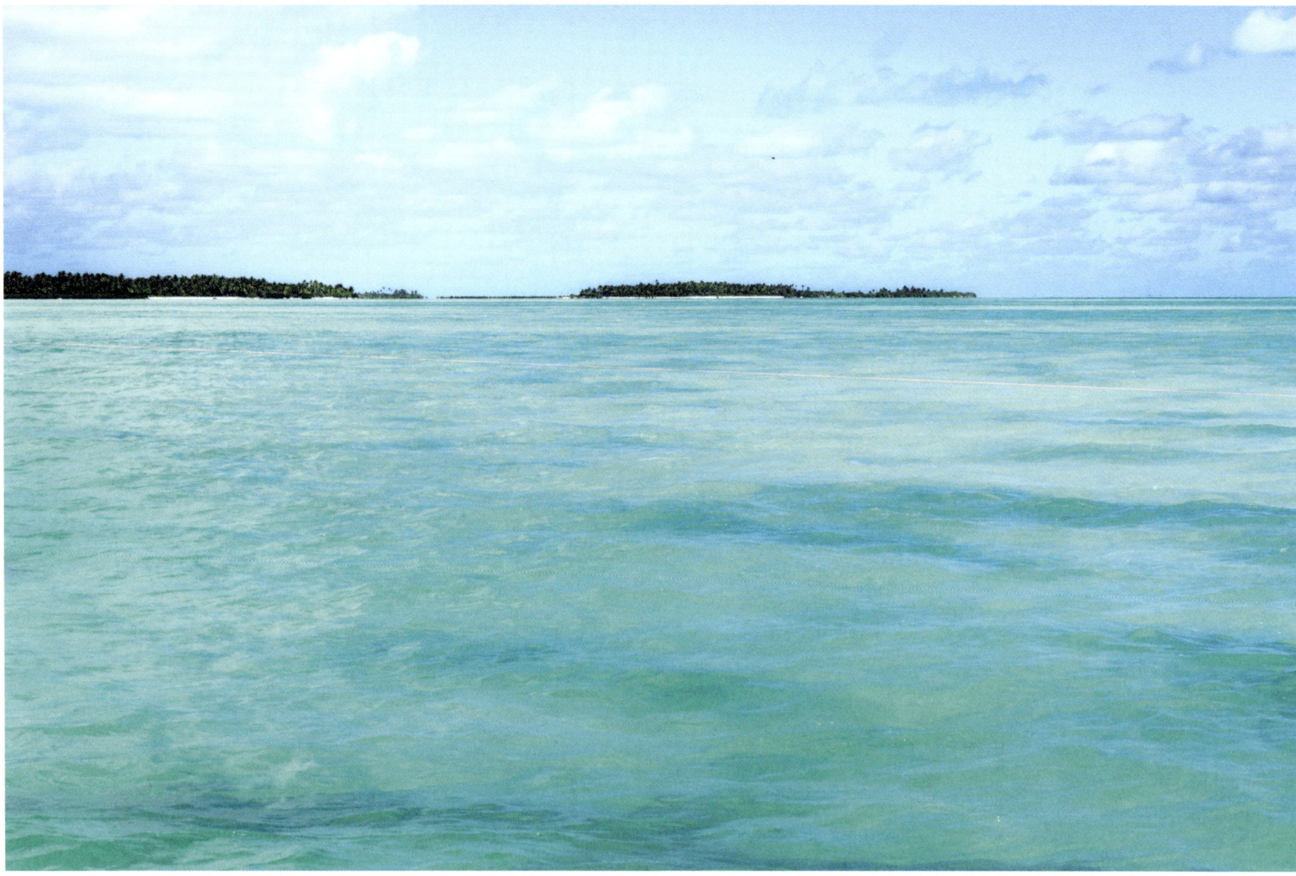

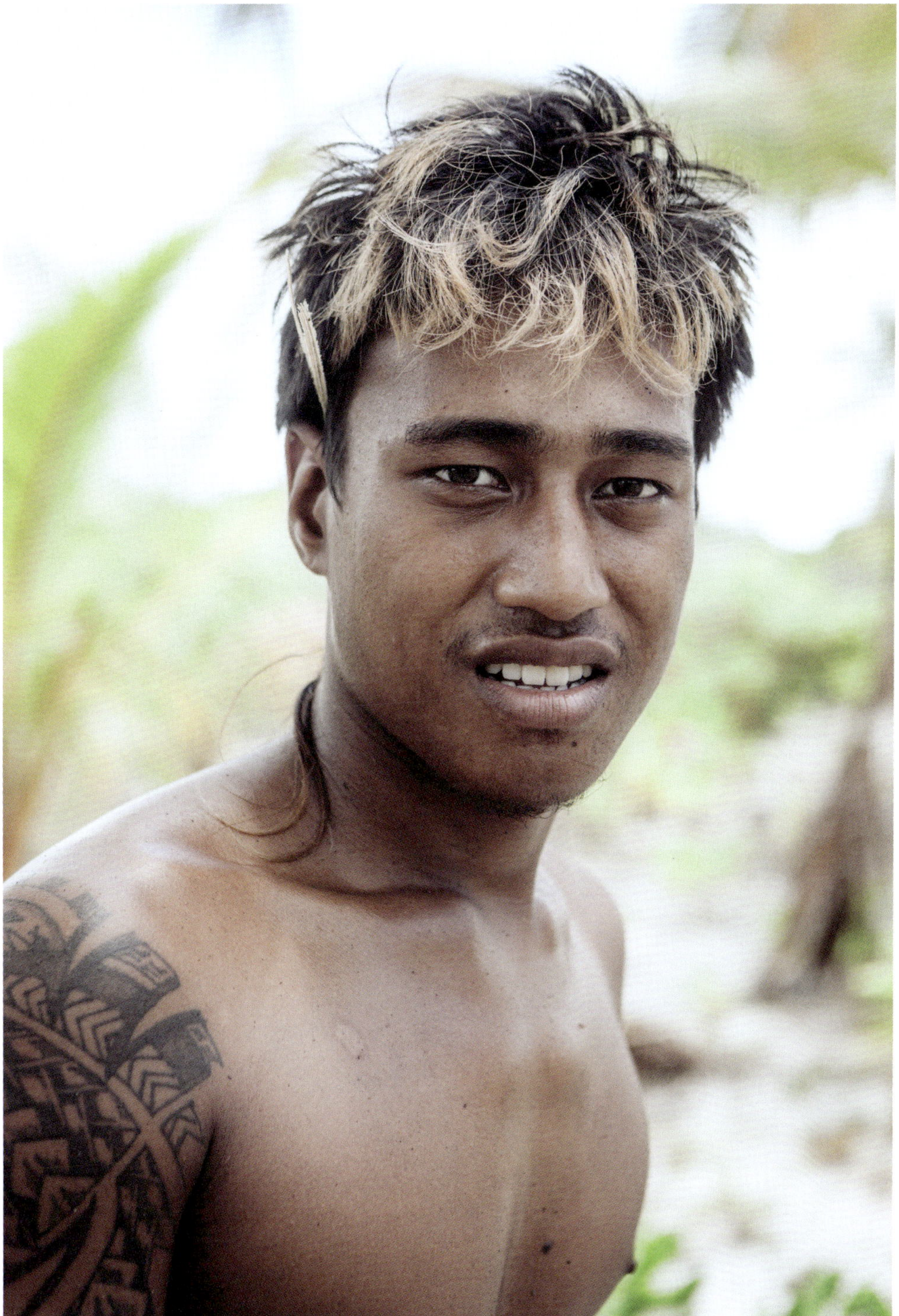

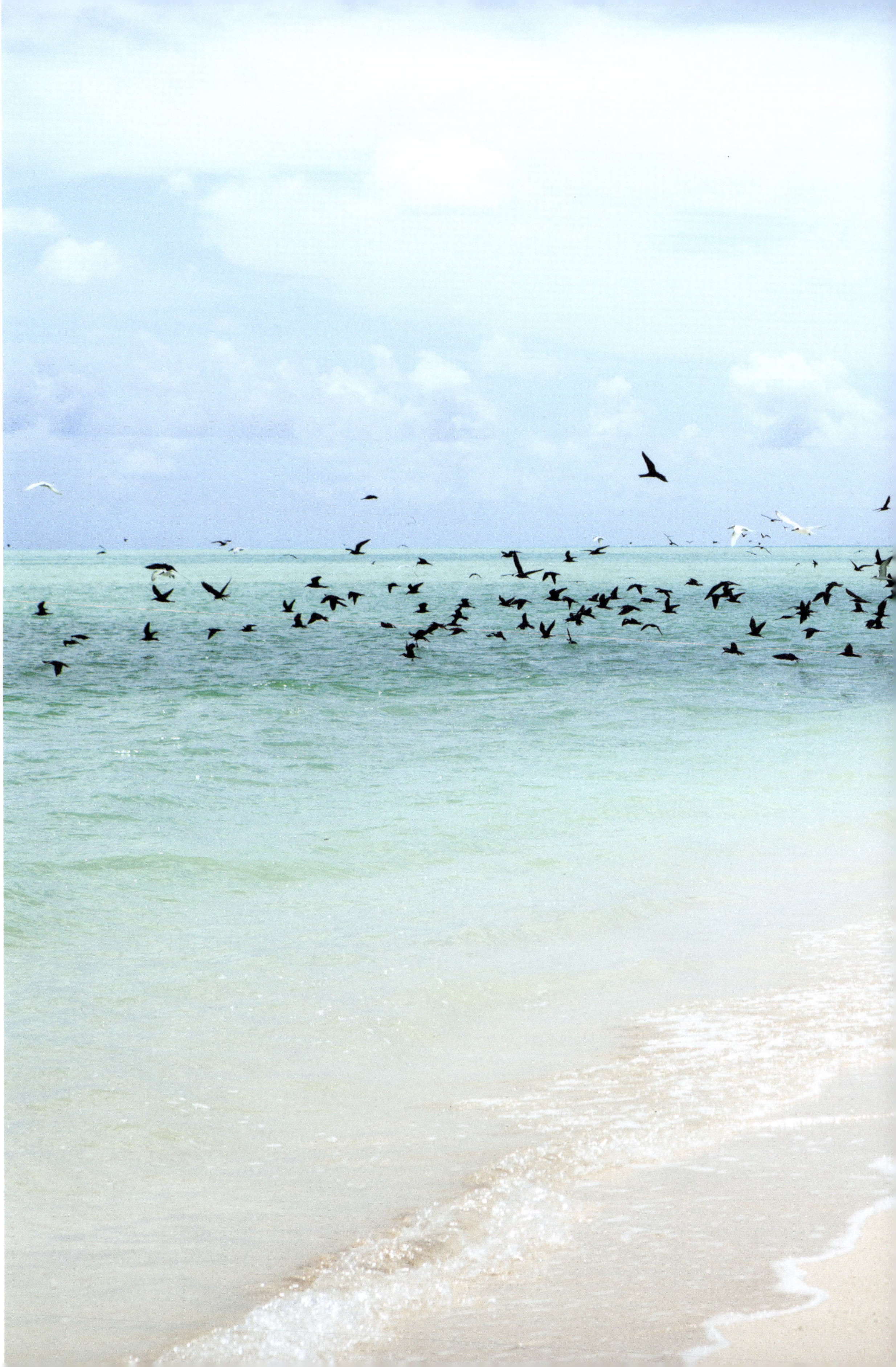

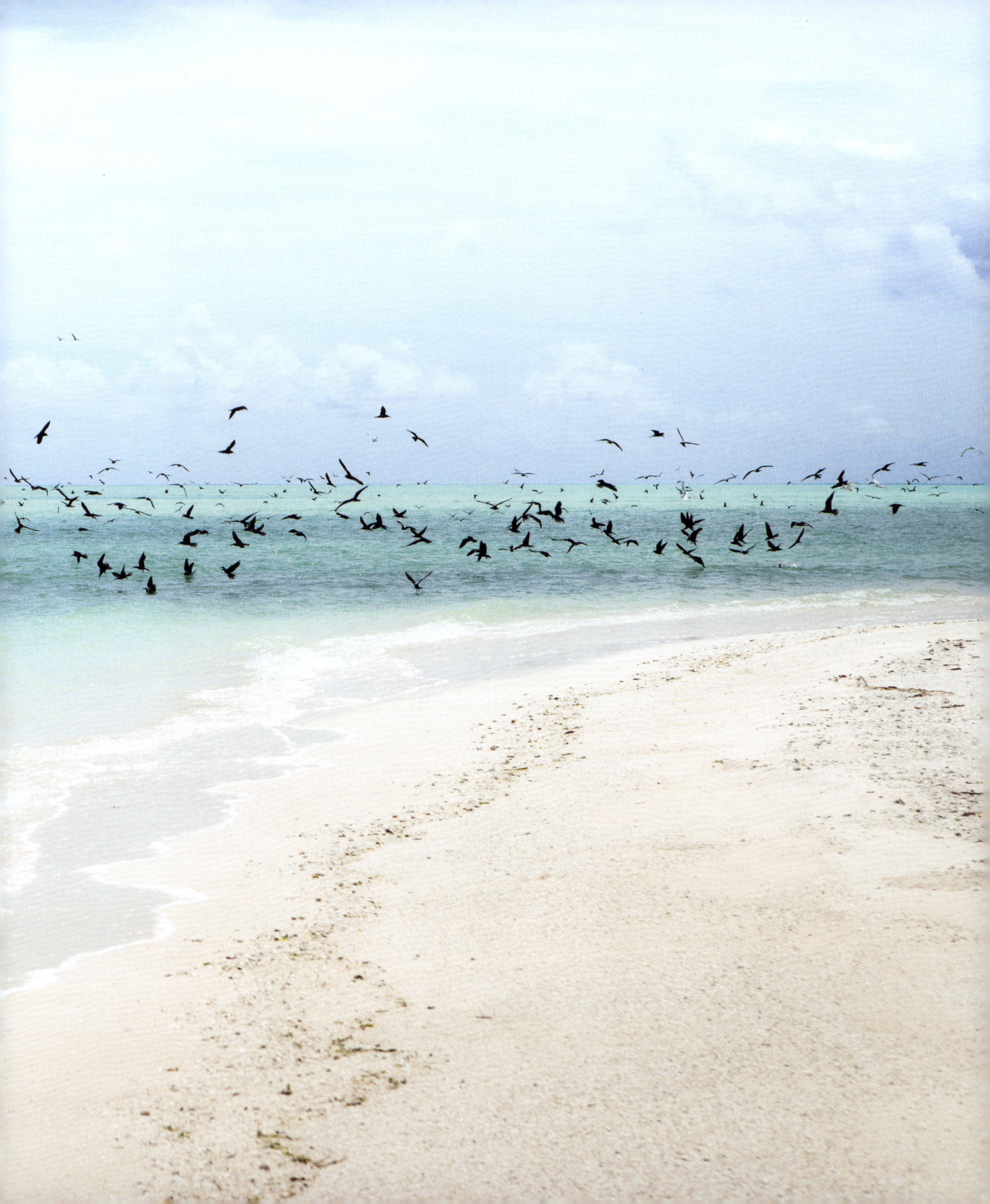

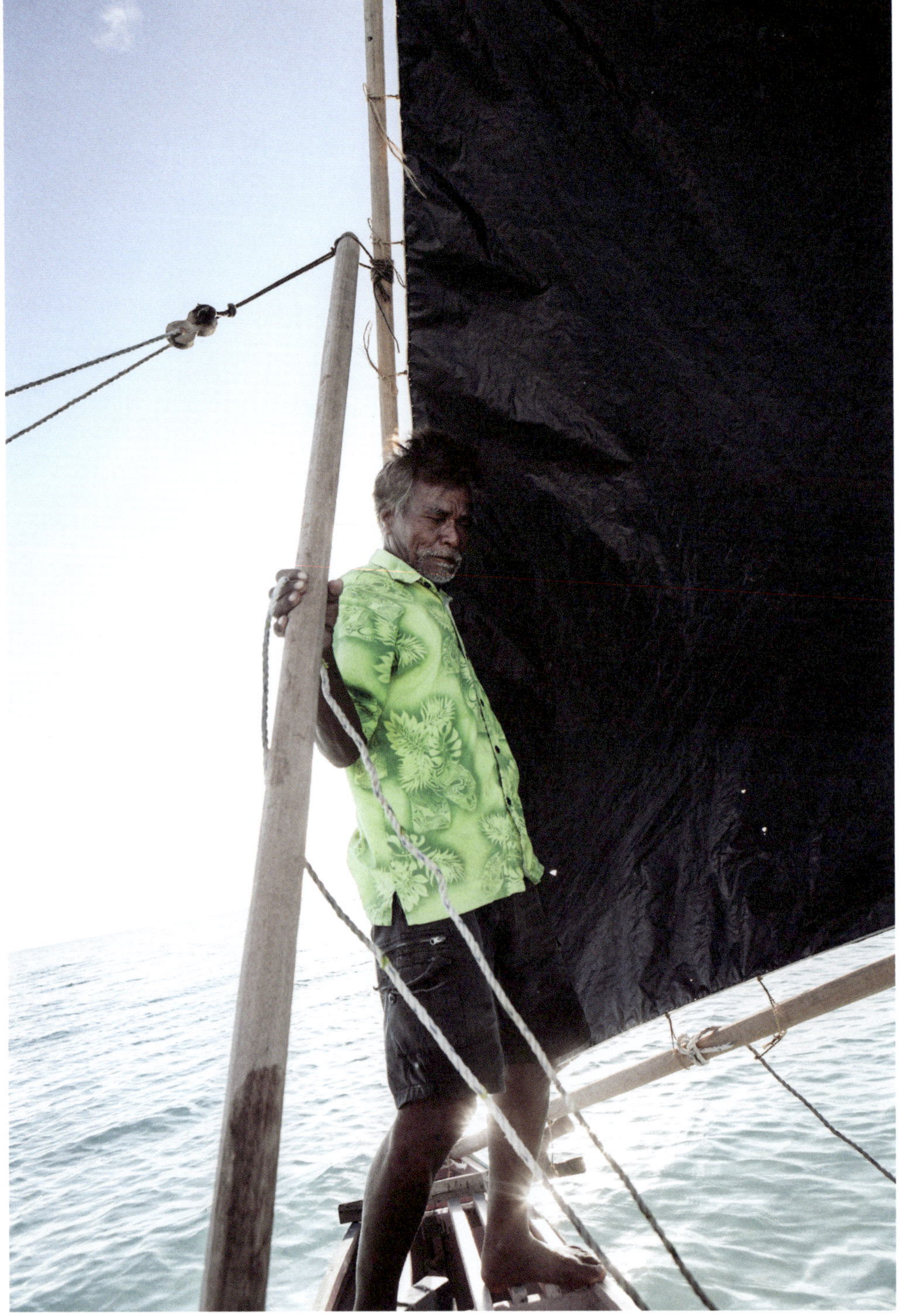

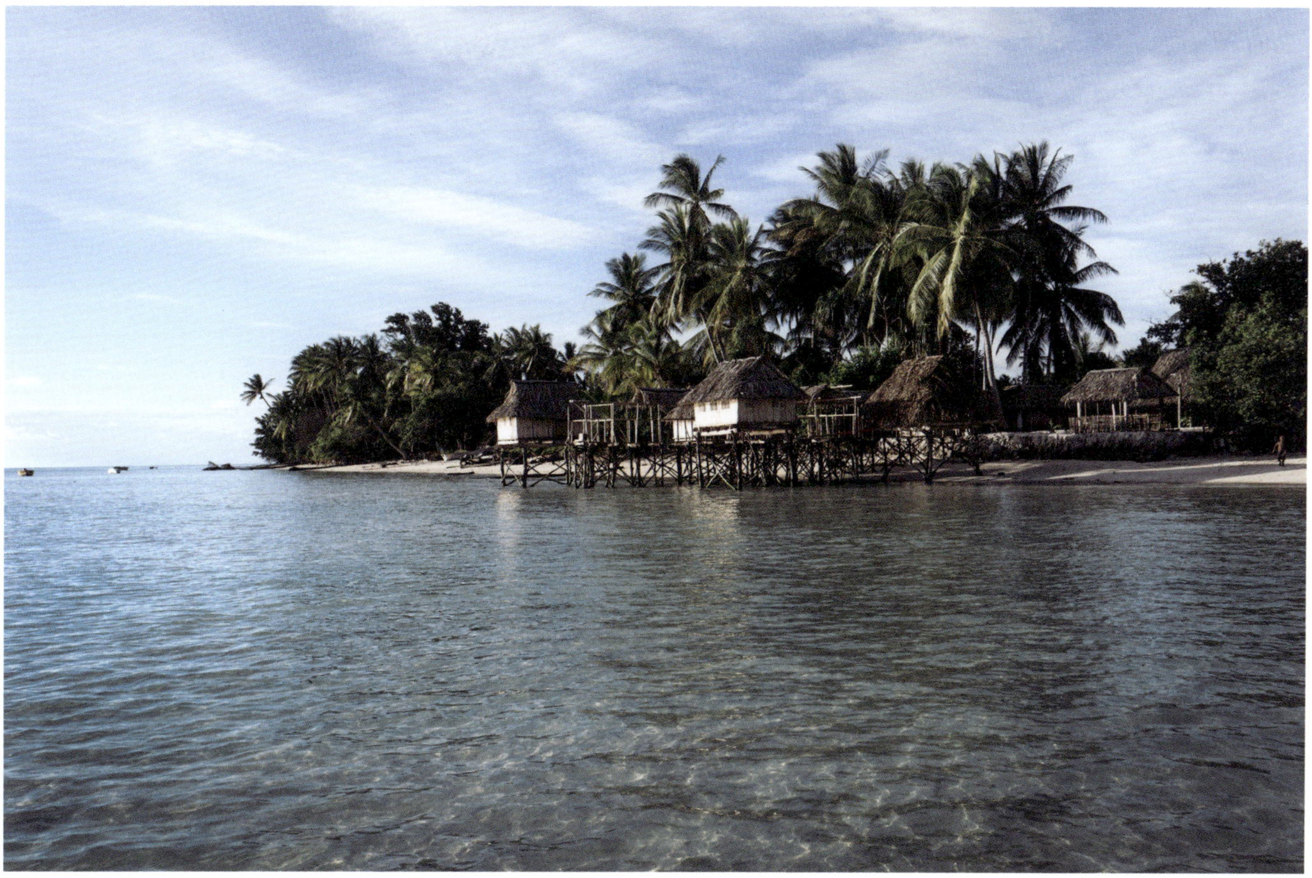

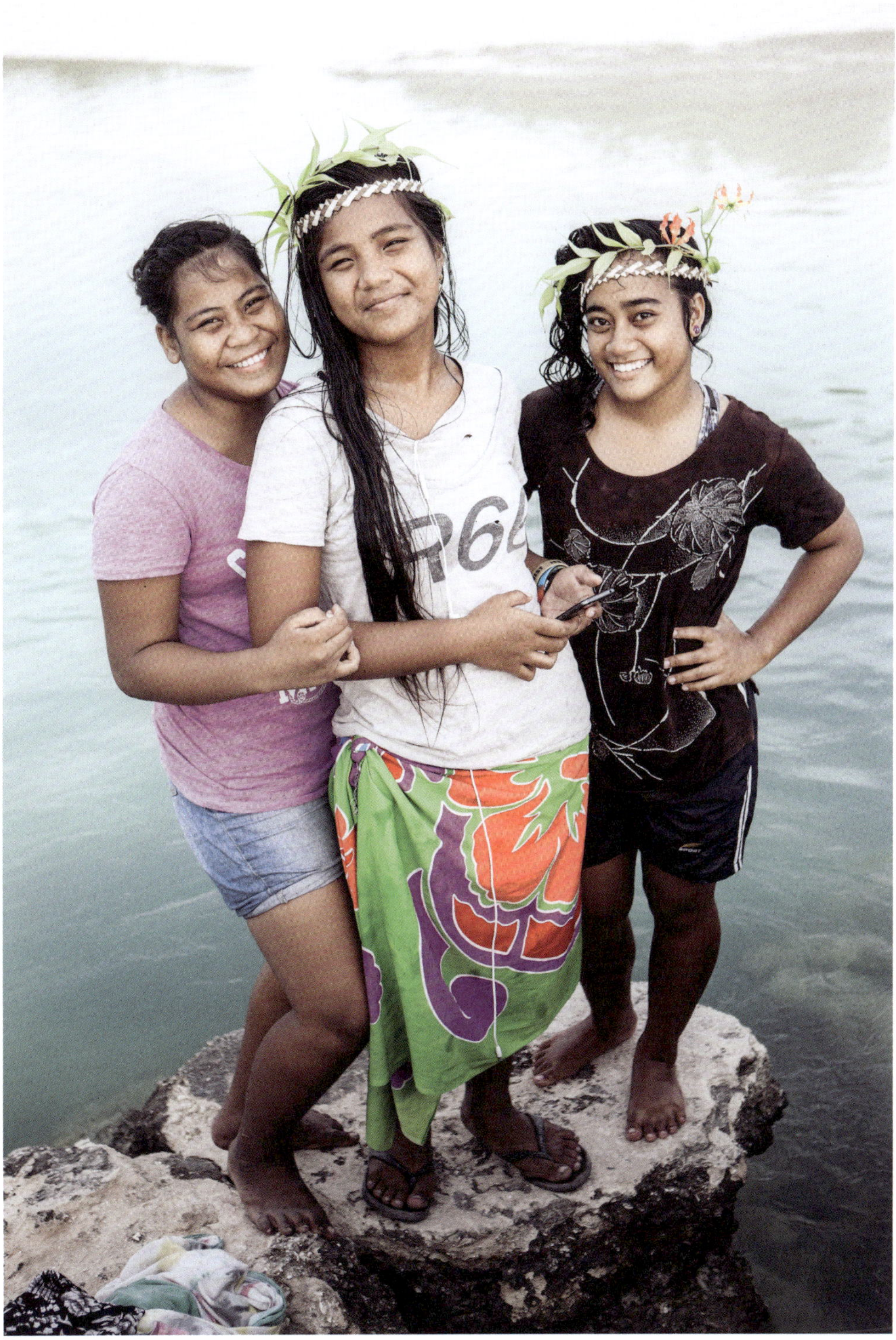

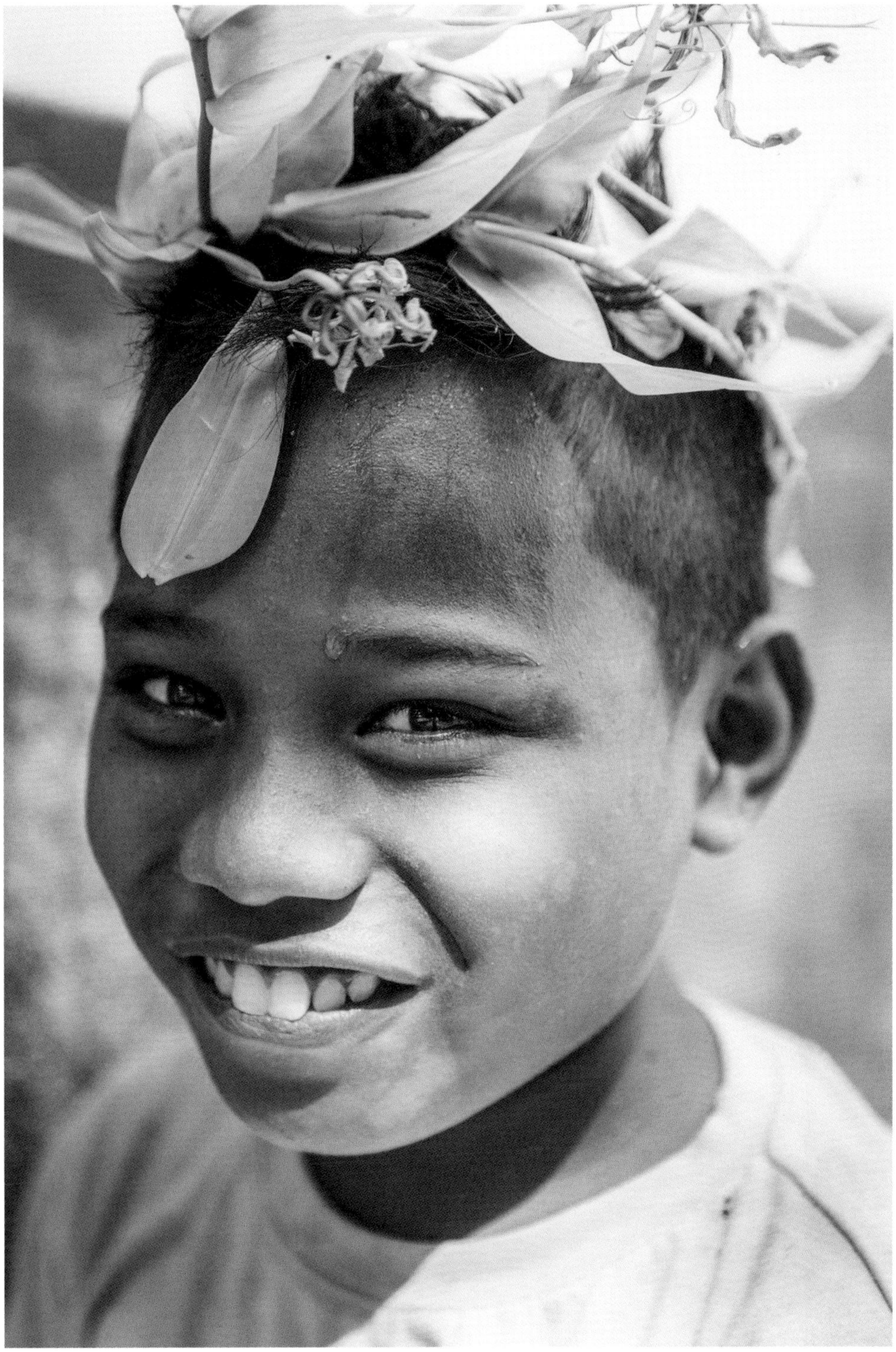

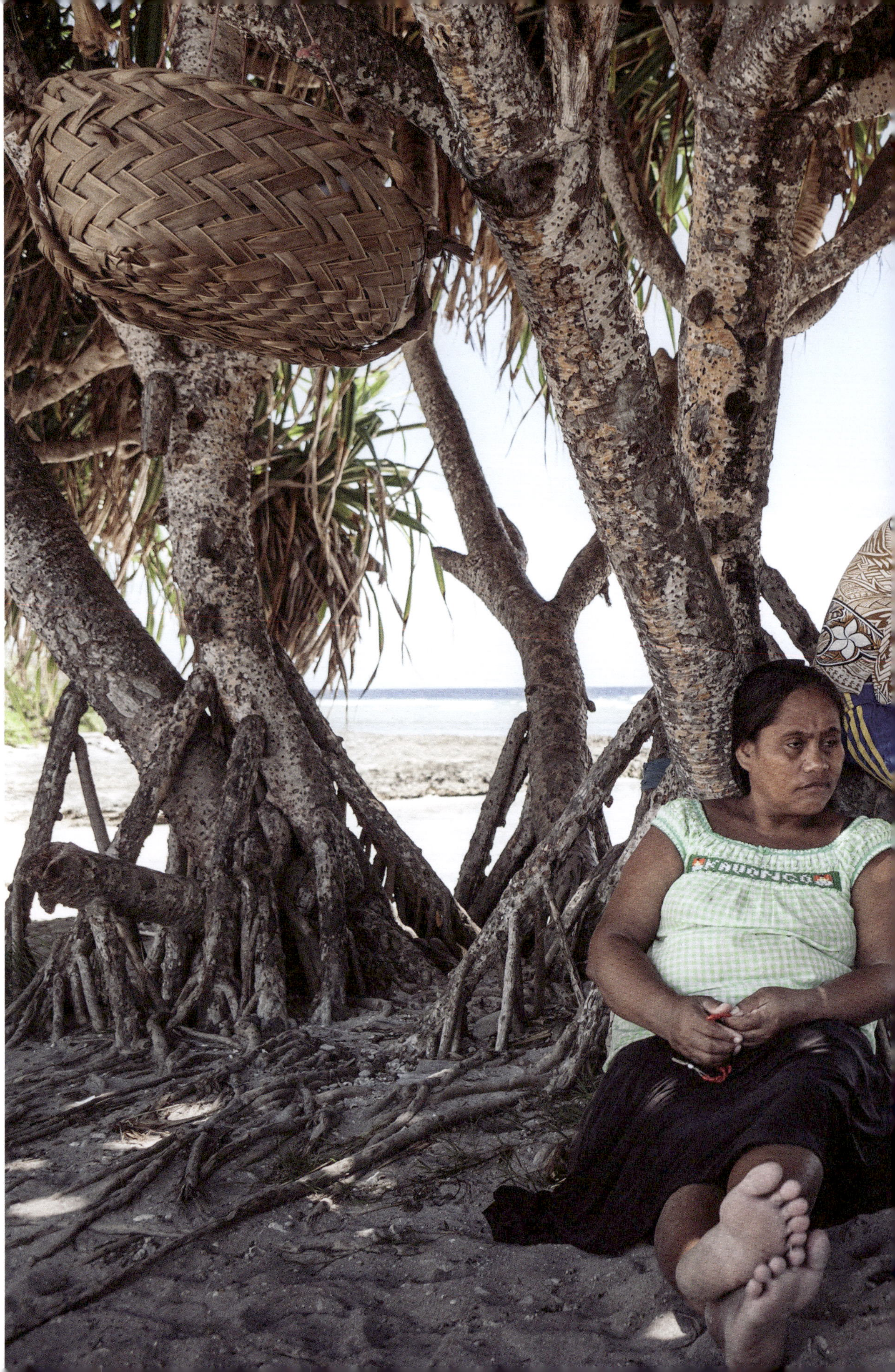

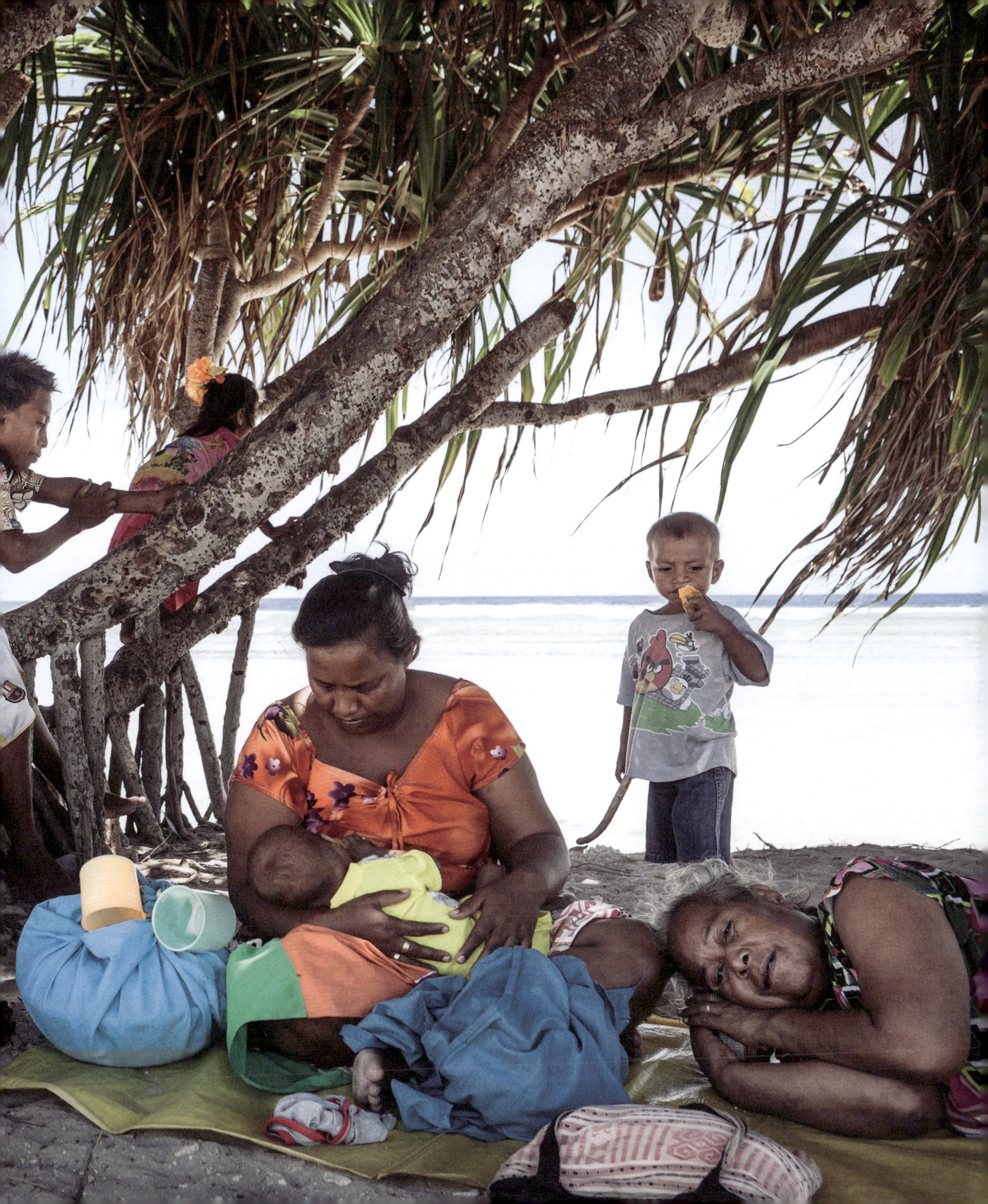

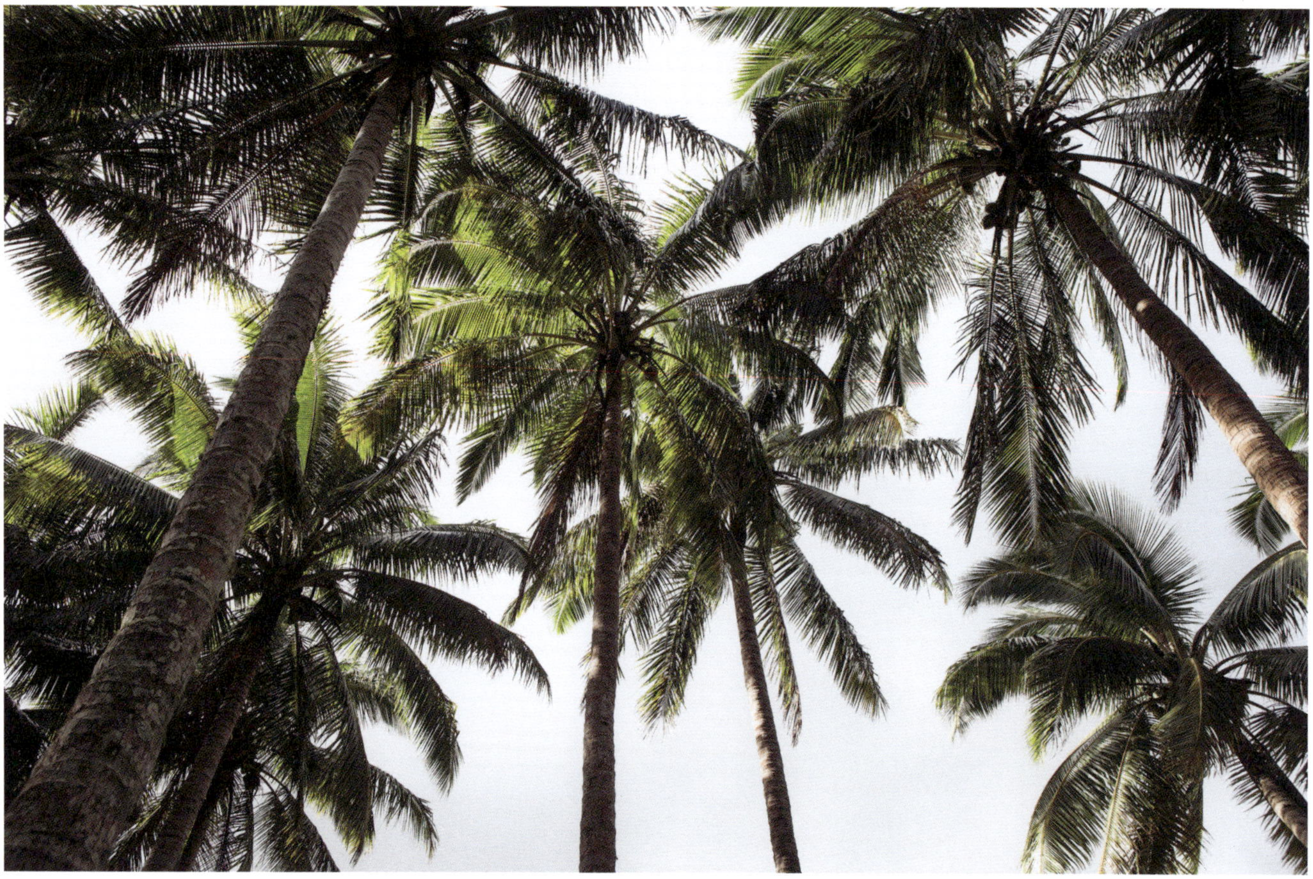

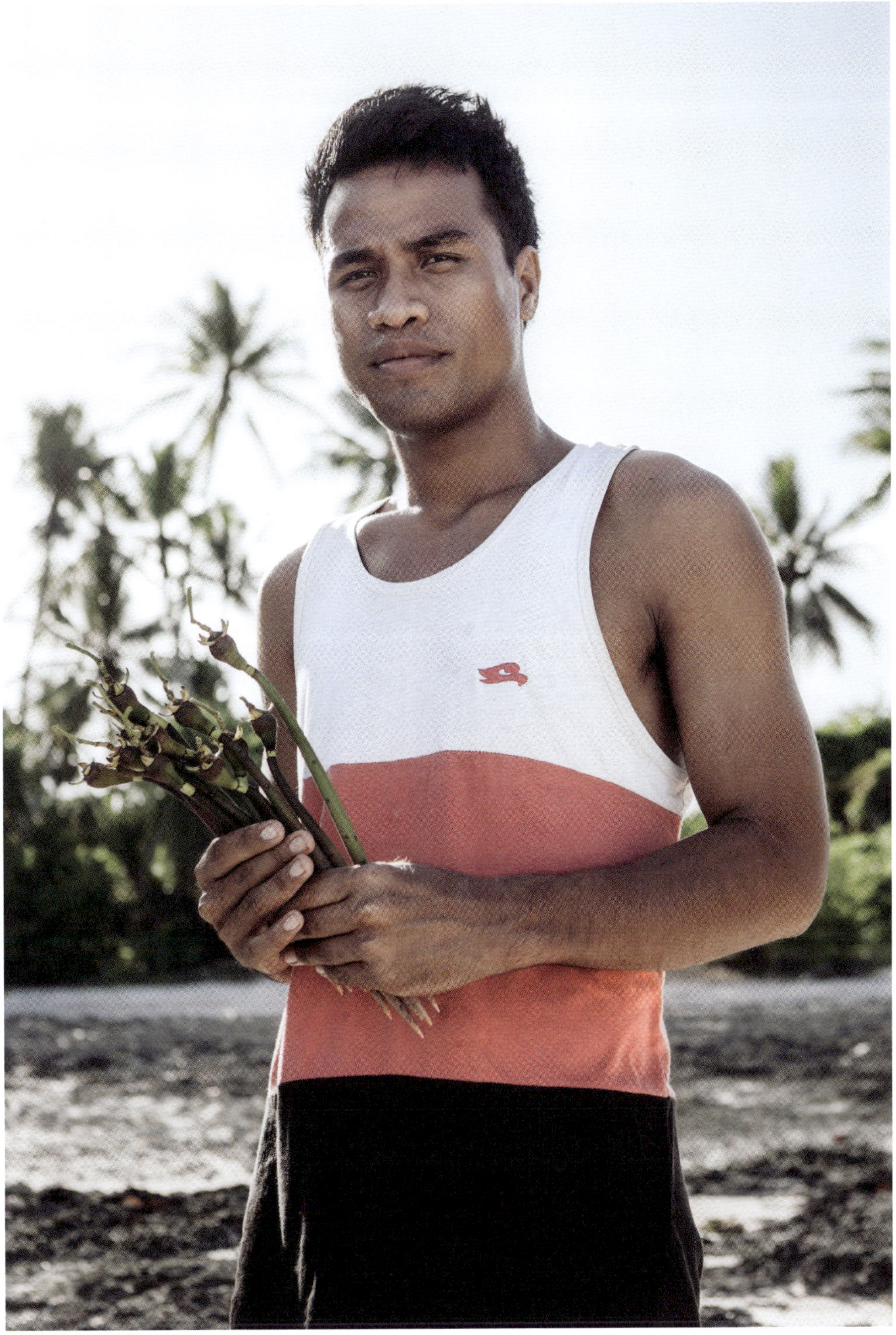

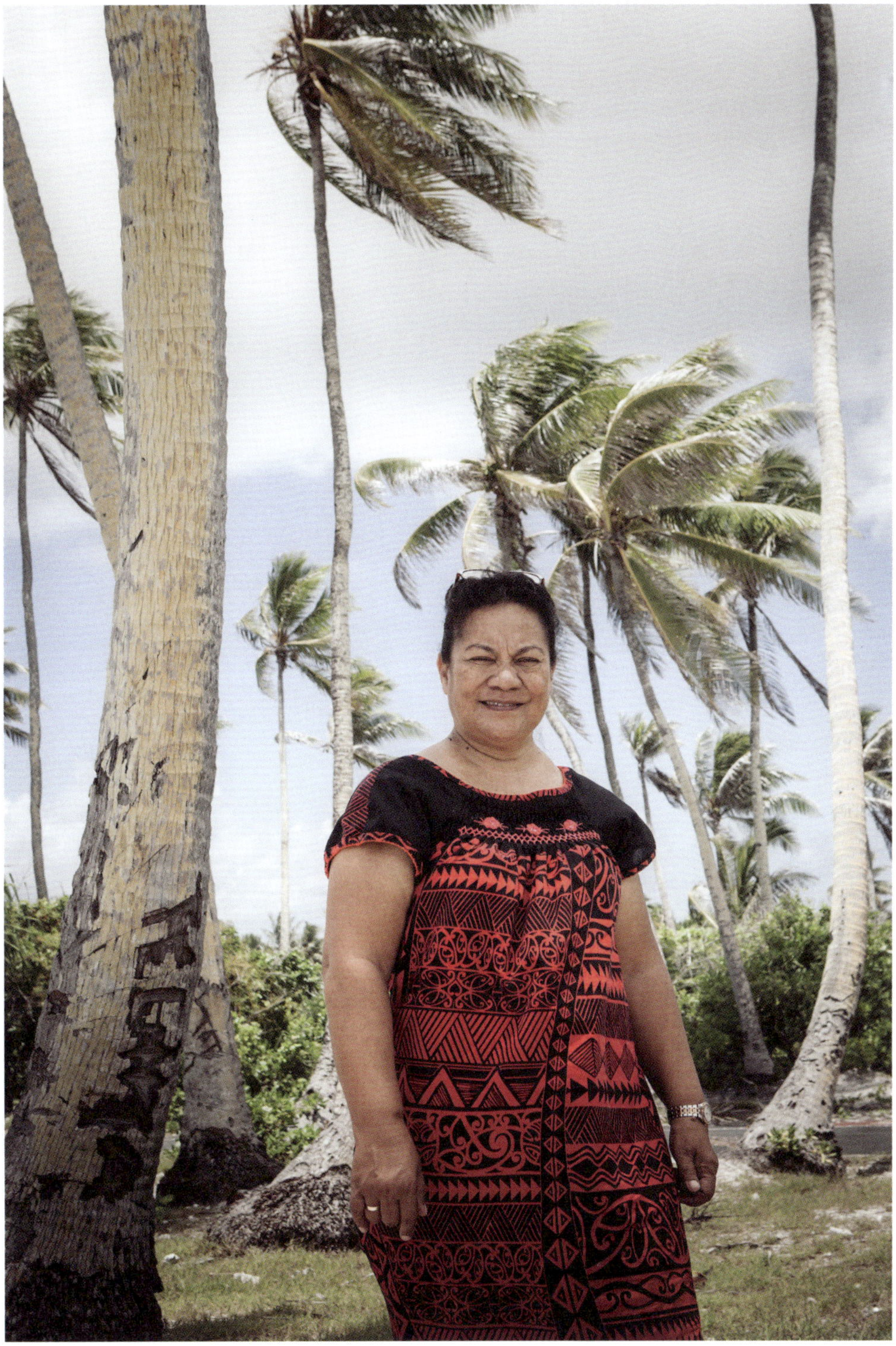

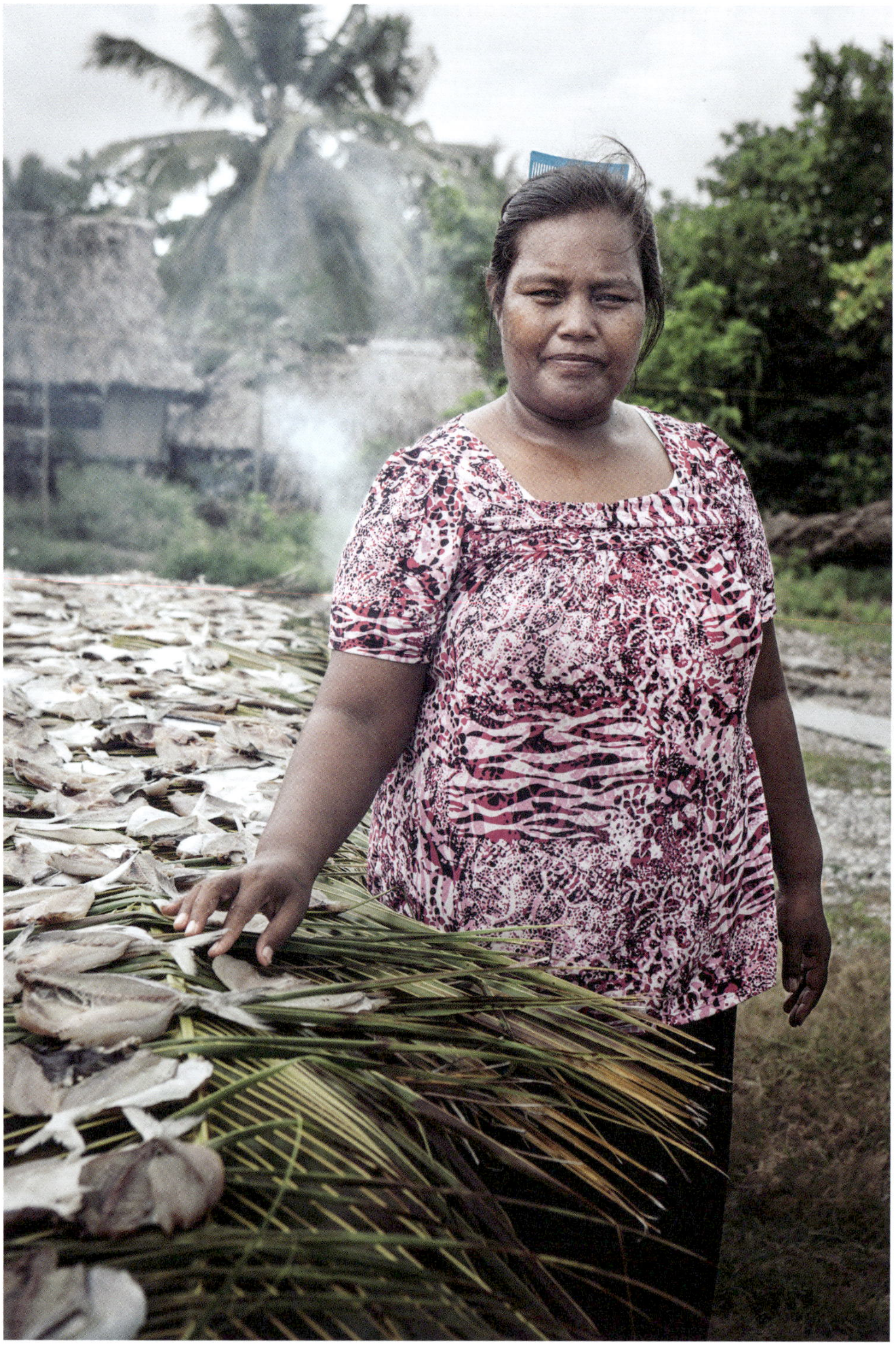

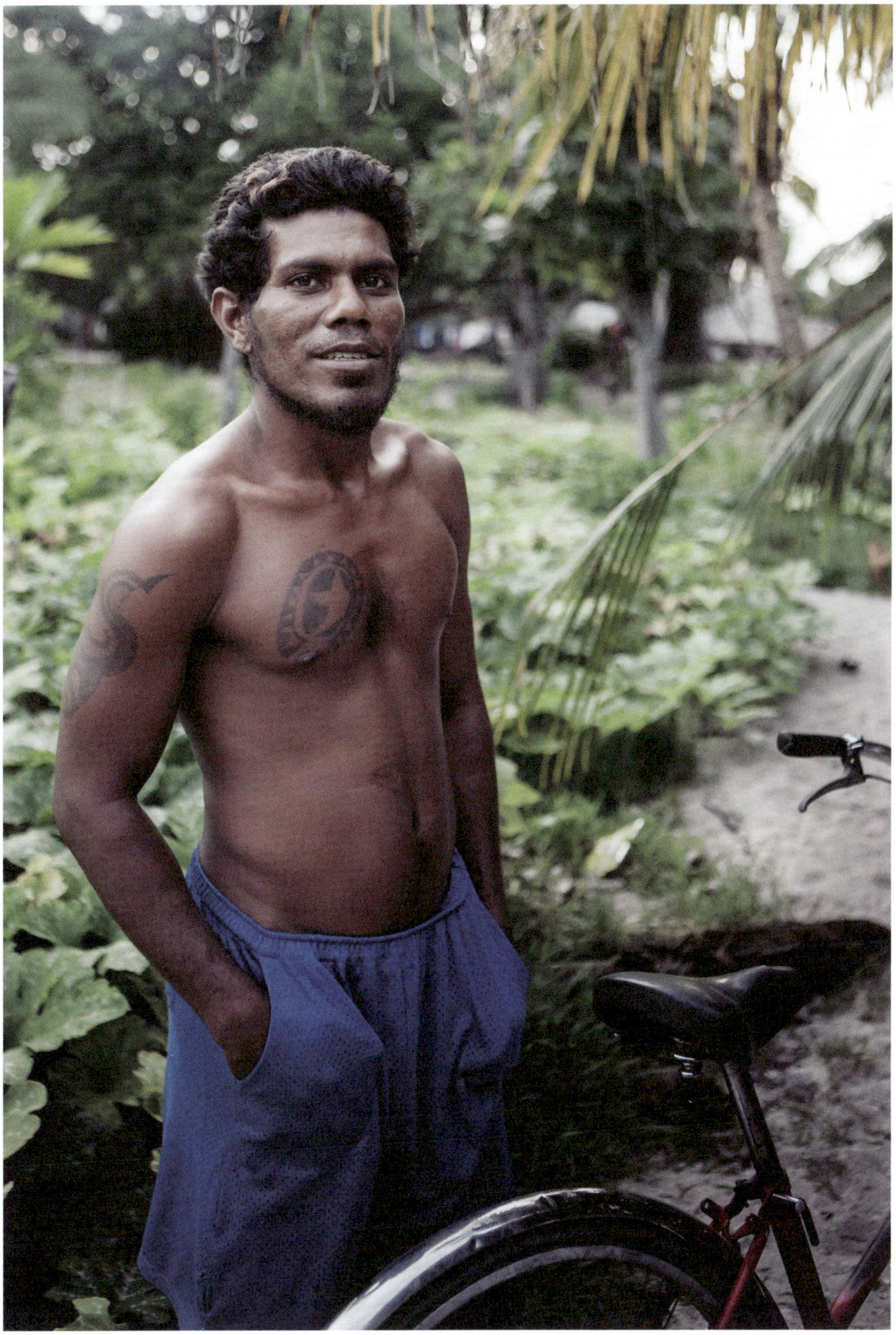

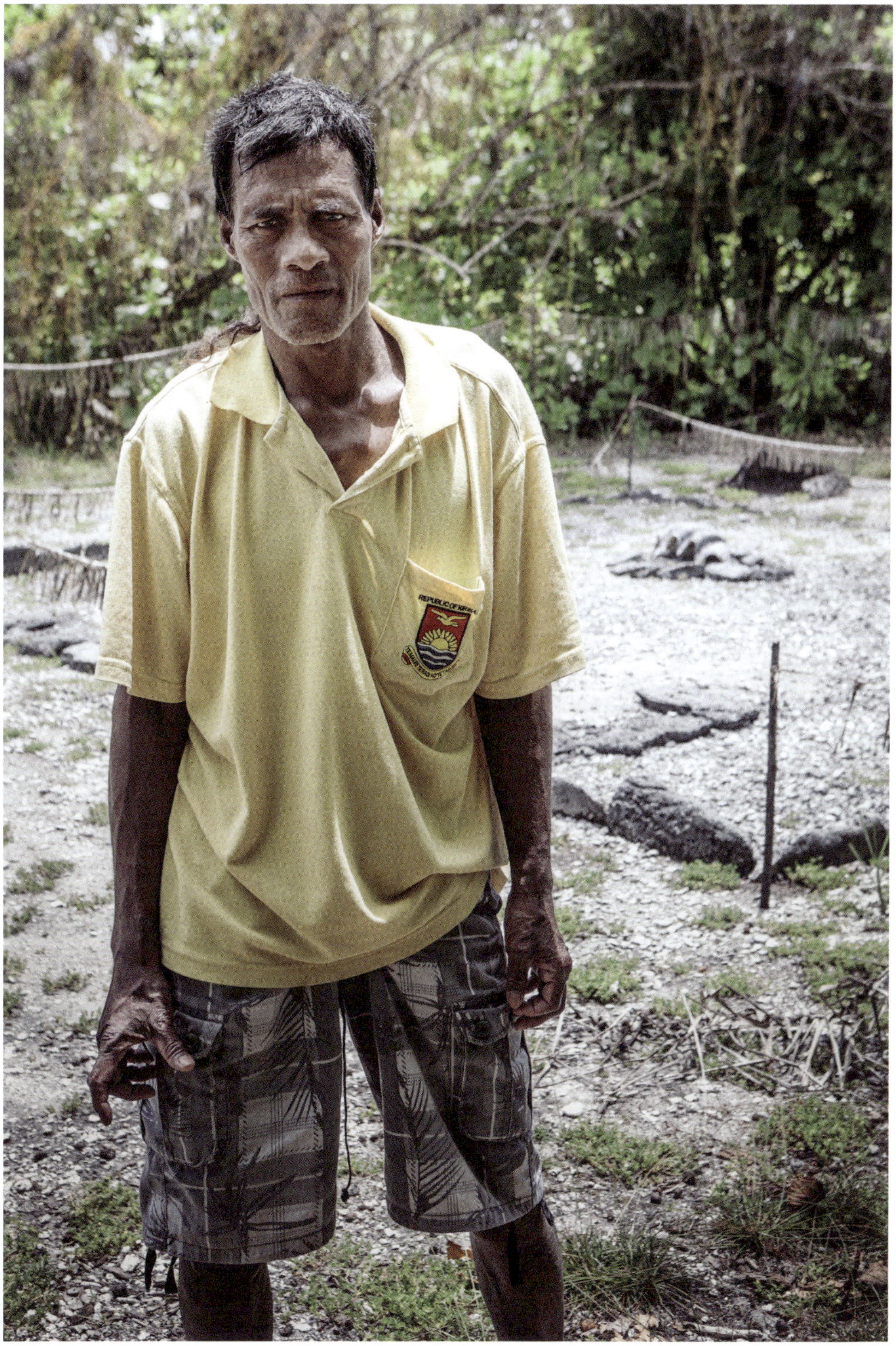

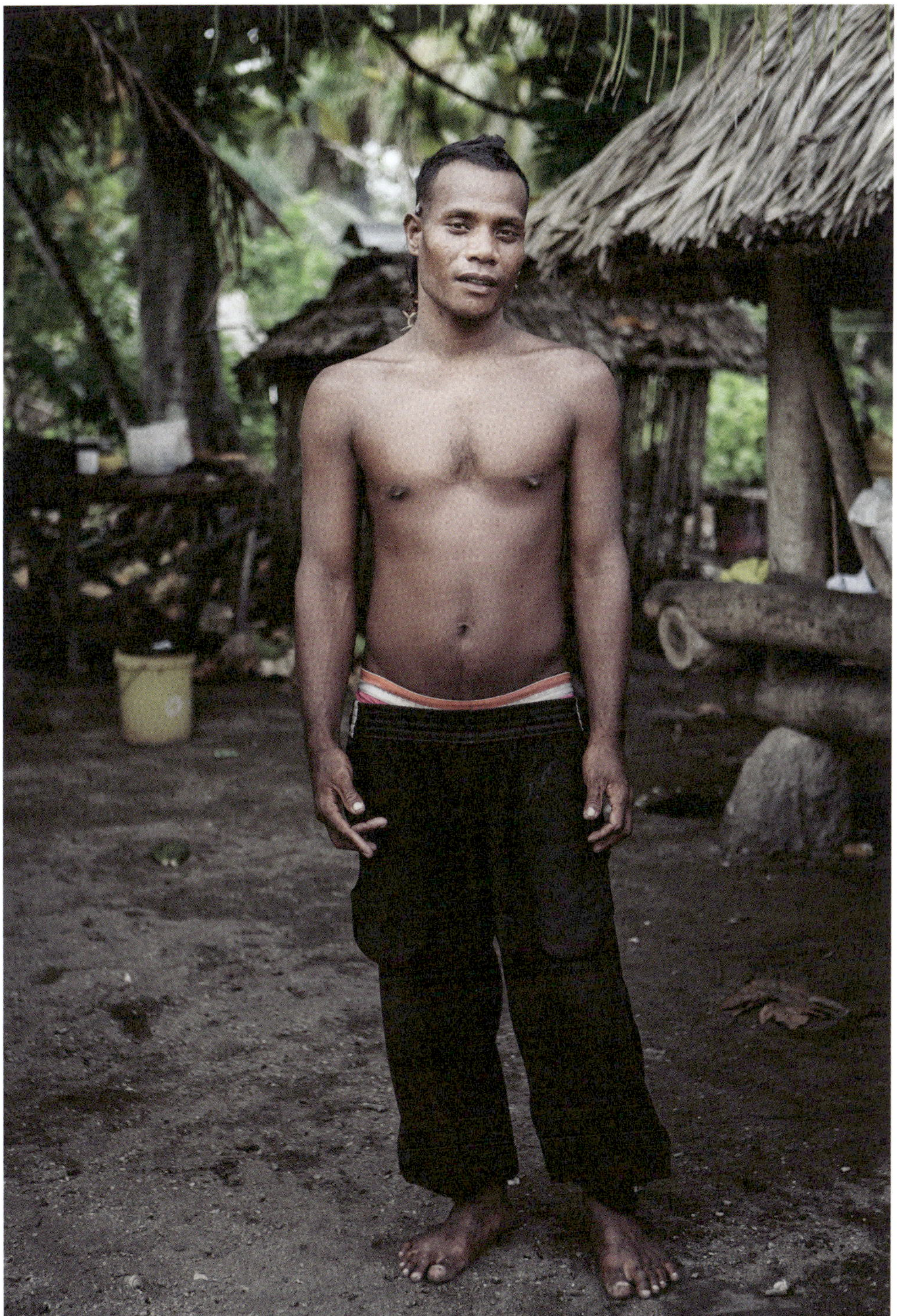

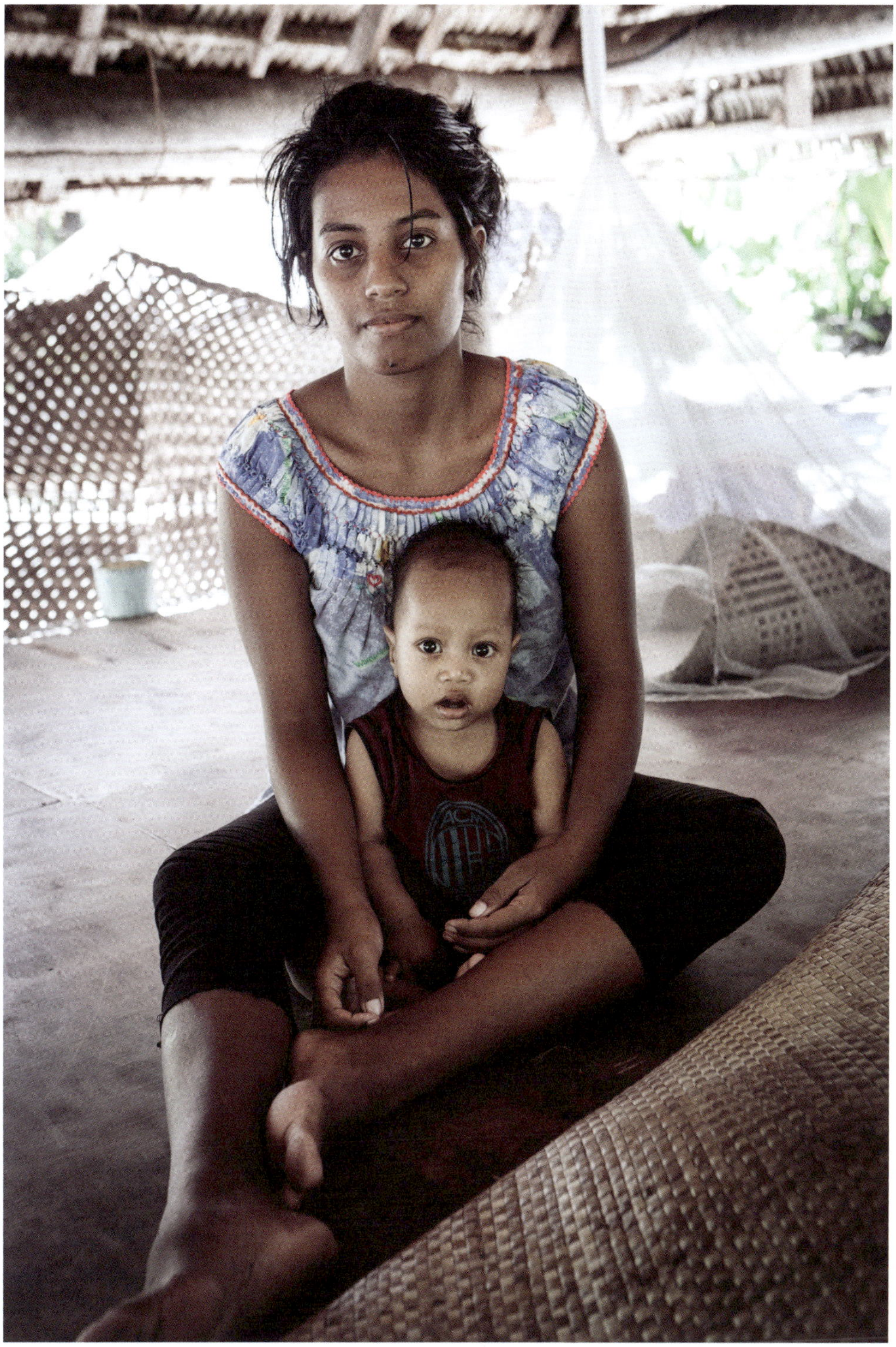

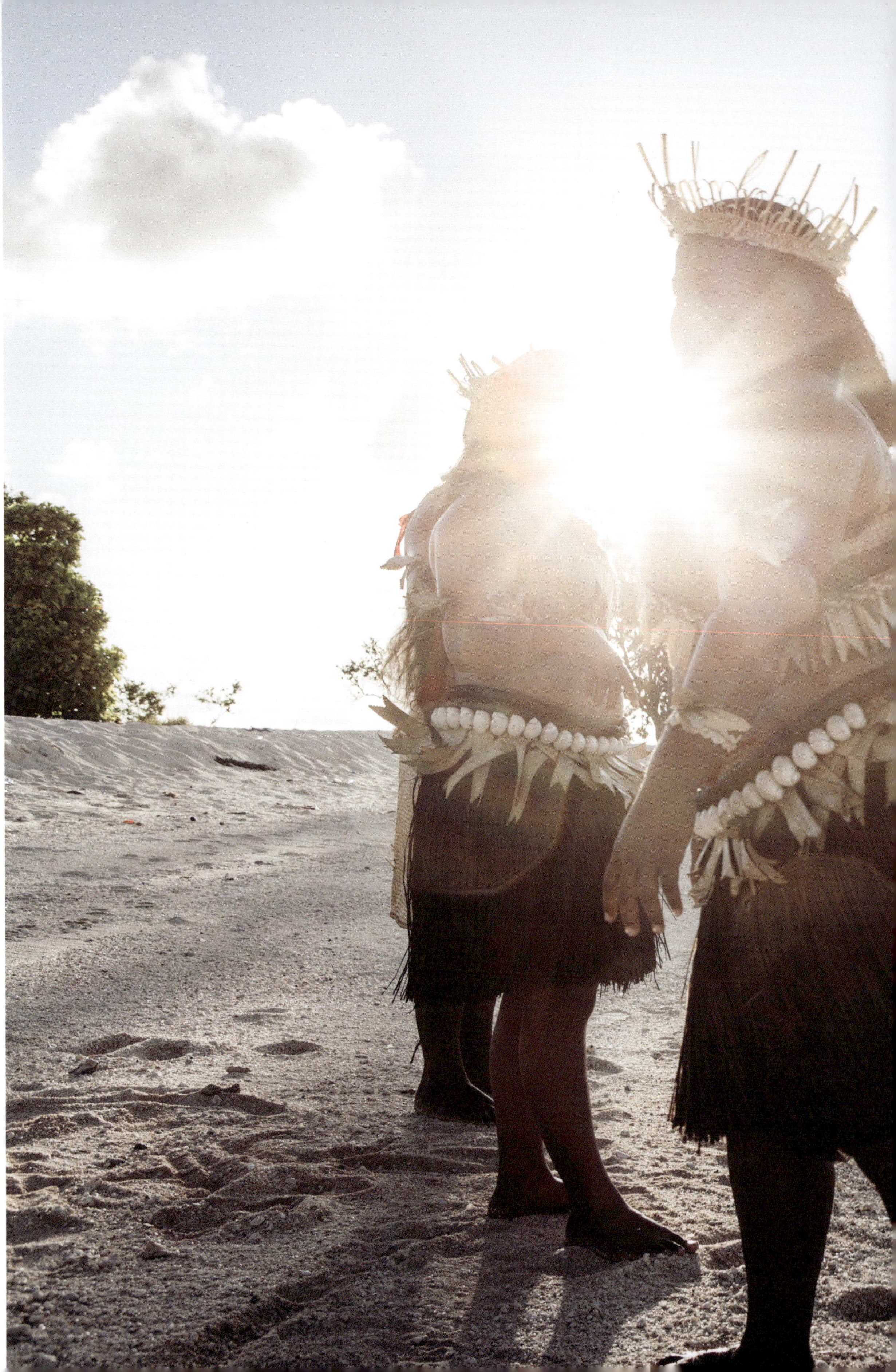

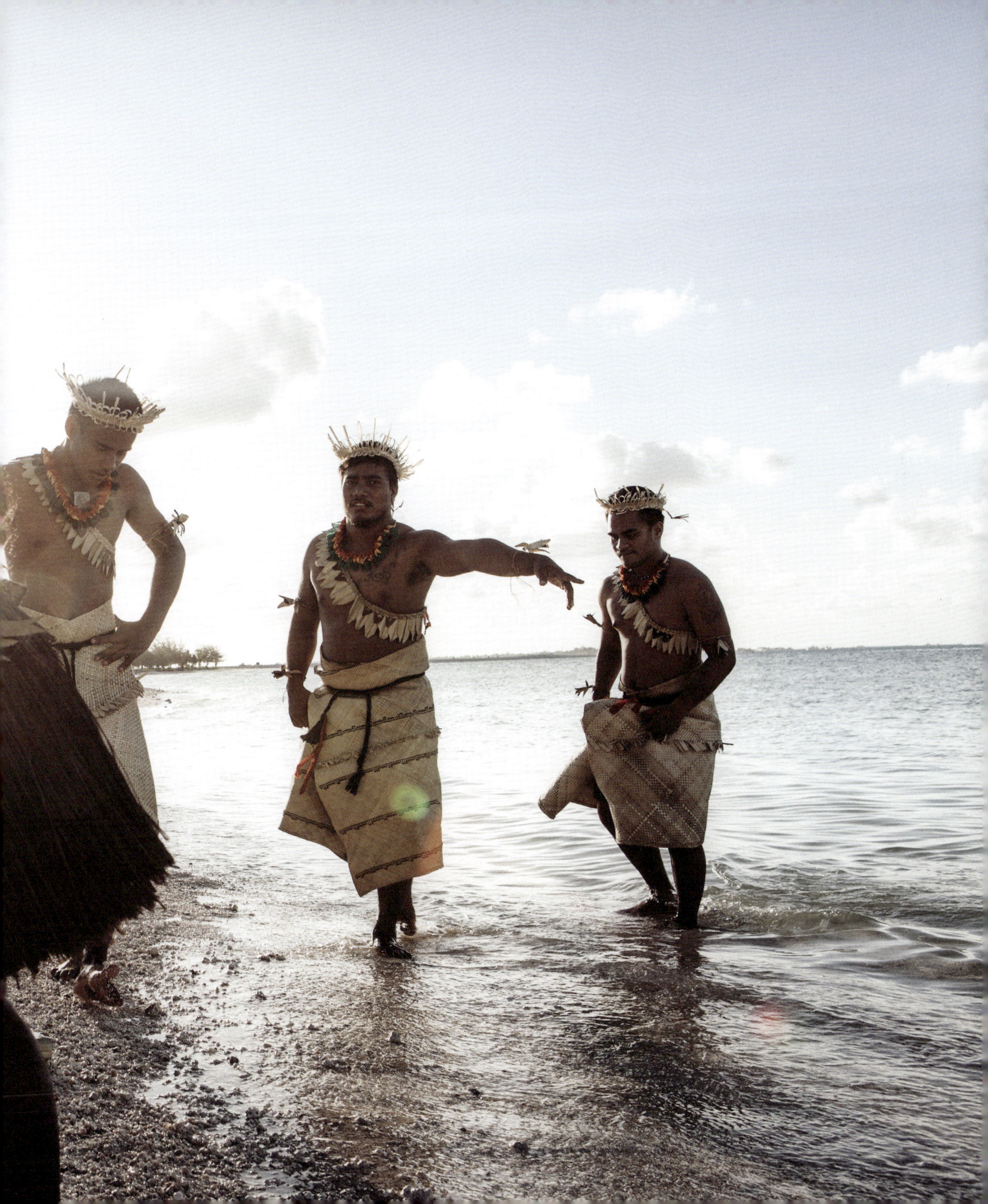

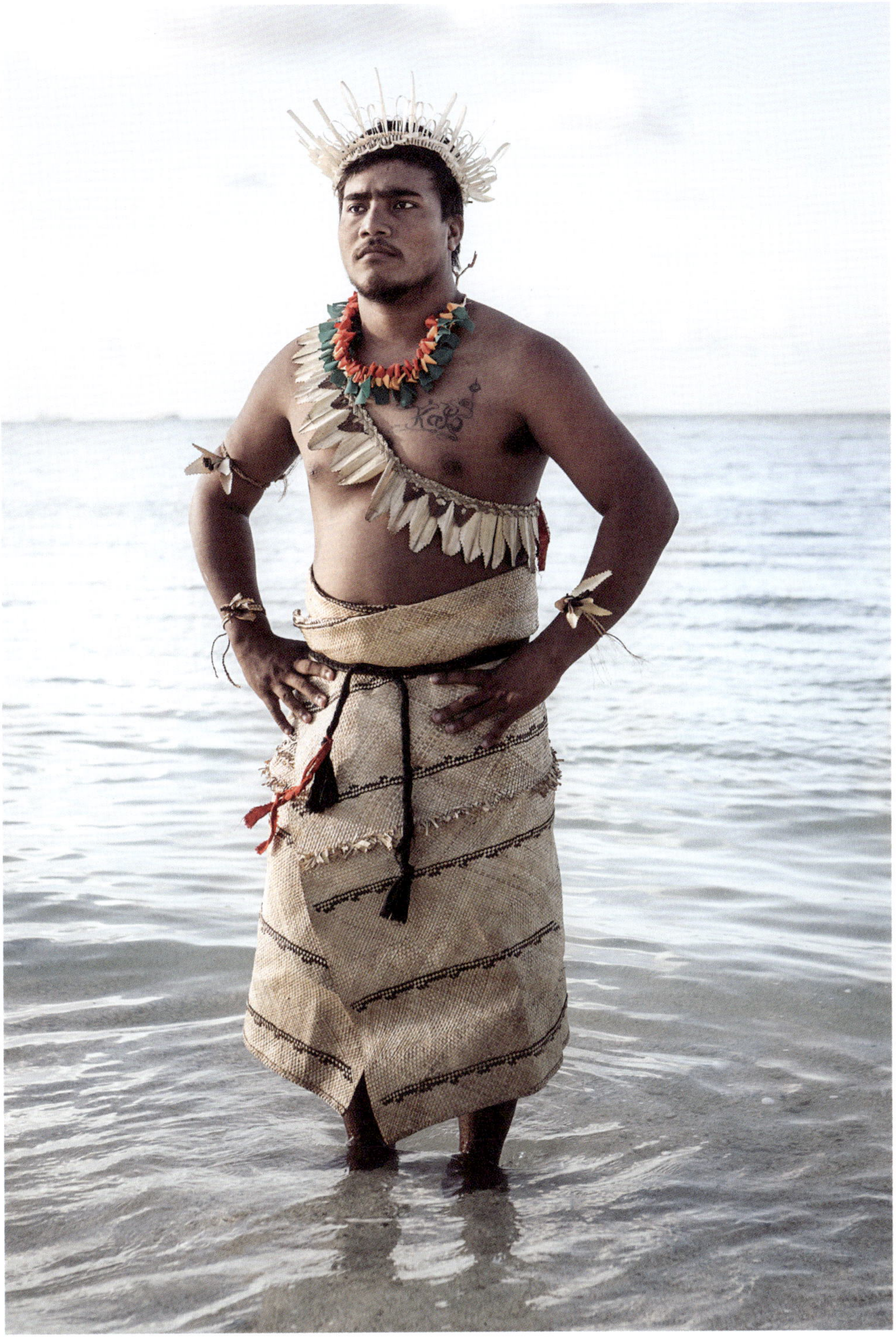

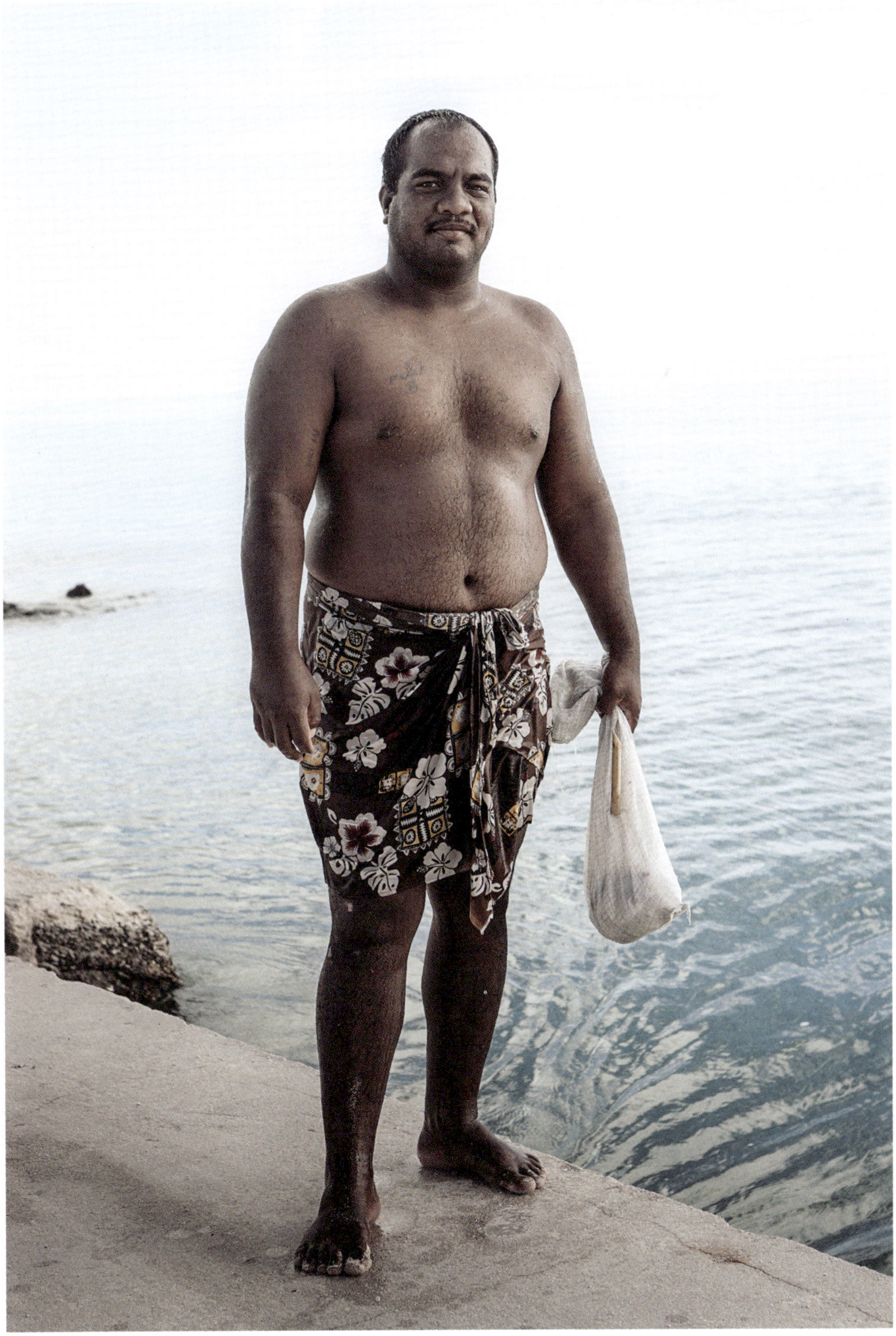

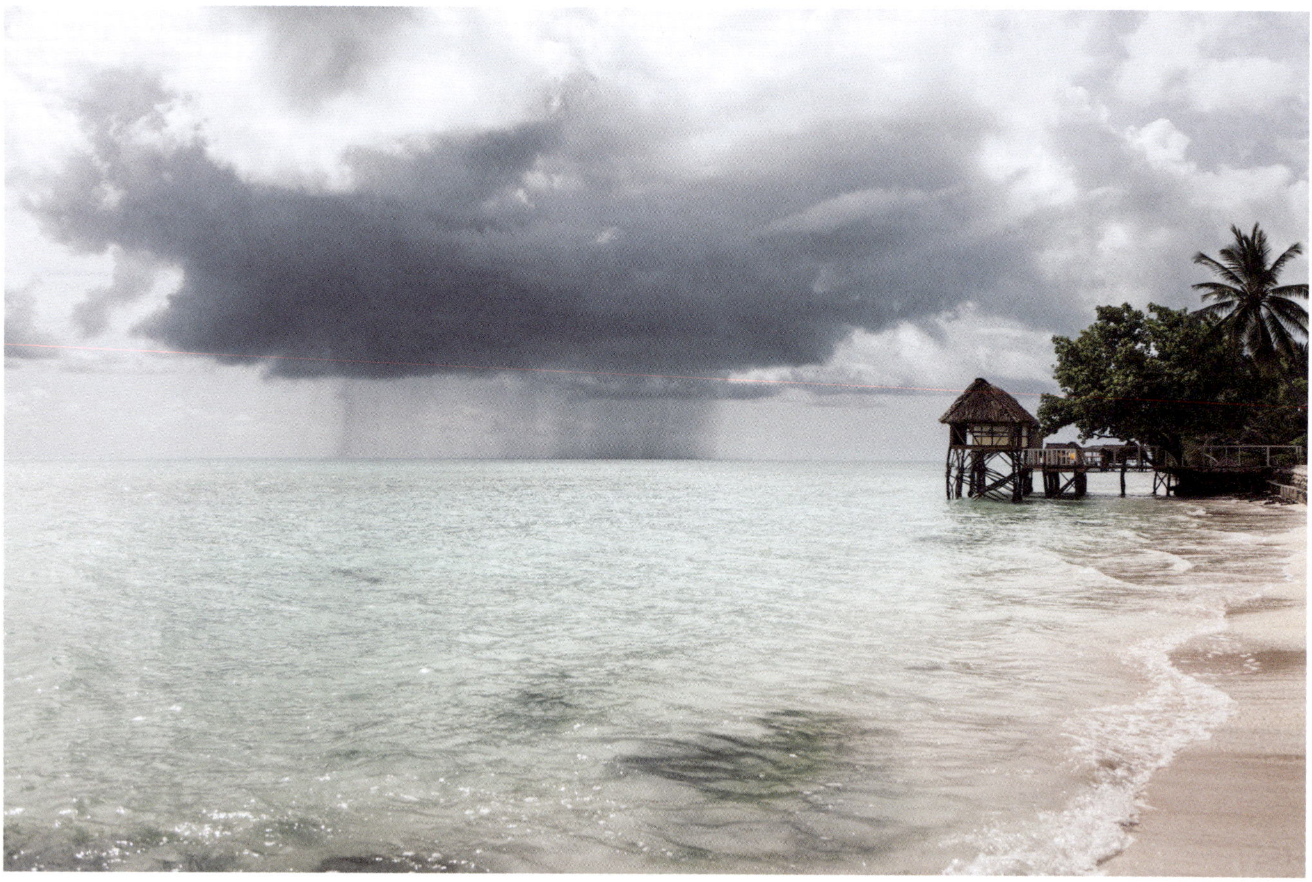

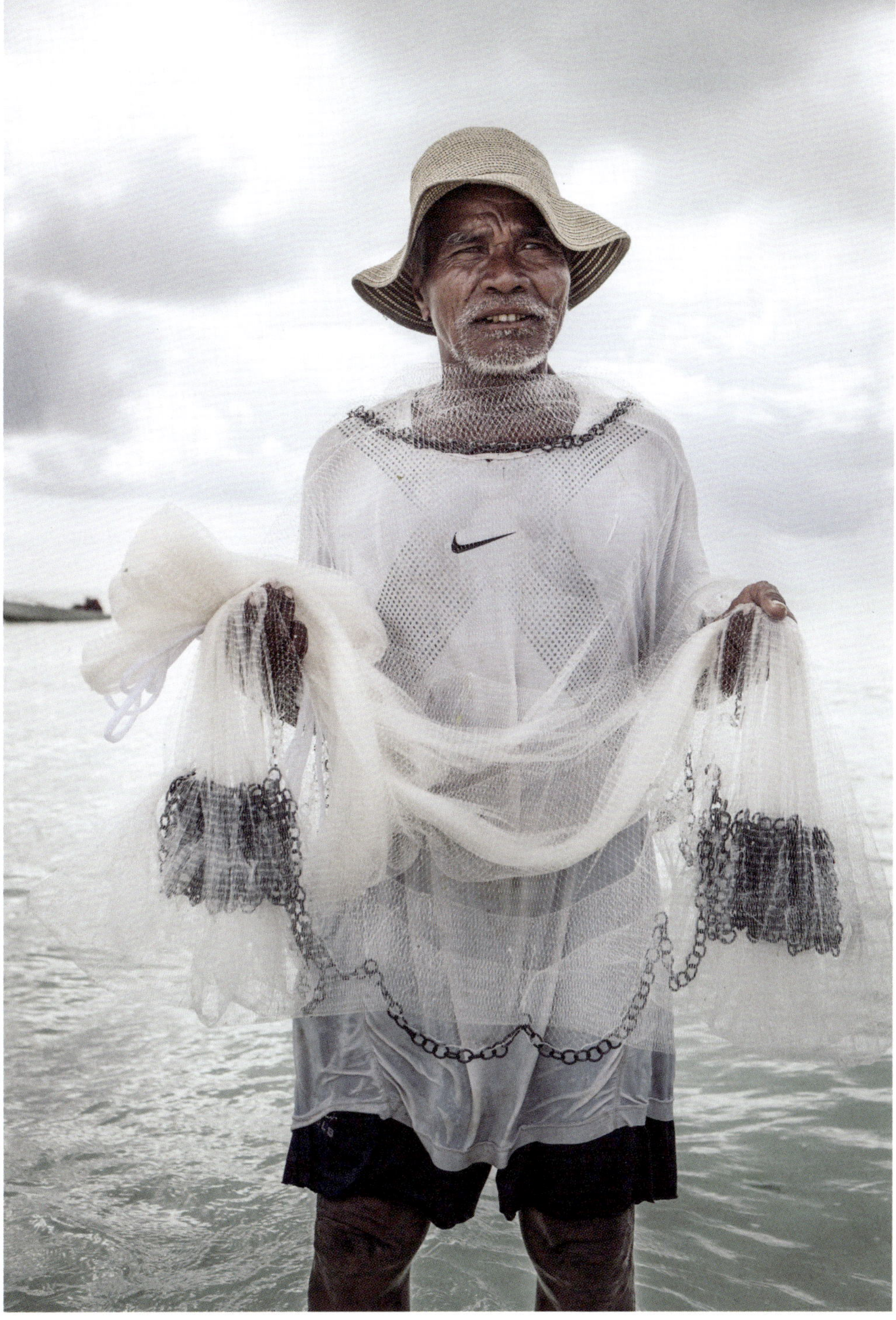

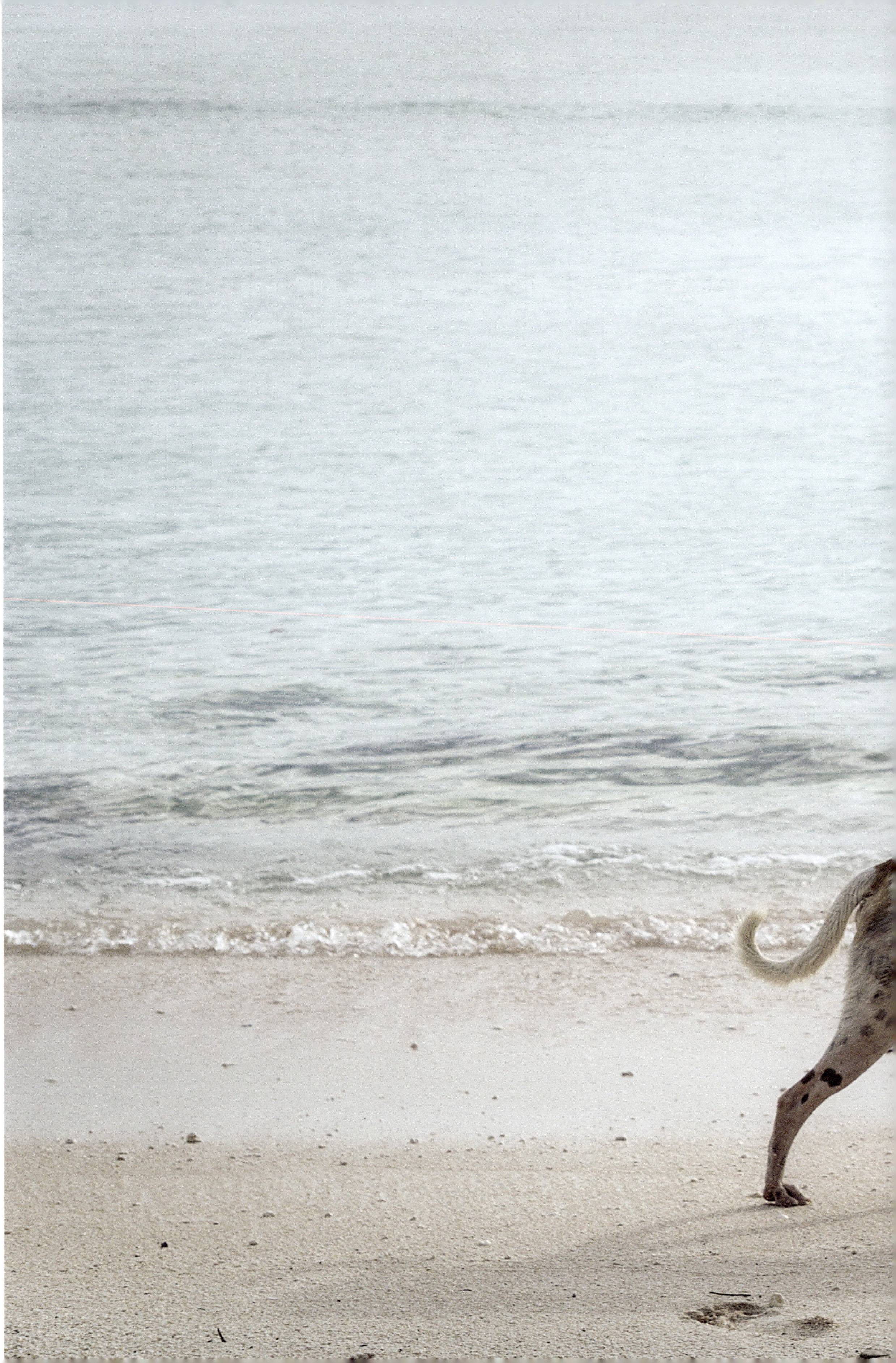

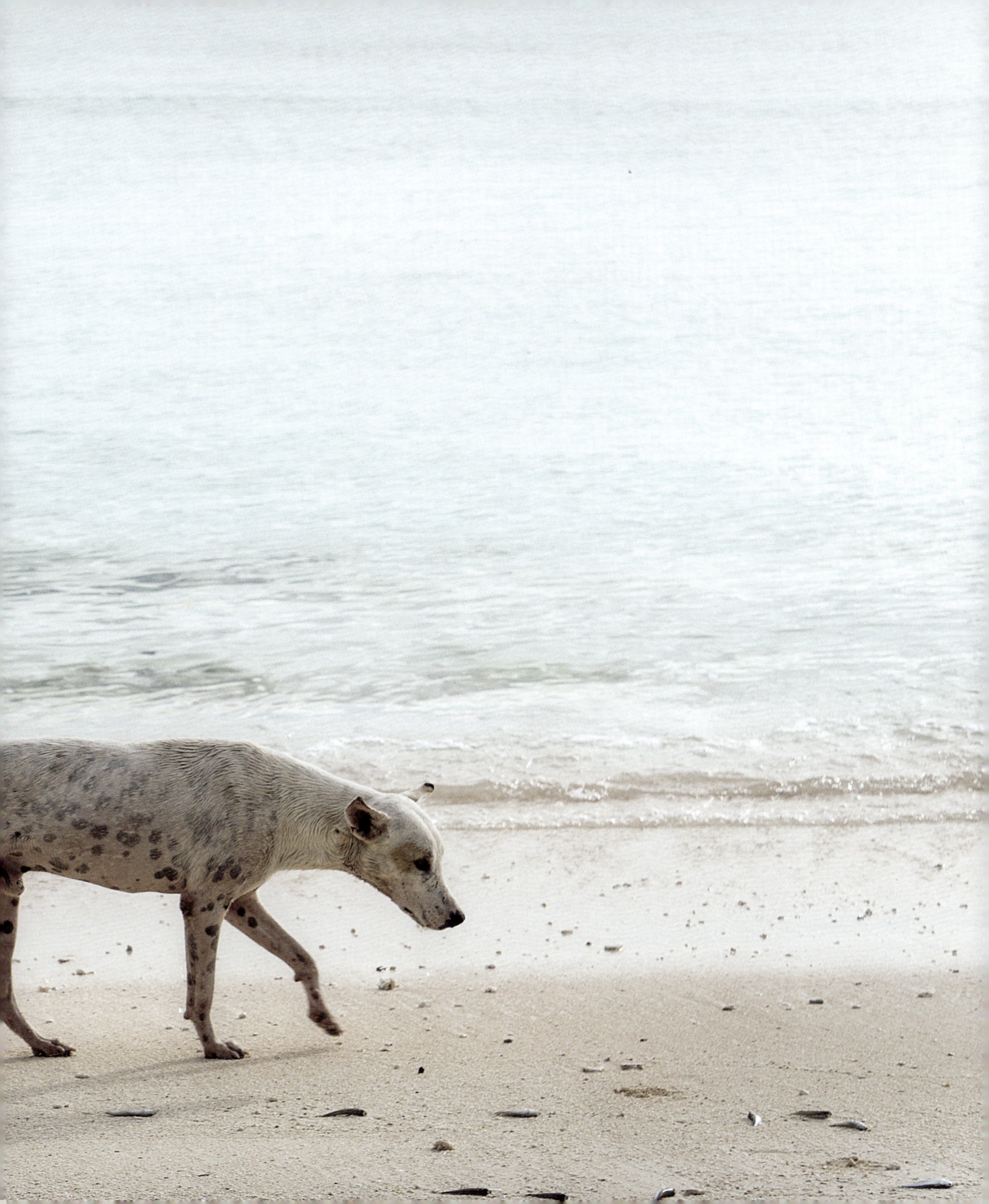

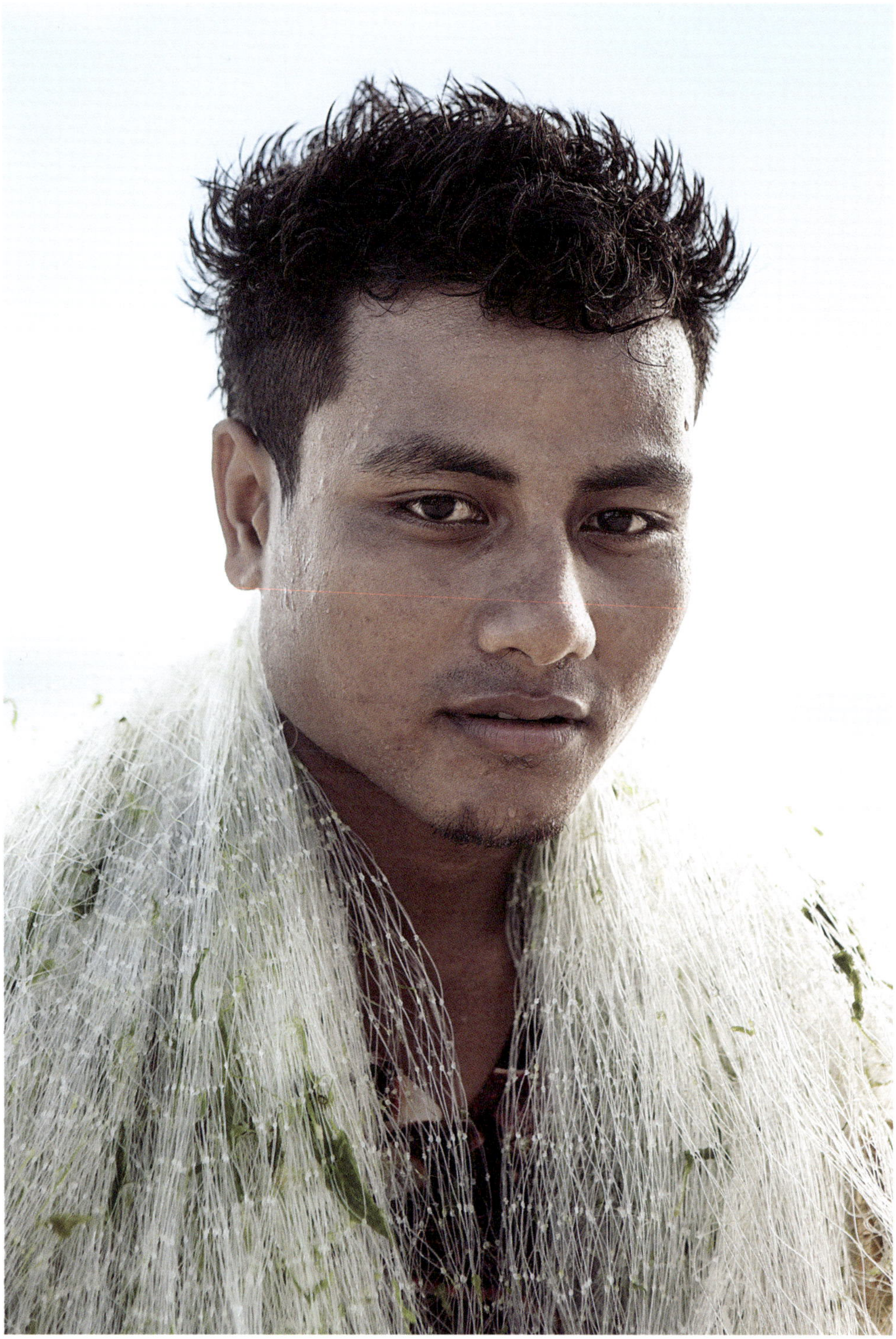

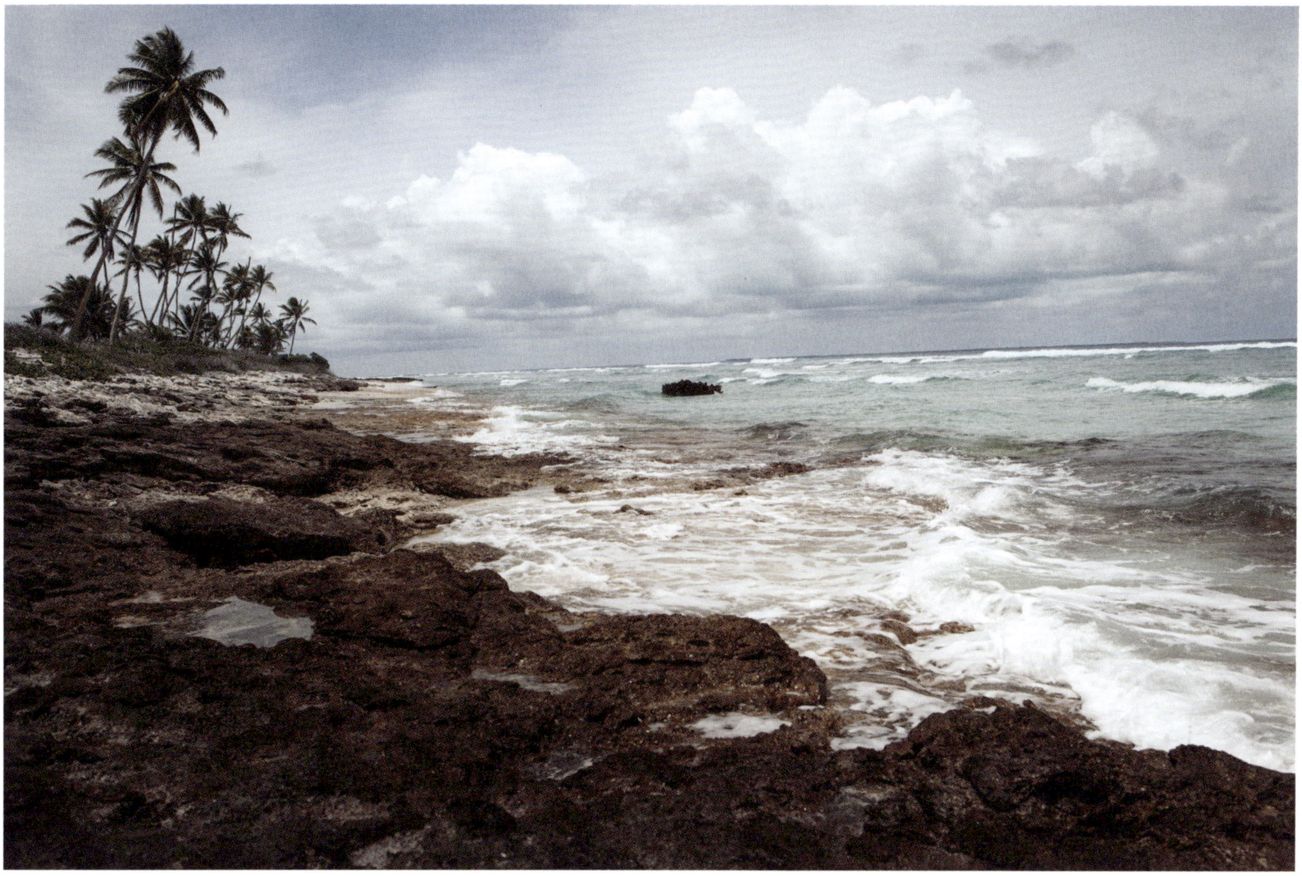

AFRICA

TANSANIA

2° 58′ 35.015″ S 36° 57′ 26.046″ E

TANZANIA

I visited the Maasai in the village of Tingatinga.

The Maasai are an ethnic group of nomadic cattle breeders, herders and warriors who live between southern Kenya and northern Tanzania. It is likely that their ancestors migrated from the Nile Valley through southern Sudan around the 15th century and settled in their current settlement region by the 18th century. They form a very small part of the population in both Kenya and Tanzania; however, they are disproportionally widely recognized, which is presumably based on their very original way of life in close proximity to several National Parks. In Kenya in particular, the displacement of the Maasai from nature reserves and the need for settlement areas for a growing population is severely restricting their nomadic way of life.

Tingatinga, the village I visited, is located in a game reserve that allows the Maasai and wild animals to coexist. The families settle there in individual bomas, a compound of mud huts and enclosures for cattle, sheep and goats. The areas are enclosed by a dense thorn hedge to protect people and livestock from the wild animals that roam at night.

Cattle are central to the culture of the Maasai. However, one of their most important sources of food is the meat of the goats and sheep they also keep. The morani, their warriors and cattle herders, subsist exclusively on saroi, a mixture of cattle blood and milk. It is an integral part of their lives and ceremonies.

The Maasai believe in a creator god, Engai, who is said to live on the summit of Oldonyo Lengai, a mountain in Tanzania, and control the weather. He is also said to have granted them dominion over all the cattle in the world in the act of creation.

According to the IPCC, Africa is the continent most threatened by climate change. This is mainly due to the impact on agriculture, which is the livelihood of around 70 percent of the African population. Increasingly extreme droughts alternate

TANSANIA

Mein Besuch galt den Massai in dem Dorf Tingatinga.

Die Massai sind eine nomadische Ethnie von Rinderzüchtern, Hirten und Kriegern, die zwischen dem südlichen Kenia und dem nördlichen Tansania leben. Als wahrscheinlich gilt, dass ihre Vorfahren um das 15. Jahrhundert aus dem Niltal durch den Südsudan zogen und sich bis zum 18. Jahrhundert in ihrem heutigen Siedlungsgebiet niederließen. Sie bilden sowohl in Kenia als auch in Tansania einen sehr kleinen Teil der Bevölkerung; ihre Bekanntheit, die vermutlich überwiegend auf ihrer sehr ursprünglichen Lebensweise in unmittelbarer Nachbarschaft zu Nationalparks beruht, steht dazu jedoch in keinem Verhältnis. Vor allem in Kenia wird durch den Bedarf an Siedlungsgebieten für eine wachsende Bevölkerung und die Verdrängung der Massai aus Naturschutzgebieten deren nomadische Lebensweise stark eingeschränkt.

Tingatinga, das von mir besuchte Dorf, liegt in einem Wildschutzgebiet, das den Massai und den Wildtieren ein Leben in Koexistenz ermöglicht. Die Familien siedeln dort in einzelnen Bomas, Verbünden aus Lehmhütten und Gehegen für Rinder, Schafe und Ziegen. Diese werden von einer dichten Dornenhecke umschlossen, um Menschen und Nutztiere vor den Wildtieren zu schützen, die nachts umherstreifen.

Zentral für die Kultur der Massai ist das Rind. Eine ihrer wichtigsten Nahrungsquellen jedoch ist das Fleisch der von ihnen ebenfalls gehaltenen Ziegen und Schafe. Die Morani, ihre Krieger und Rinderhirten, ernähren sich dagegen ausschließlich von Saroi, einem Gemisch aus Rinderblut und -milch. Es ist fester Bestandteil ihres Lebens und ihrer Zeremonien.

Die Massai glauben an einen Schöpfergott, Engai; dieser soll auf dem Gipfel des Oldonyo Lengai, einem Berg in Tansania, wohnen, und über das Wetter bestimmen. Zudem habe er ihnen im Schöpfungsakt die Herrschaft über alle auf der Welt vorhandenen Rinder zugesprochen.

with heavy rainfall events, leading to a progressive loss of topsoil. According to the Potsdam Institute for Climate Impact Research, these are not exceptional weather phenomena, but consequences of climate change. The air is getting warmer and can hold more moisture, so that as it discharges, extreme precipitation is increasing. These weather events exacerbate the situation in African countries, where there are enough crises even without manmade climate change, warns the UN Refugee Agency.

Biodiversity is also suffering massively from the increasing climate crisis. According to estimates, a third of animal and plant species in Africa will have disappeared within 60 years.

Afrika ist laut Einschätzung des IPCC der Kontinent, dem durch den Klimawandel das meiste Unheil droht. Dies liegt vor allem an den Auswirkungen auf die Landwirtschaft, von der rund 70 Prozent der afrikanischen Bevölkerung leben. Immer extremere Dürren wechseln sich mit Starkregenereignissen ab, was zu einem fortschreitenden Verlust an Mutterboden führt. Dies seien keine außergewöhnlichen Wetterphänomene, sondern Folgen des Klimawandels, so das Potsdam-Institut für Klimafolgenforschung. Die Luft wird wärmer und kann mehr Feuchtigkeit halten, somit nehmen bei deren Entladung Extremniederschläge zu. Diese Wetterereignisse verschärfen die Situation in den afrikanischen Ländern, in denen es auch ohne die von Menschen verursachte Klimaveränderung genug Krisen gibt, warnt die UN-Flüchtlingshilfe.

Auch die Artenvielfalt leidet massiv unter der zunehmenden Klimakrise. Schätzungen zufolge werden in 60 Jahren ein Drittel der Tier- und Pflanzenarten Afrikas verschwunden sein.

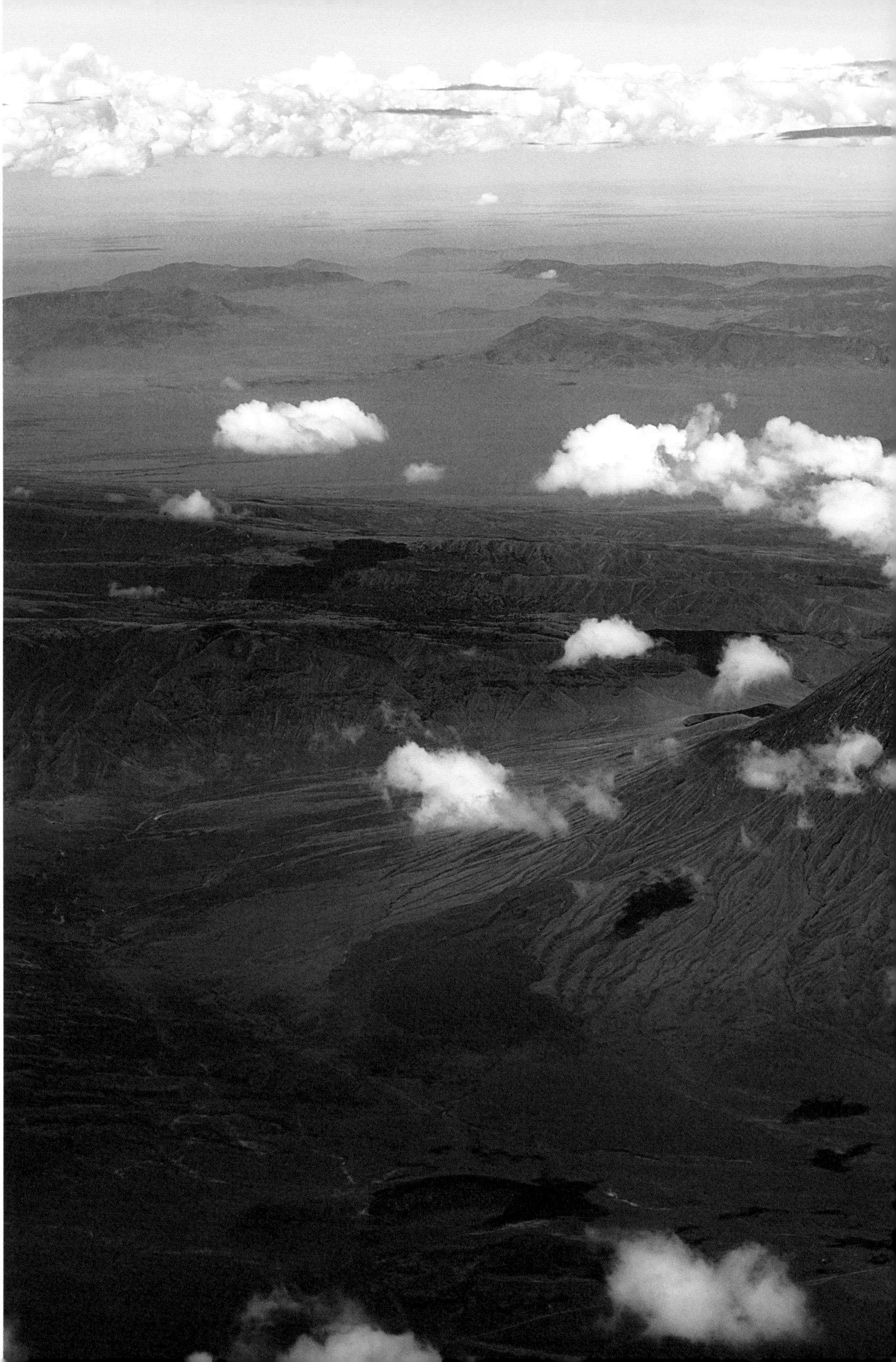

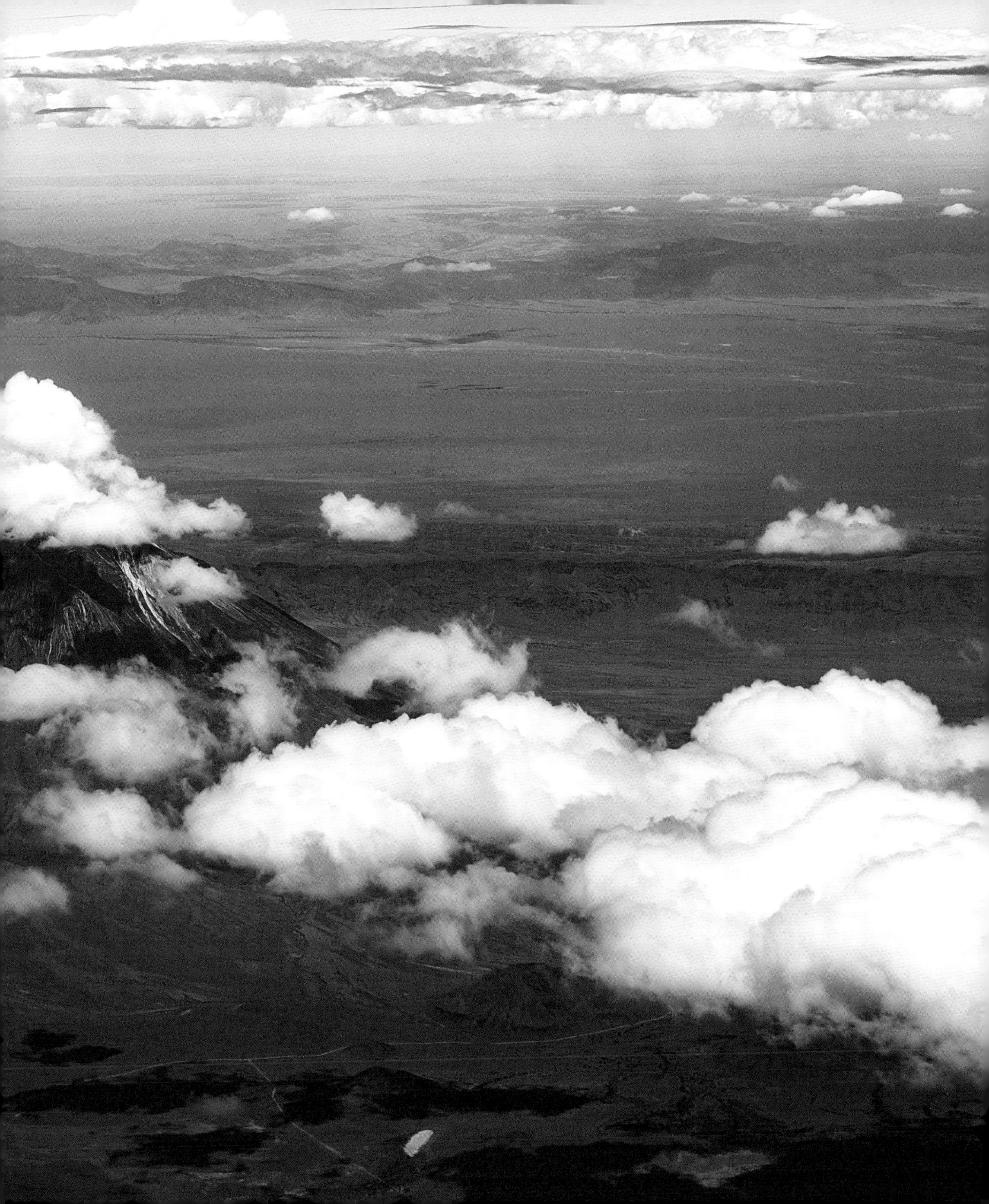

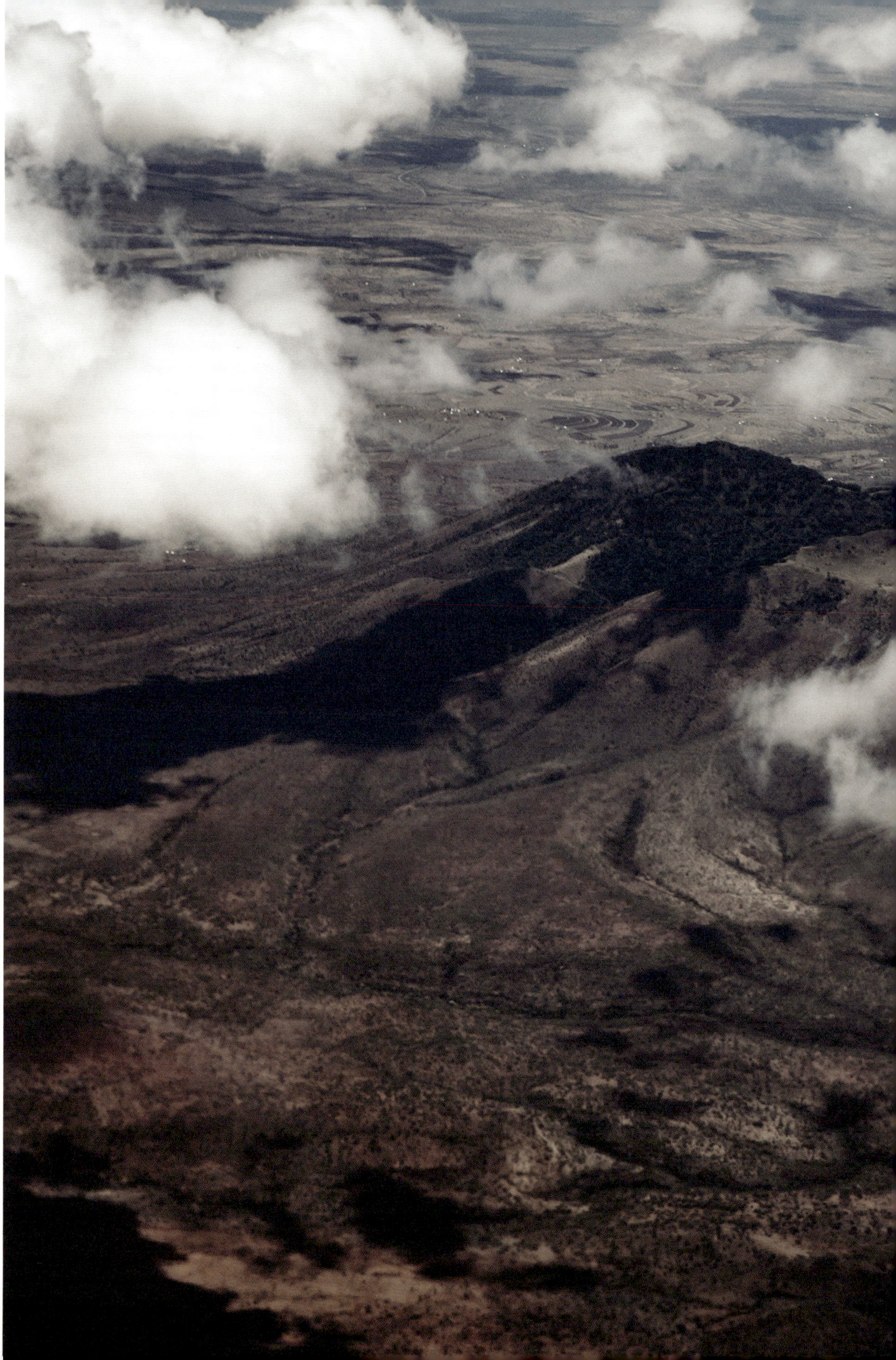

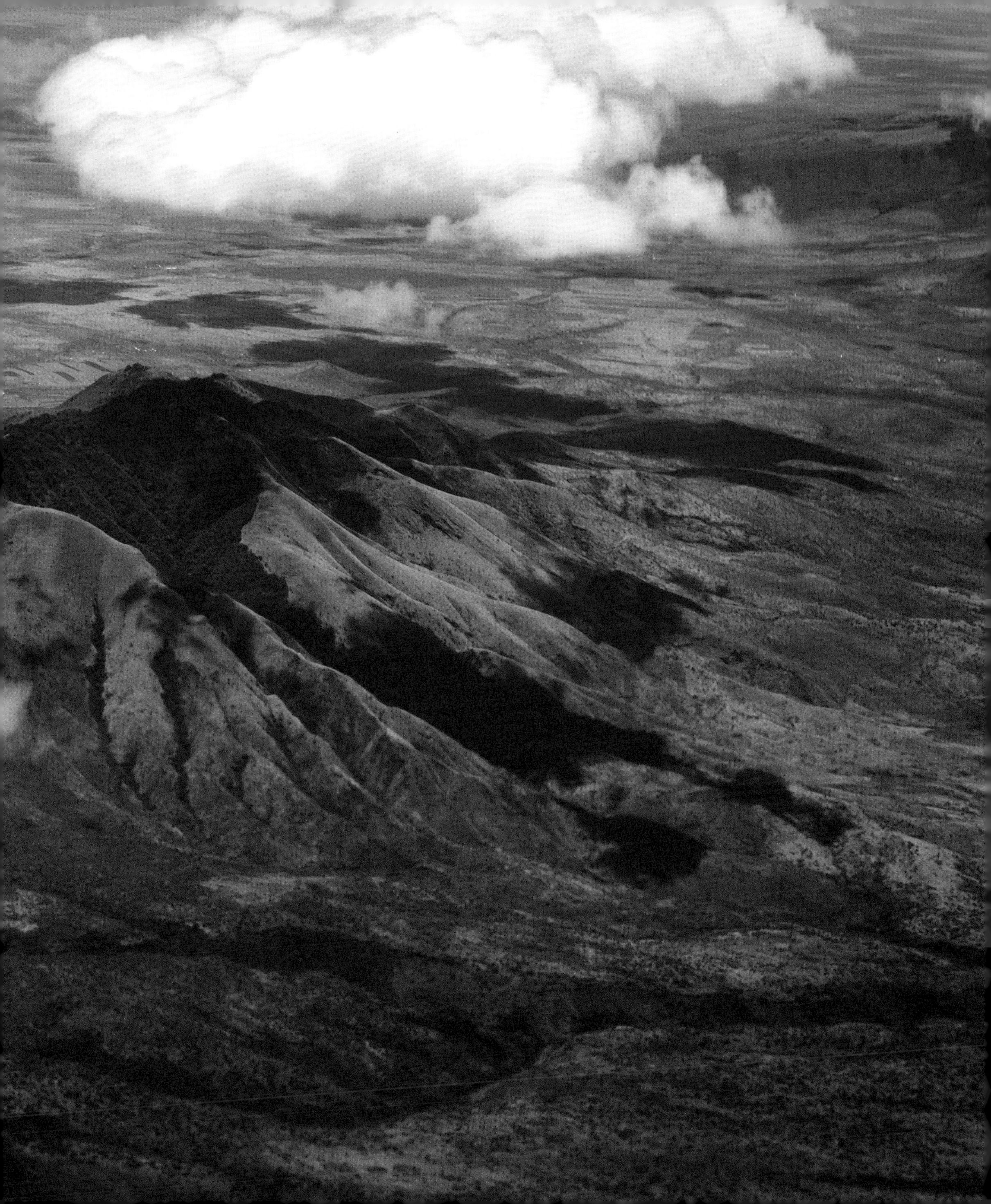

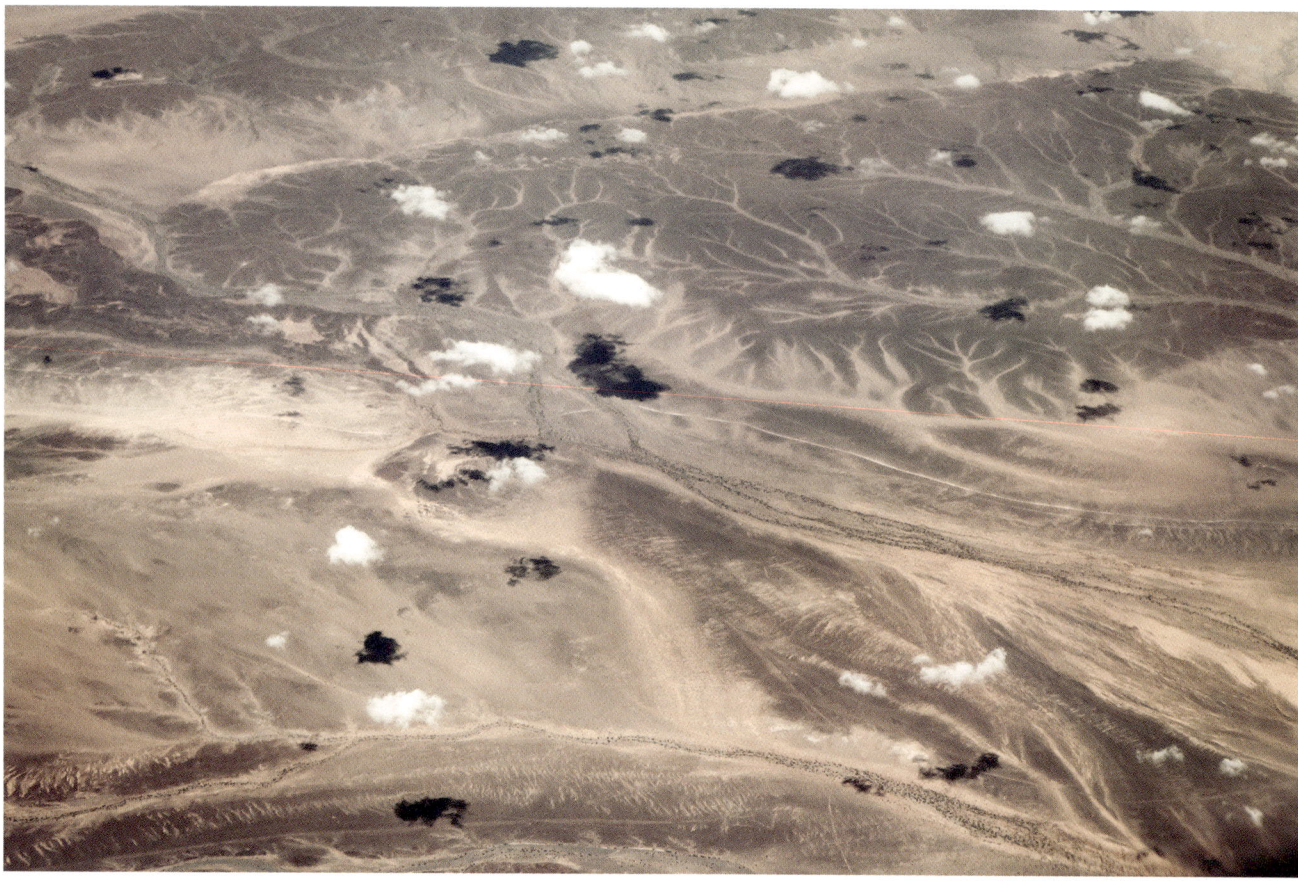

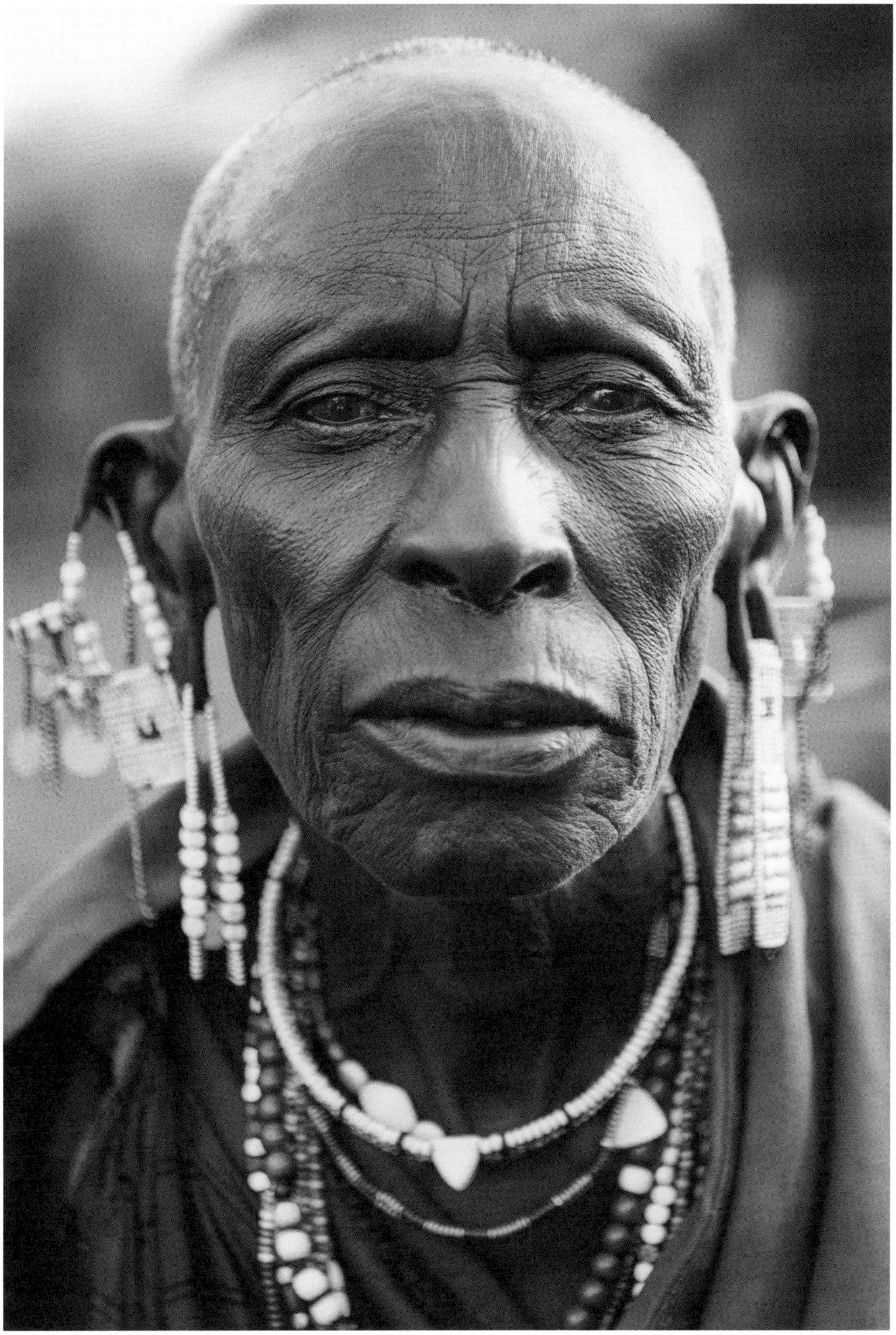

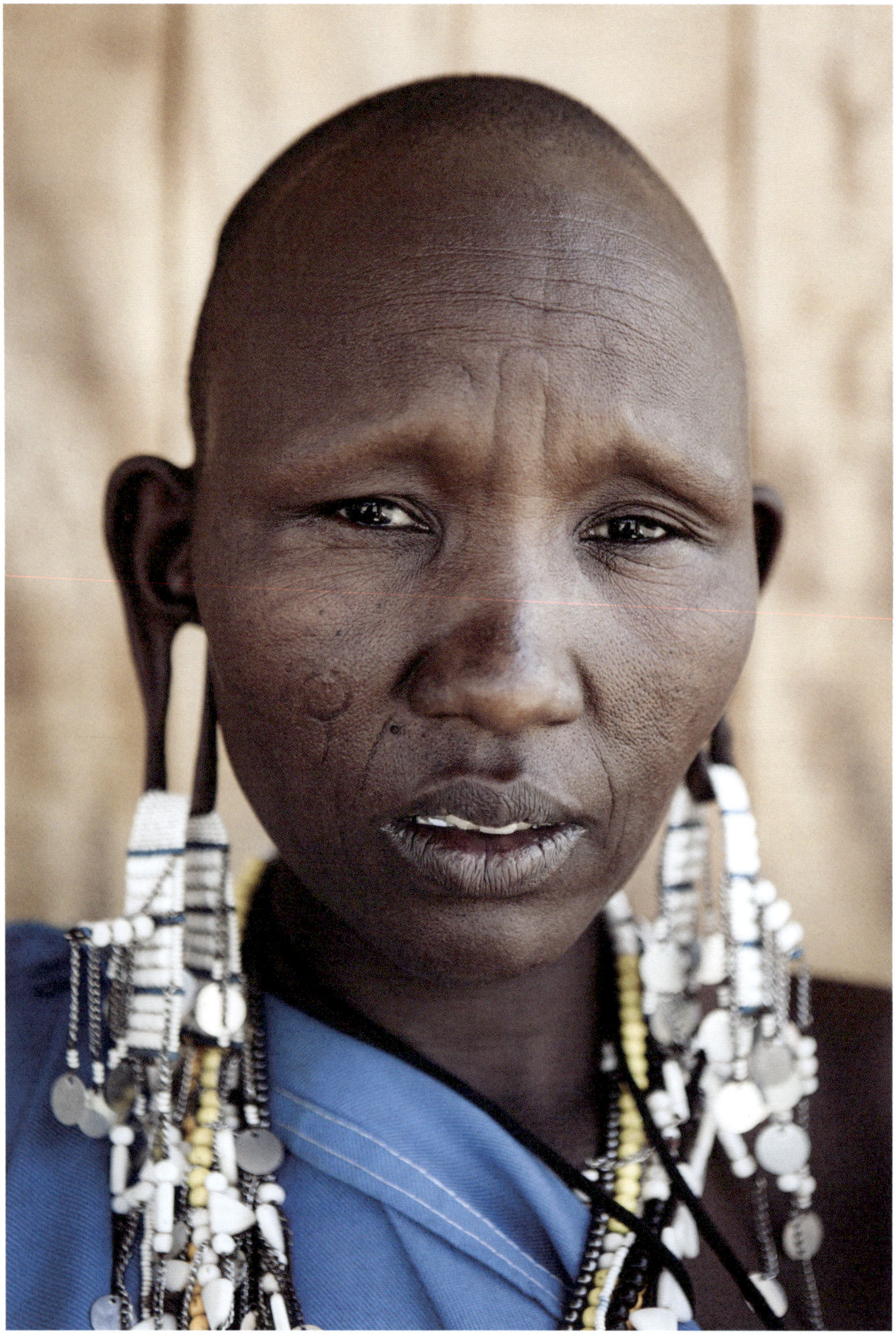

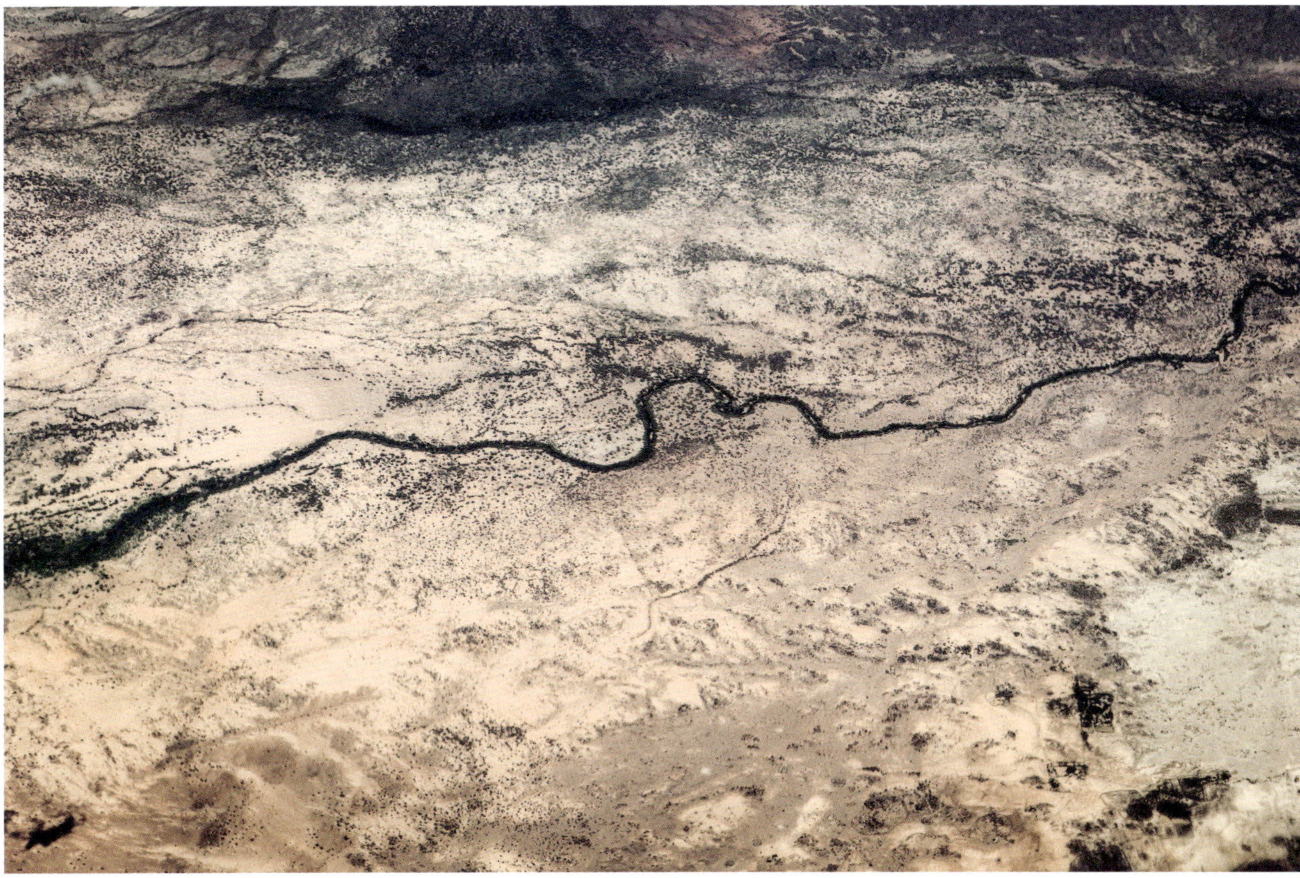

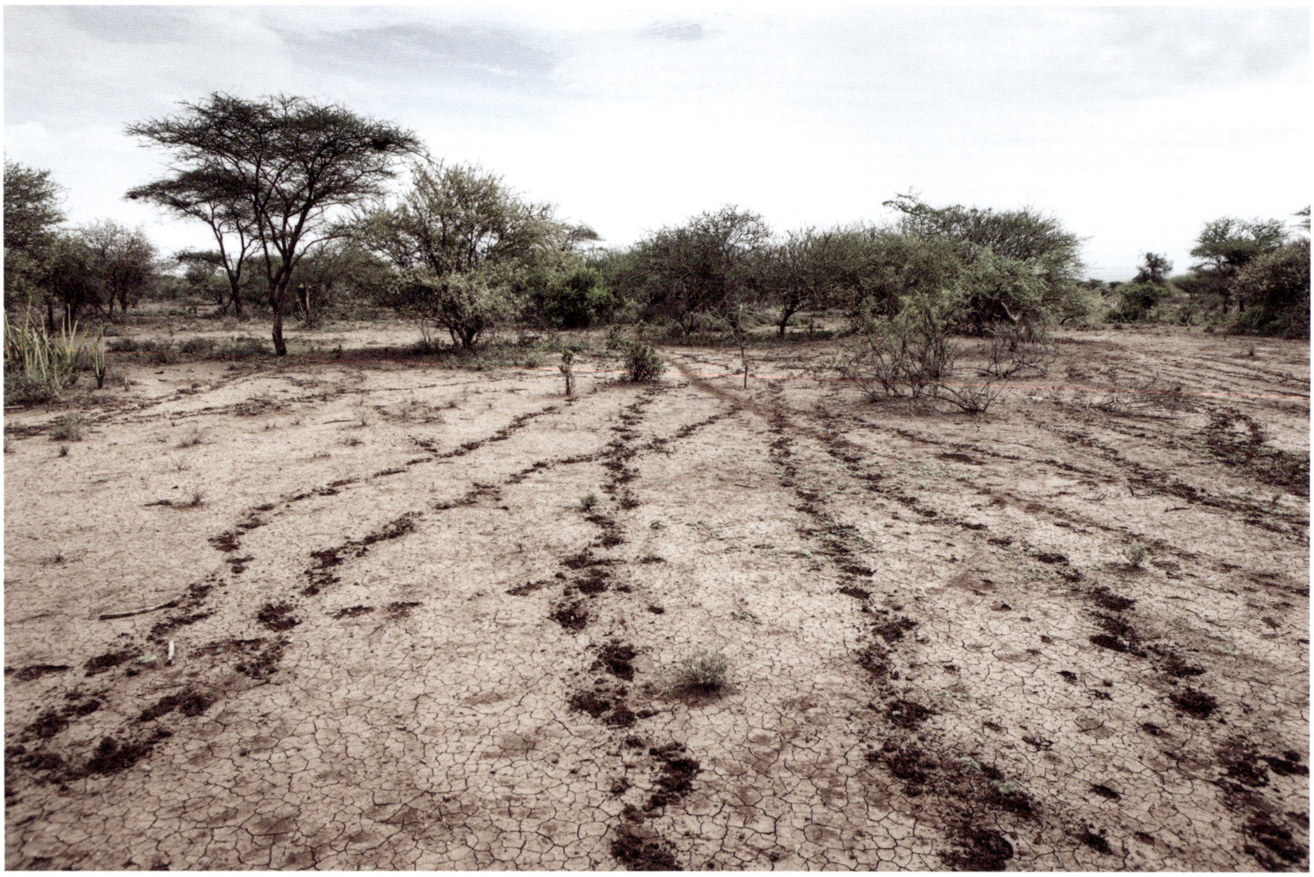

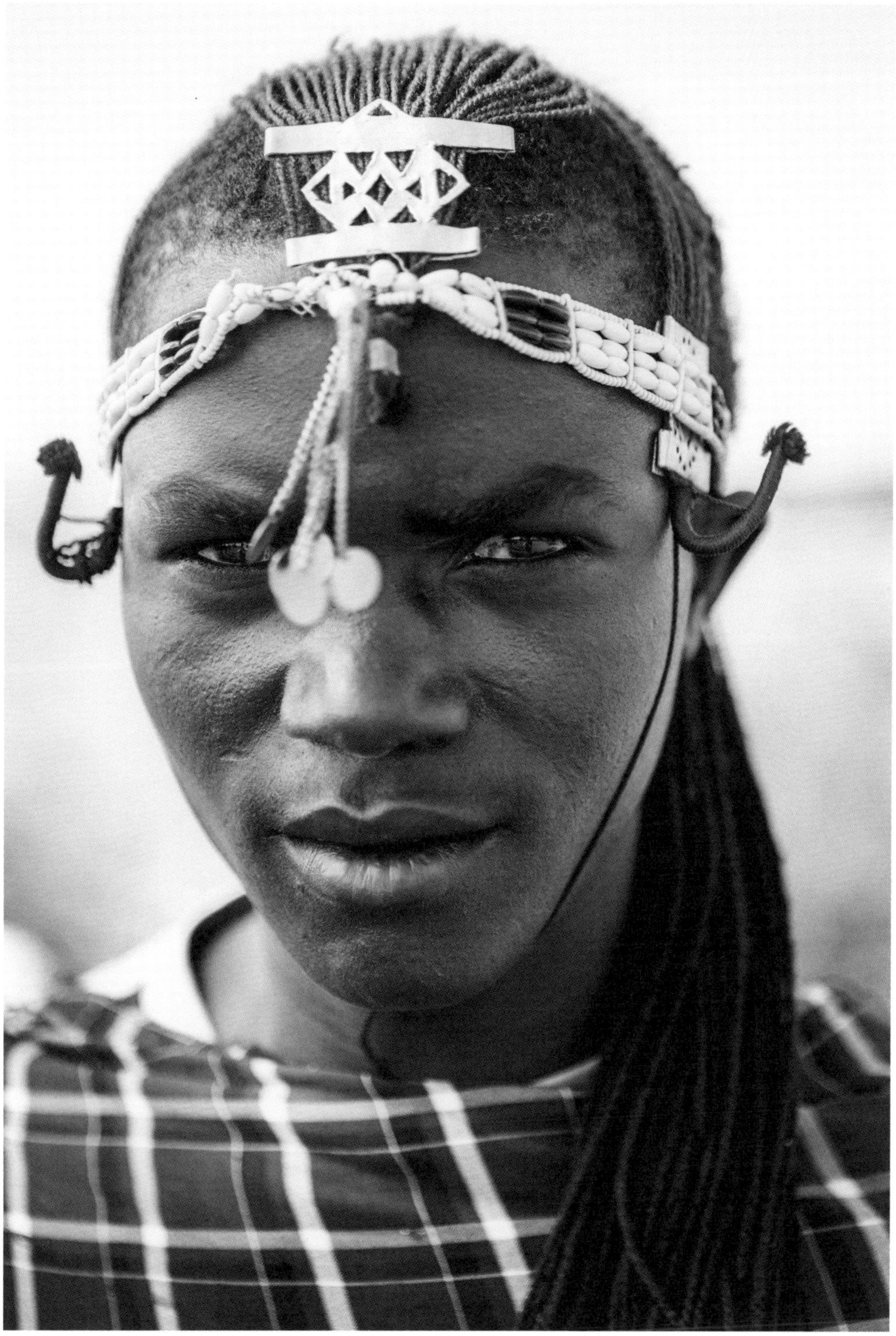

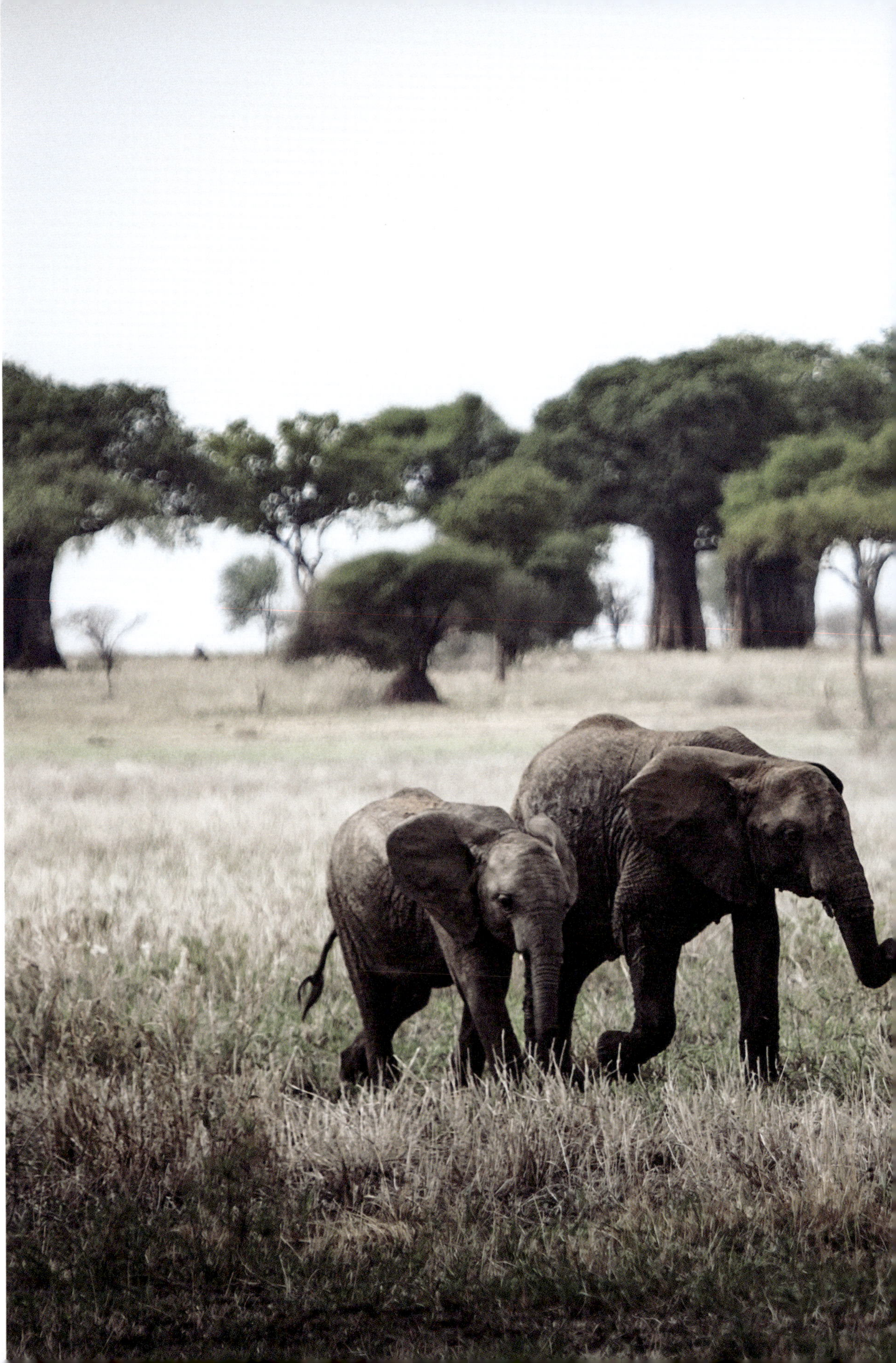

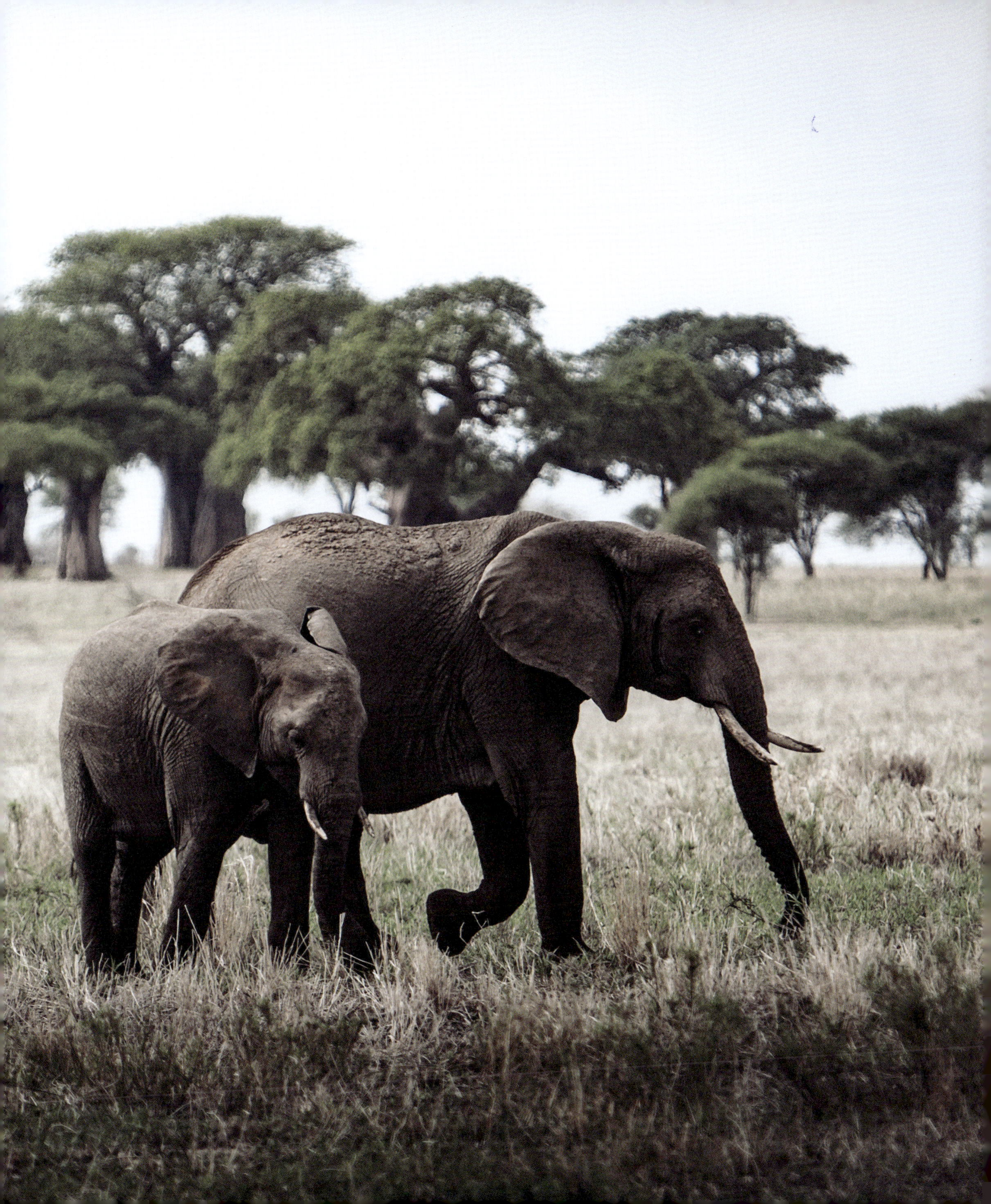

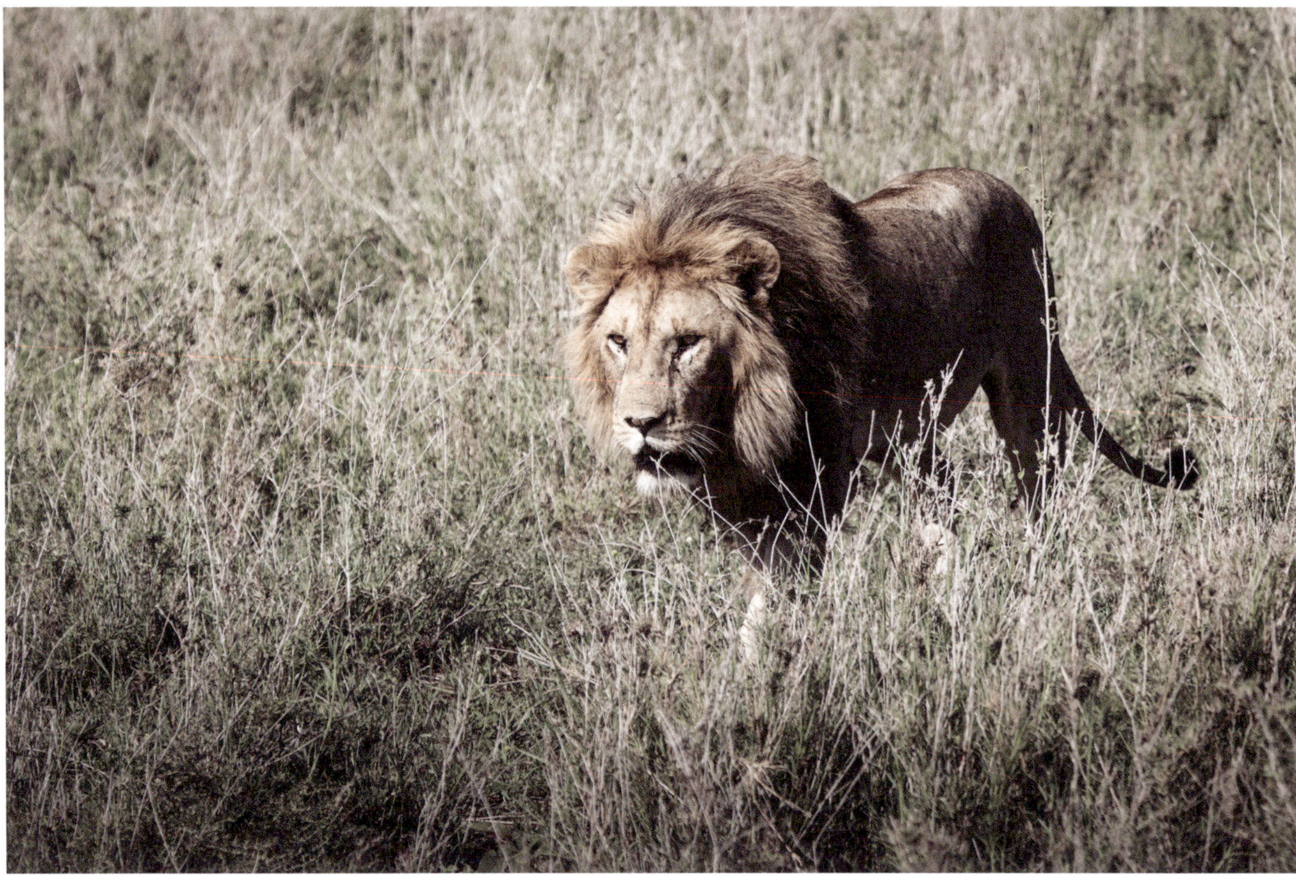

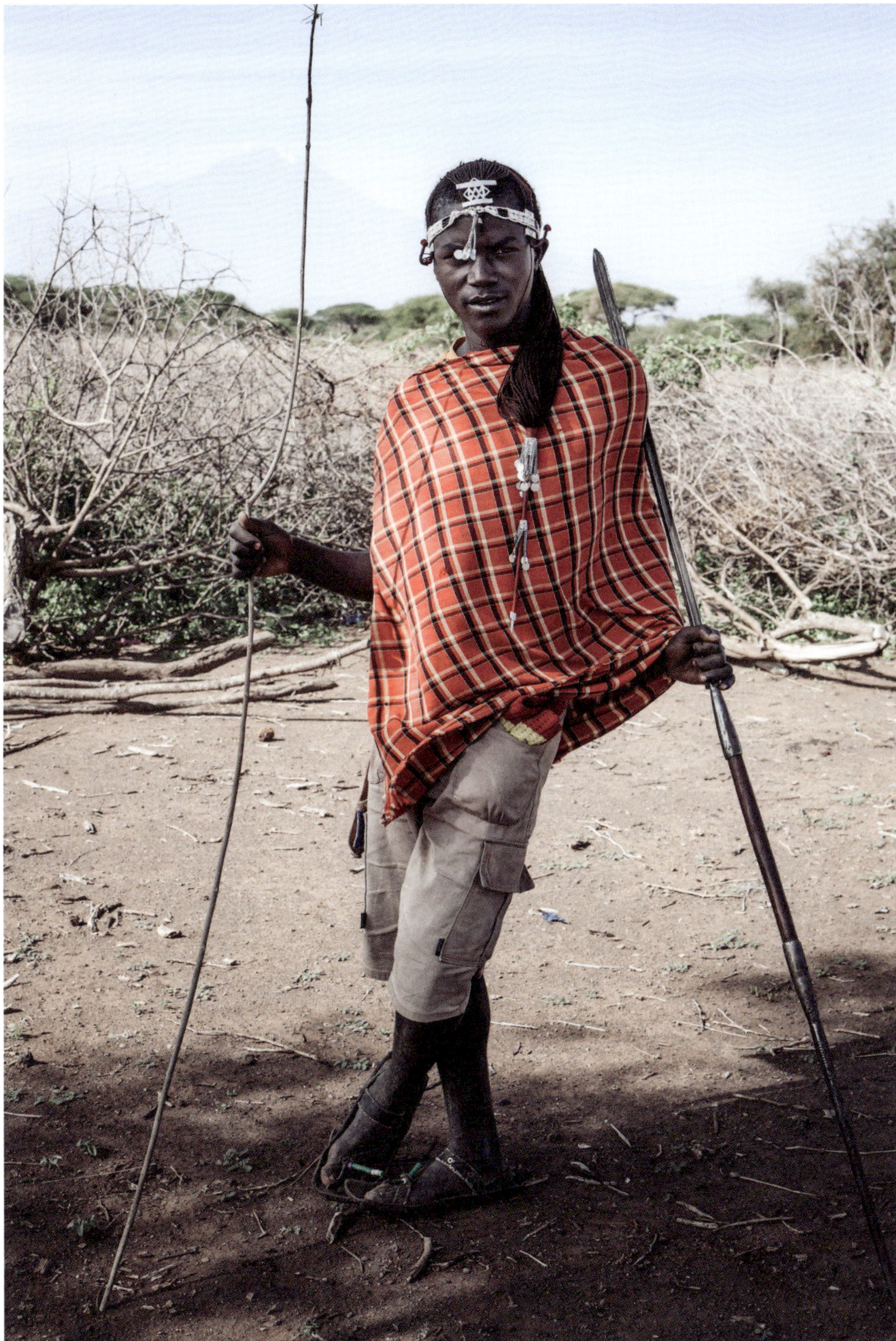

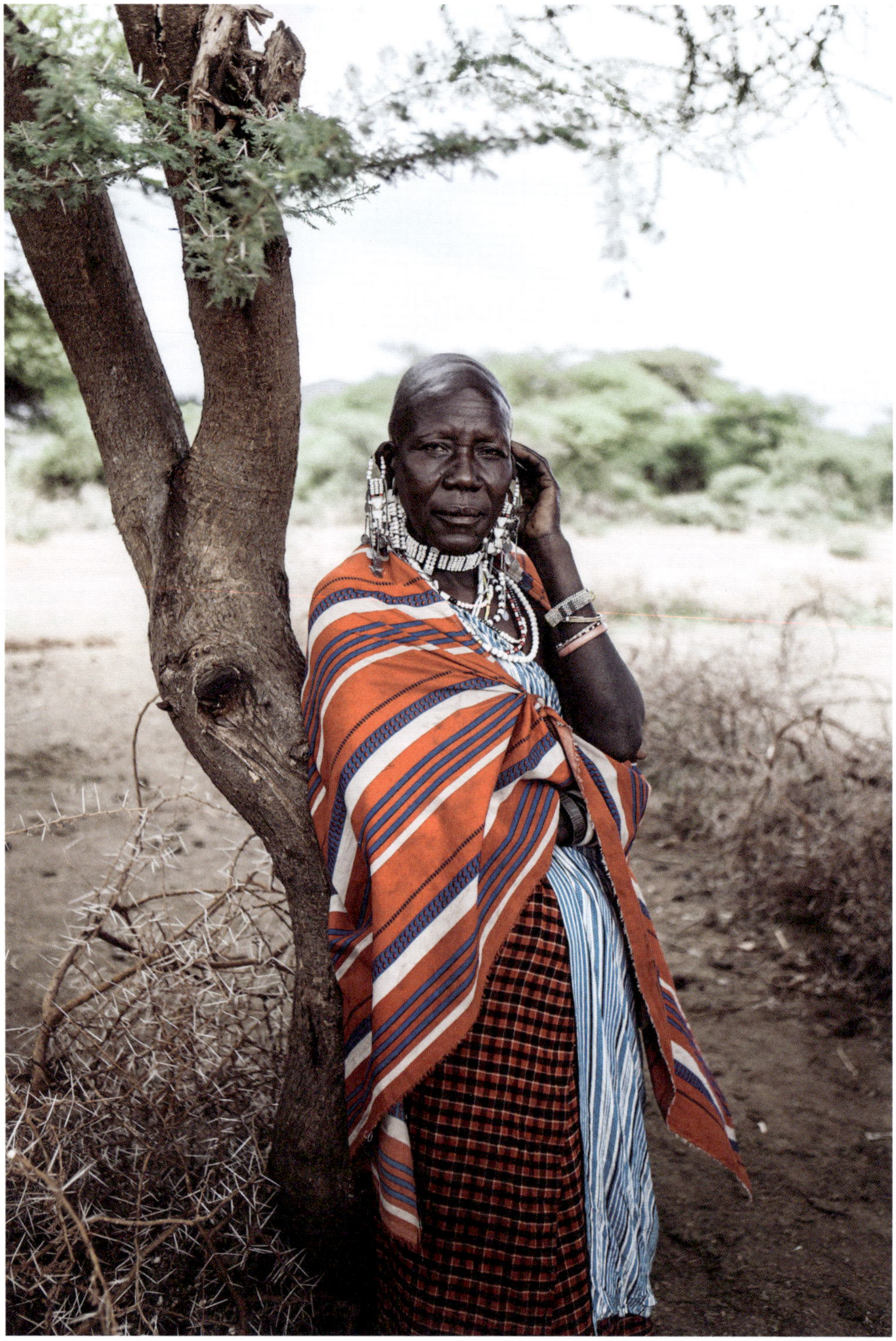

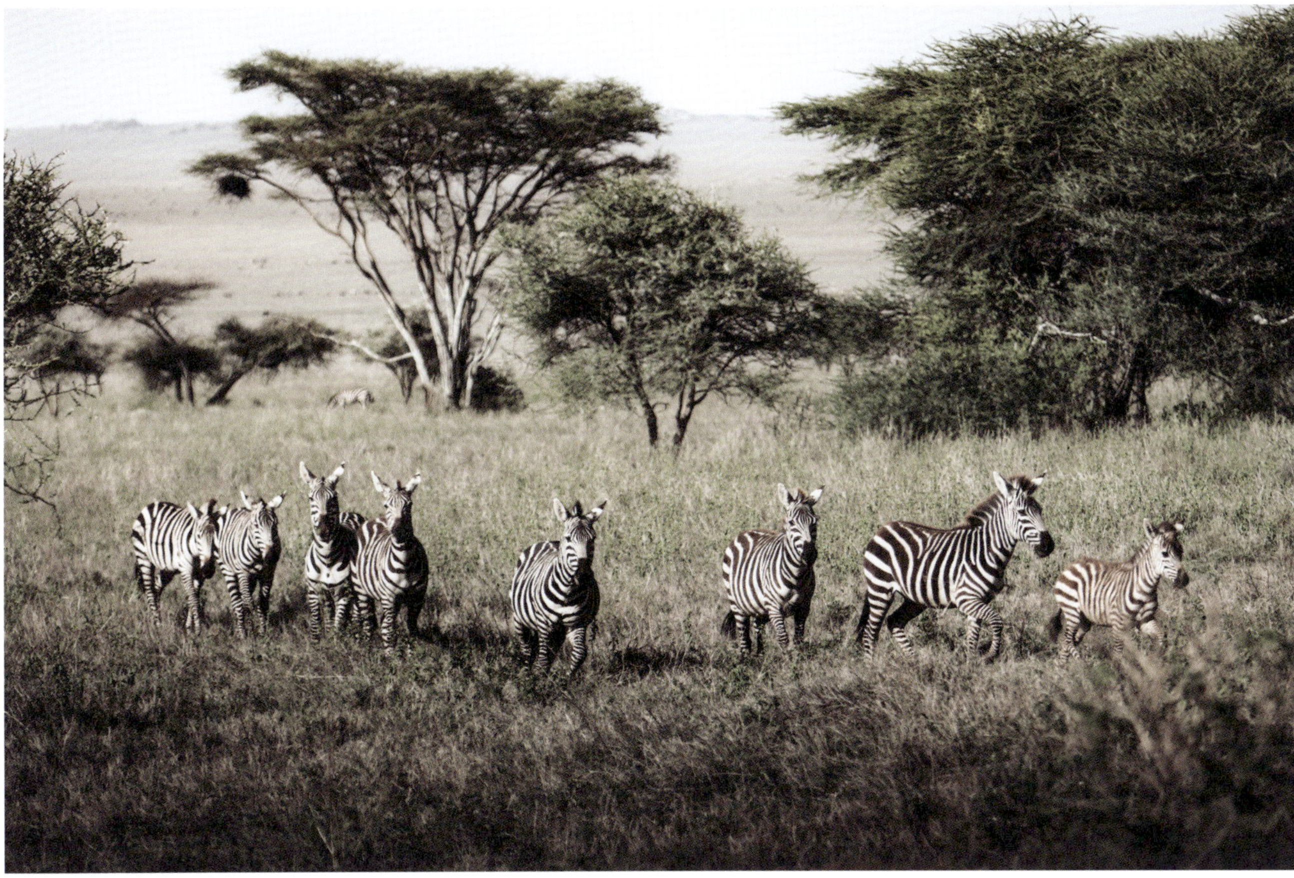

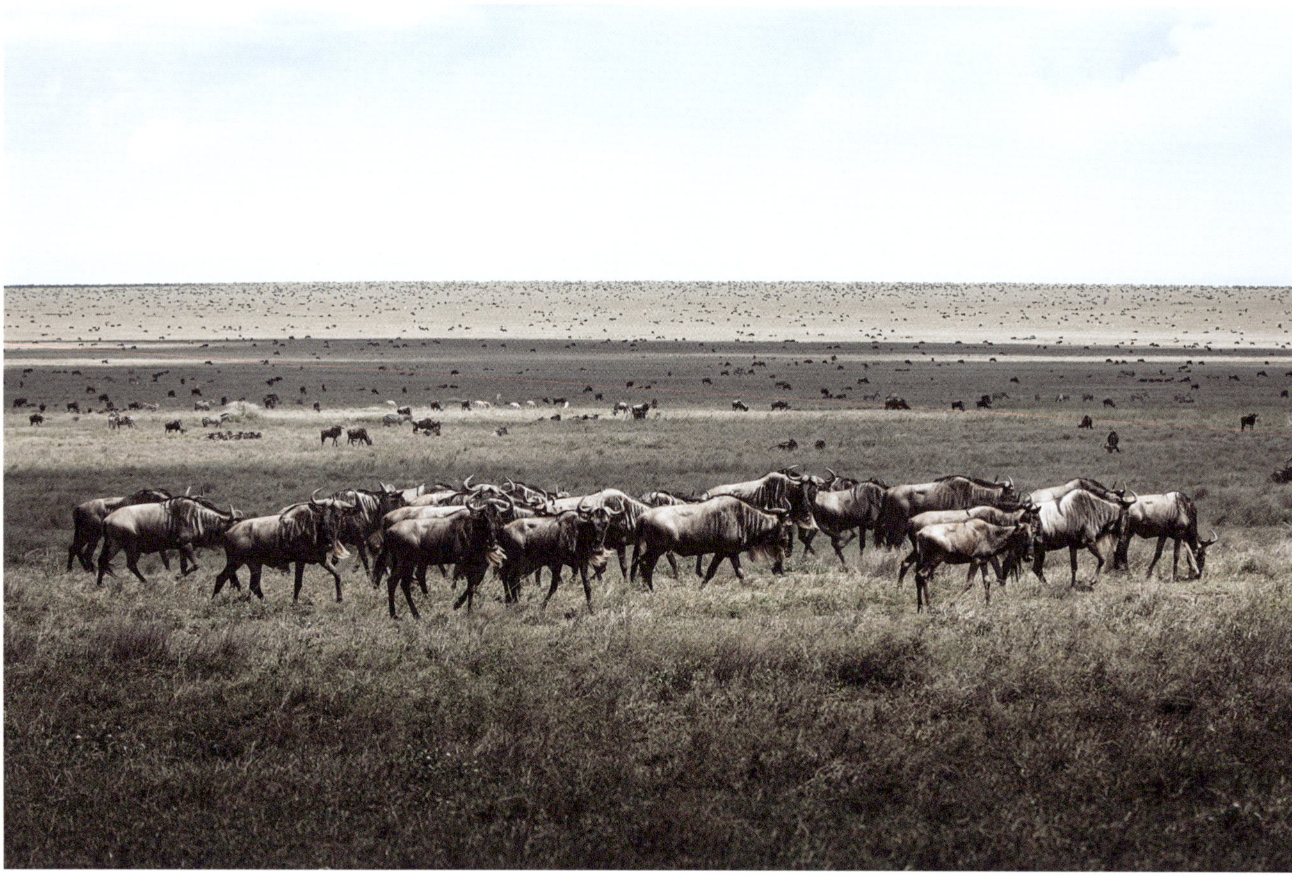

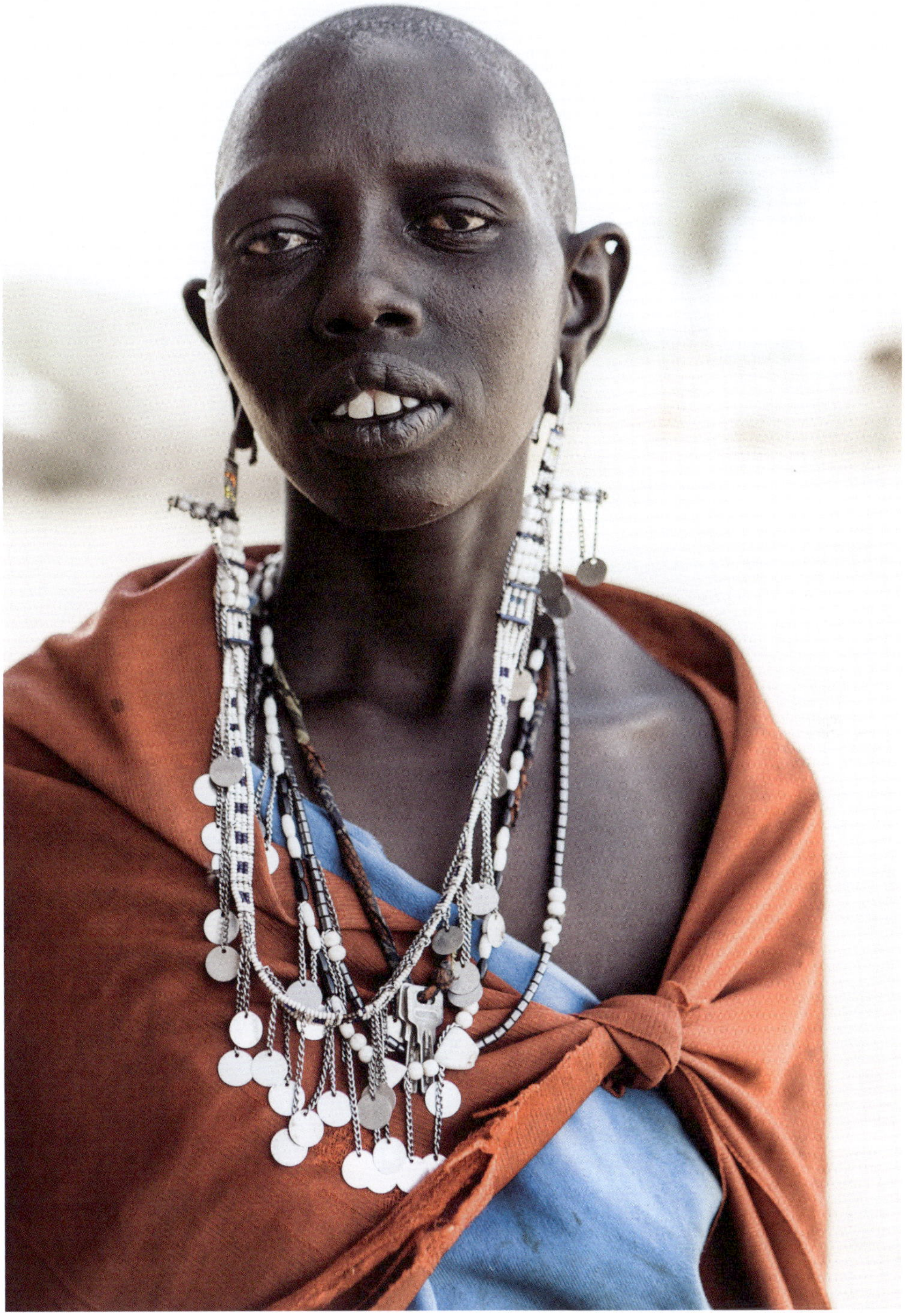

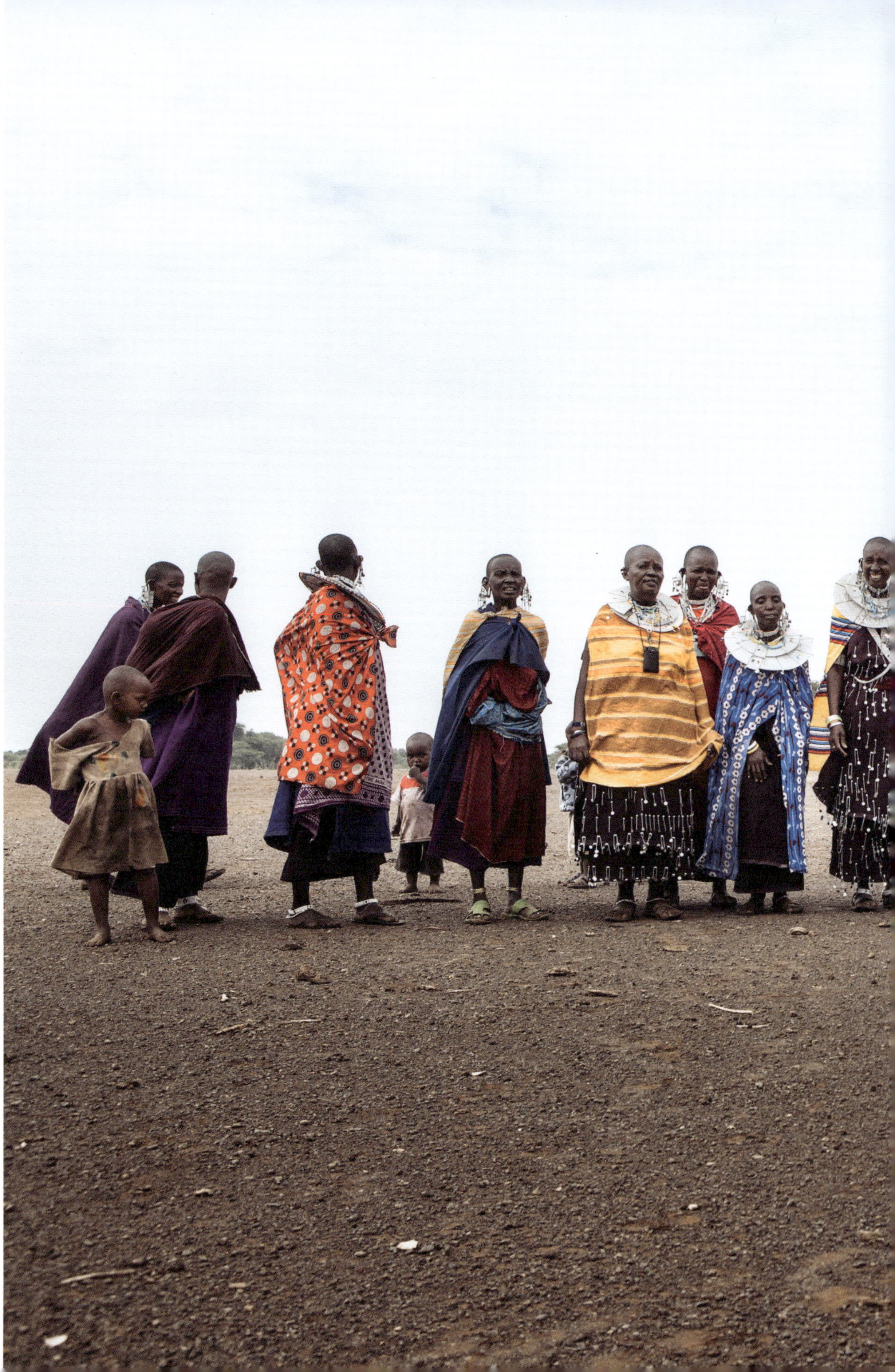

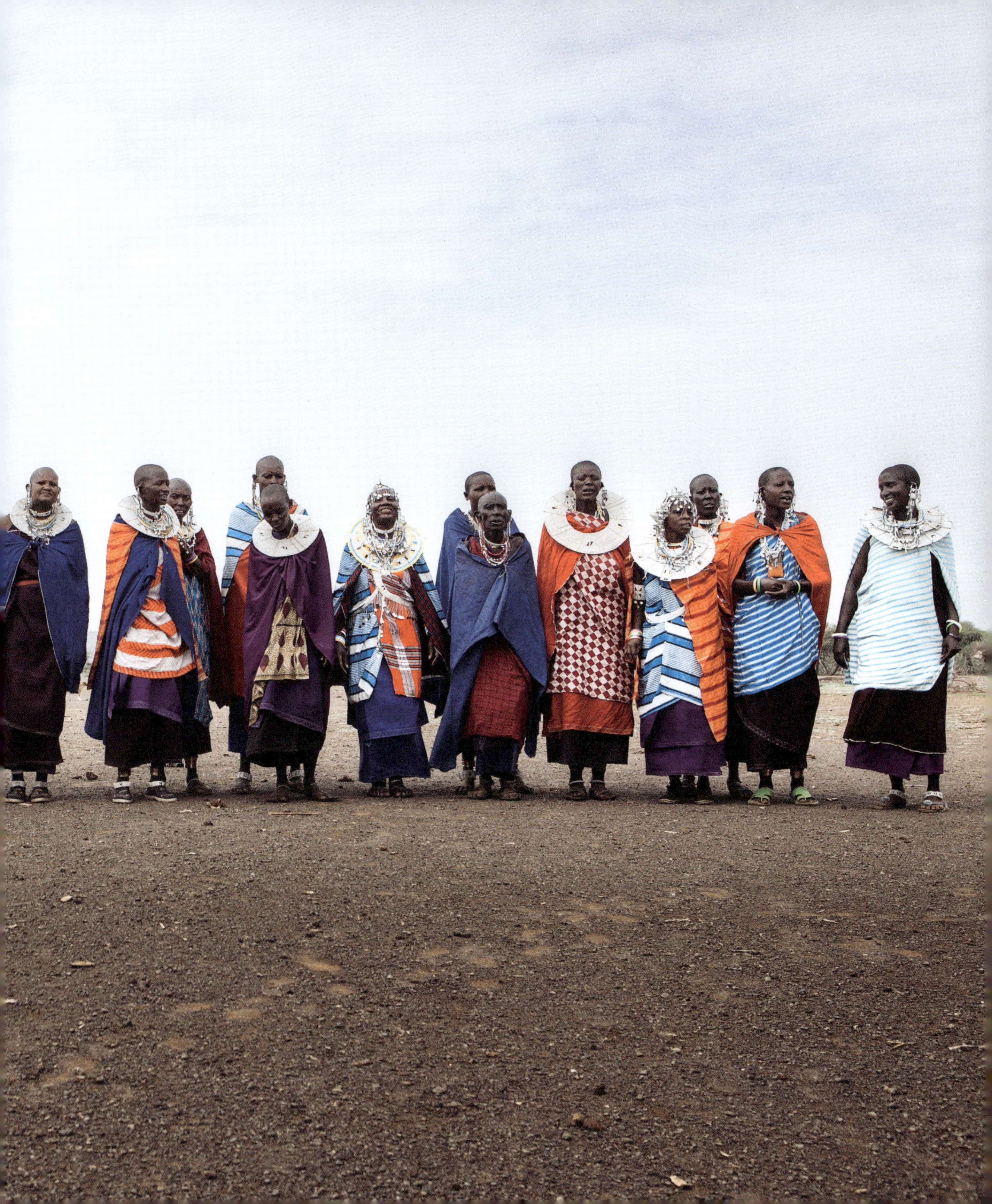

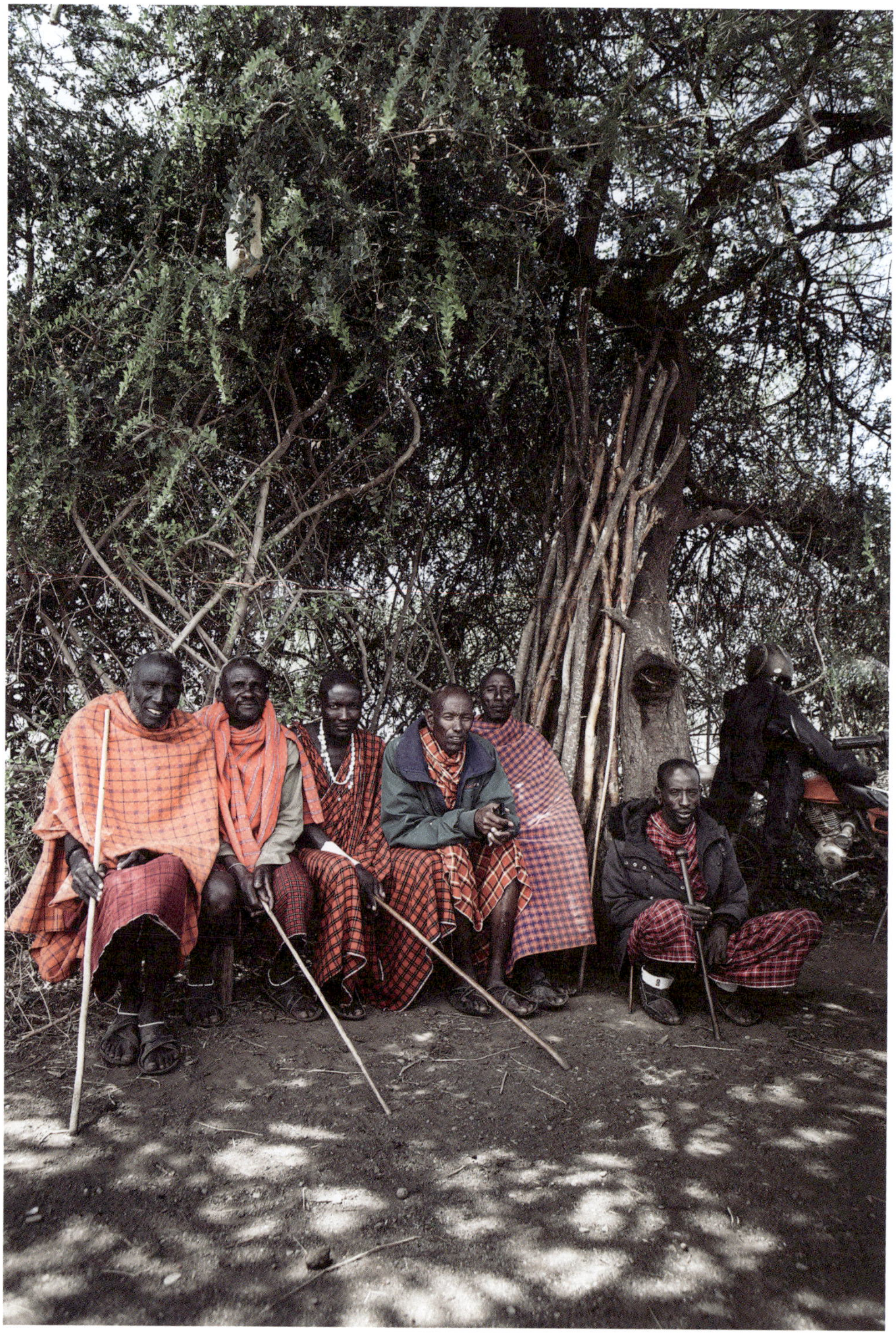

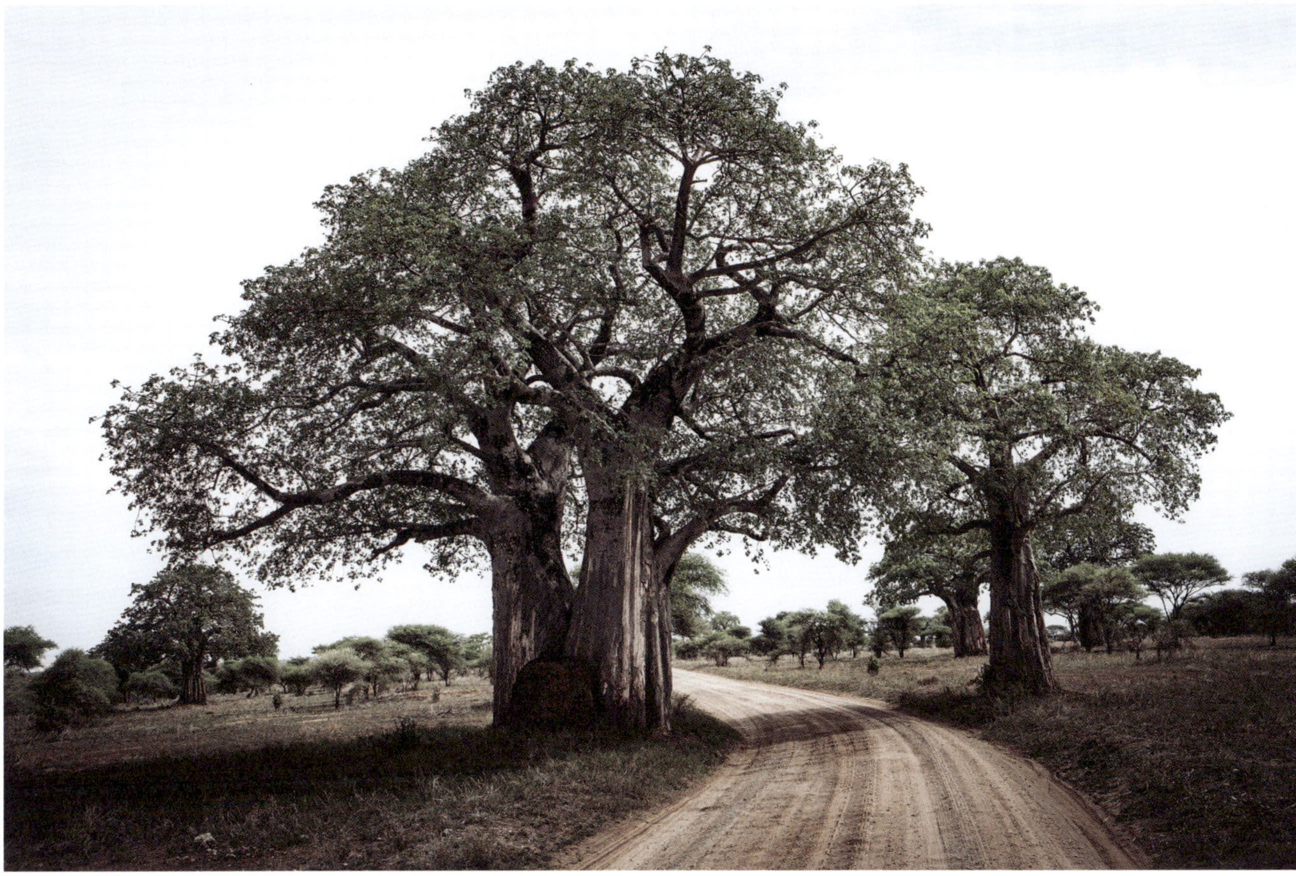

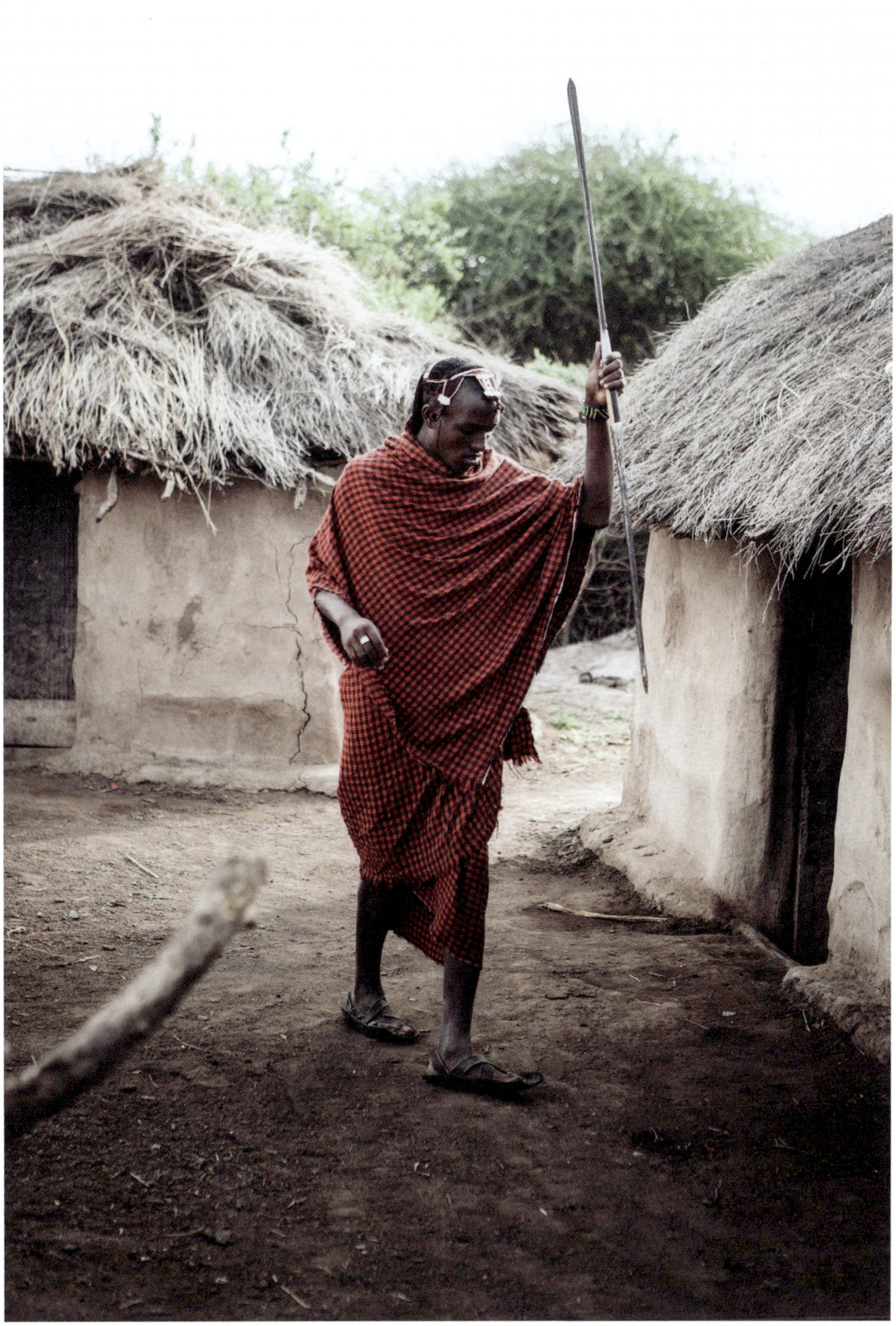

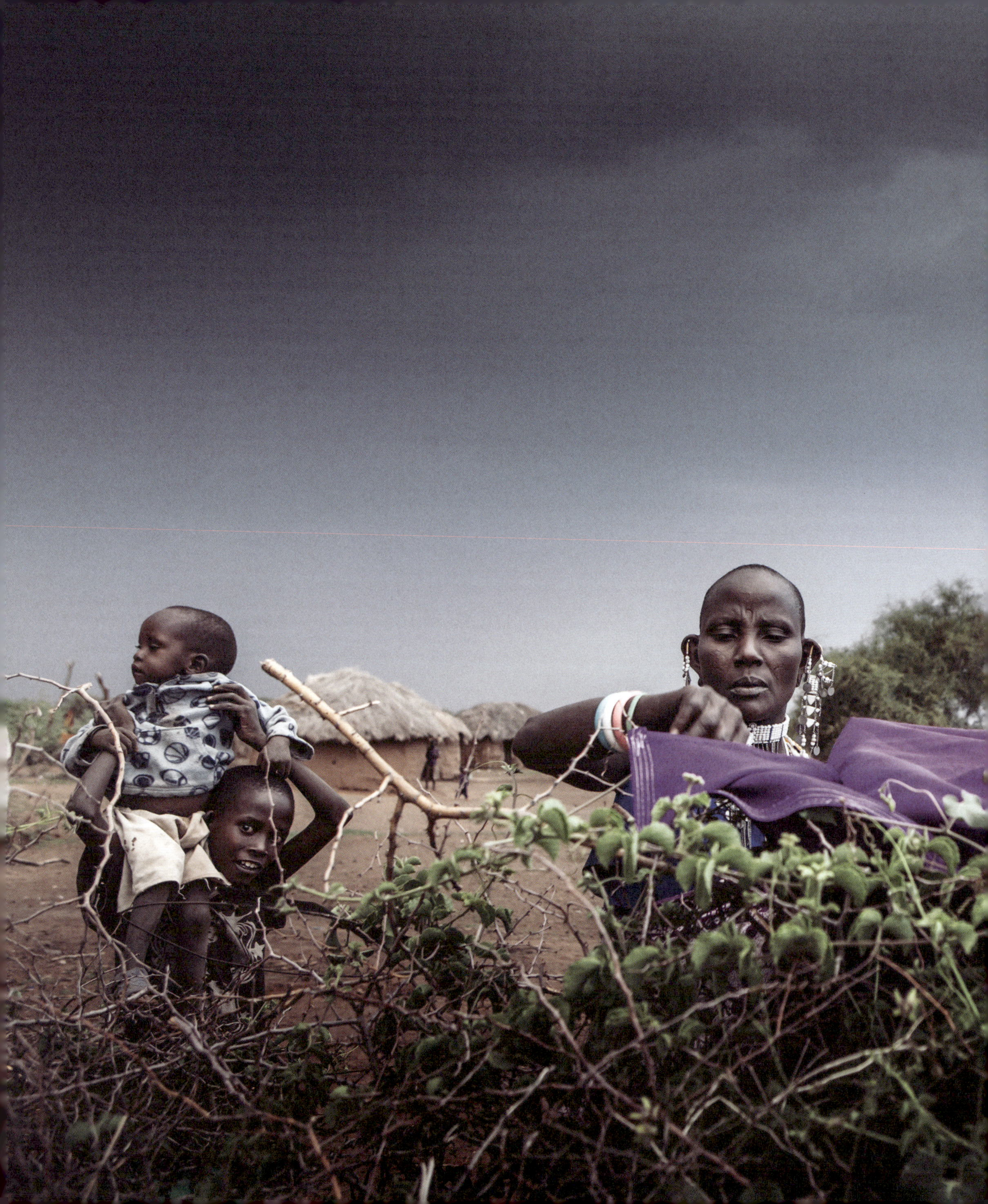

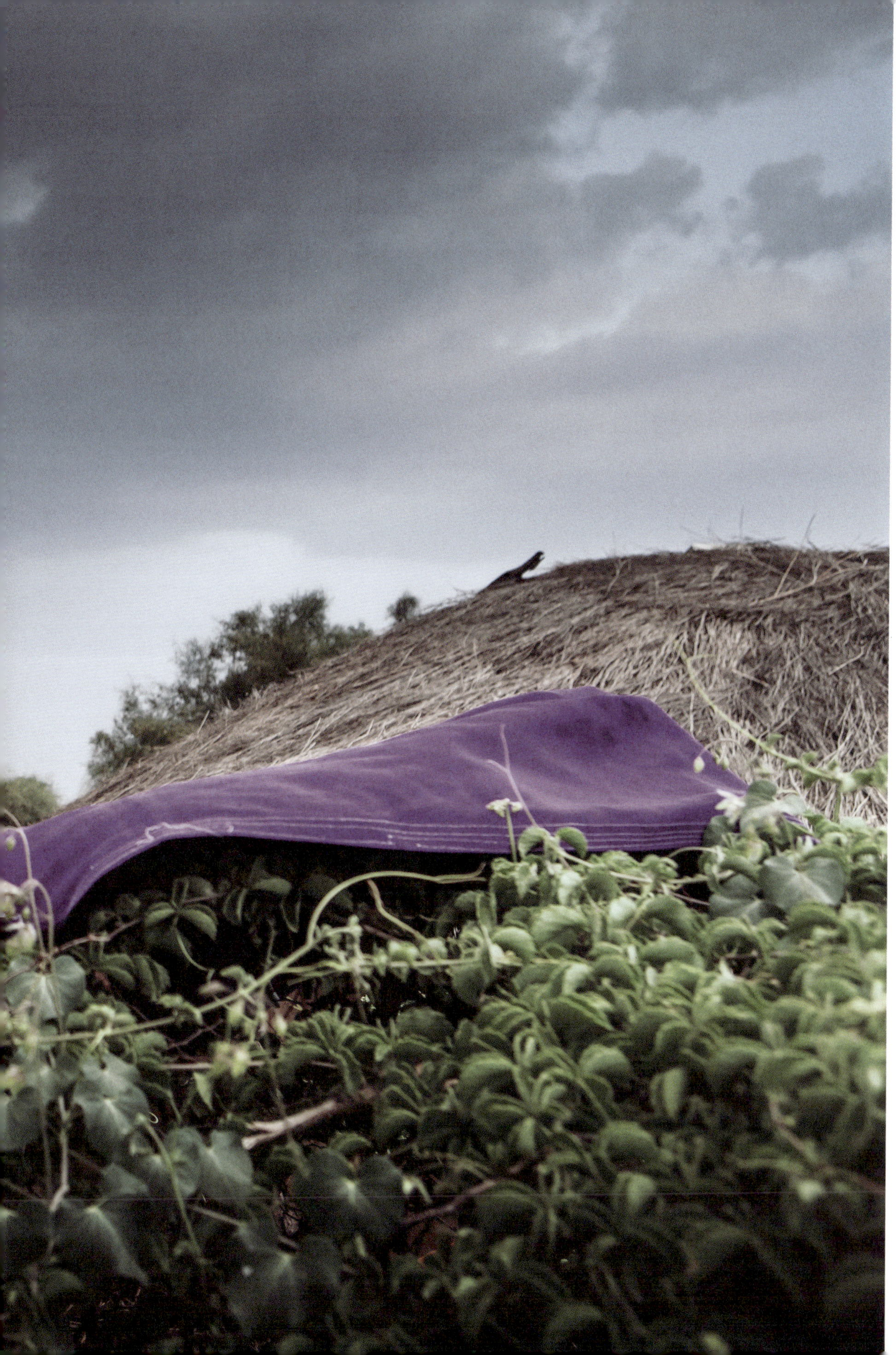

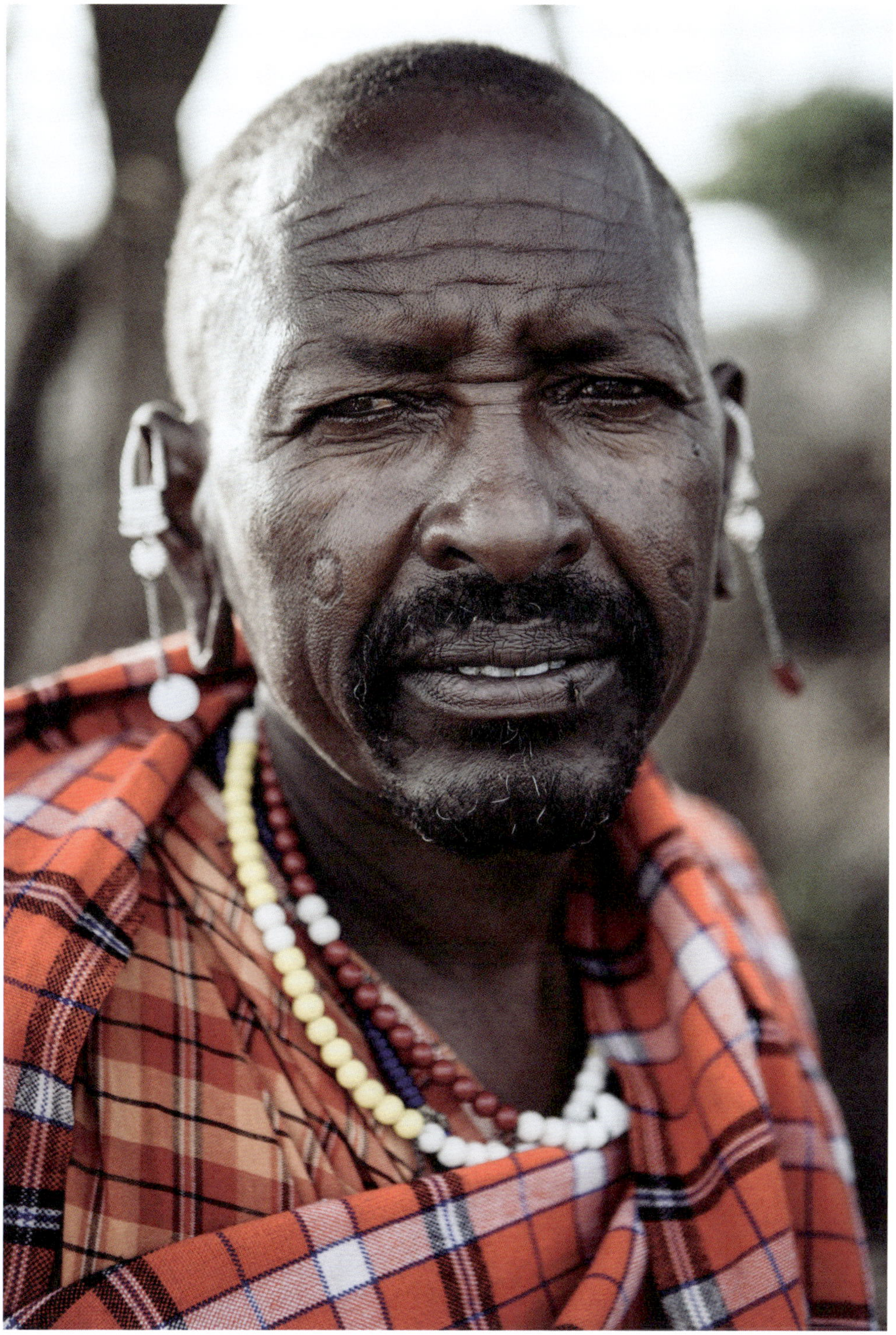

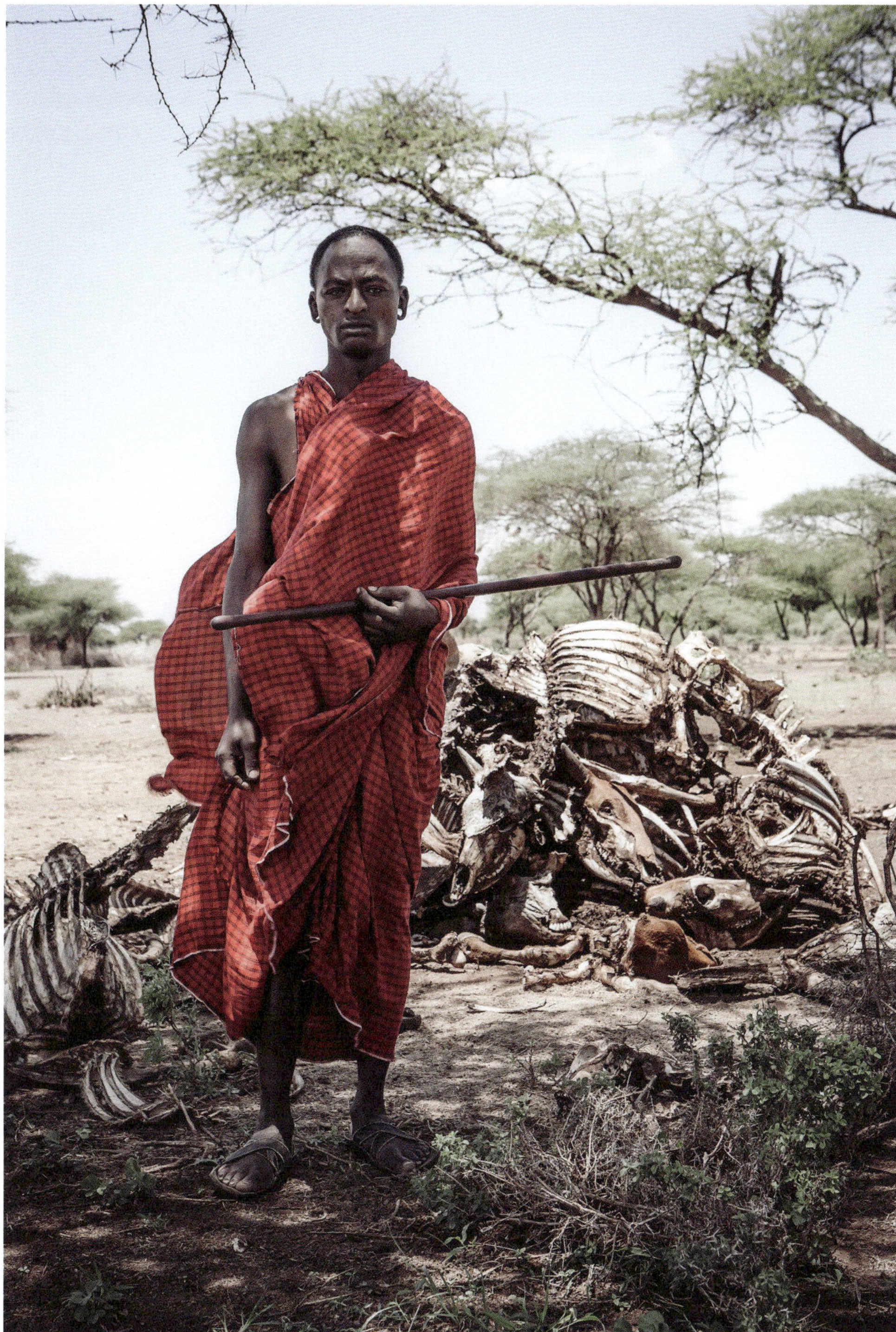

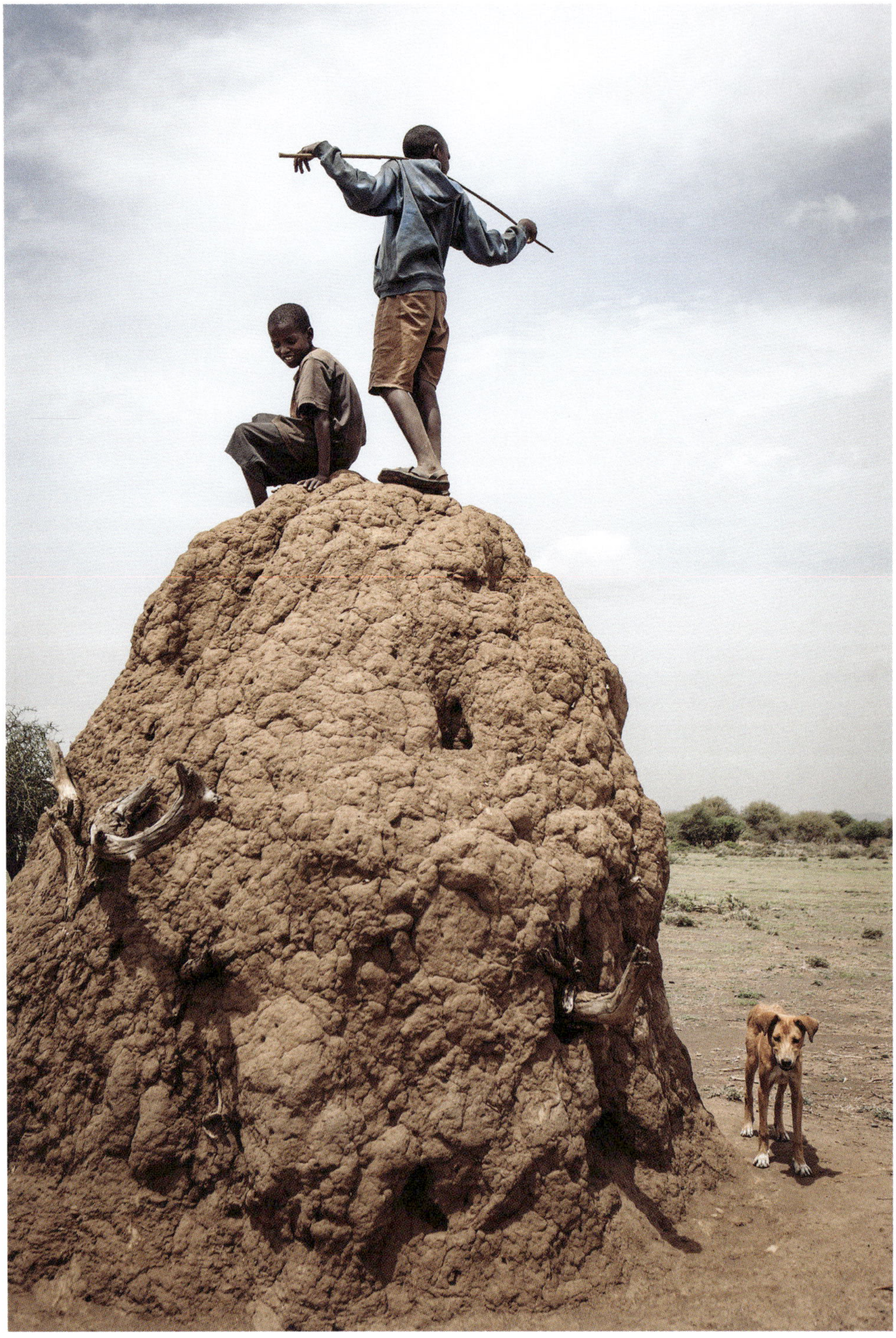

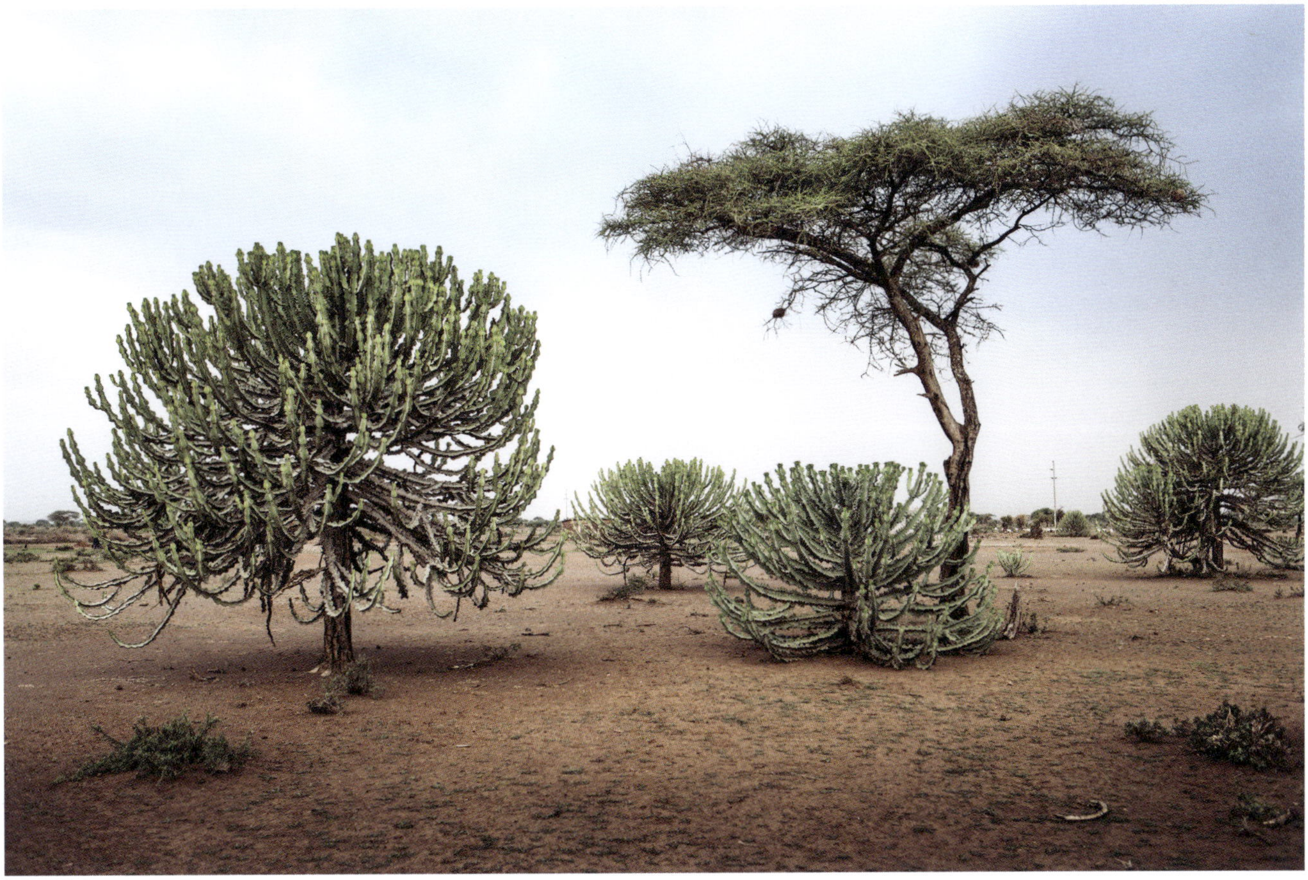

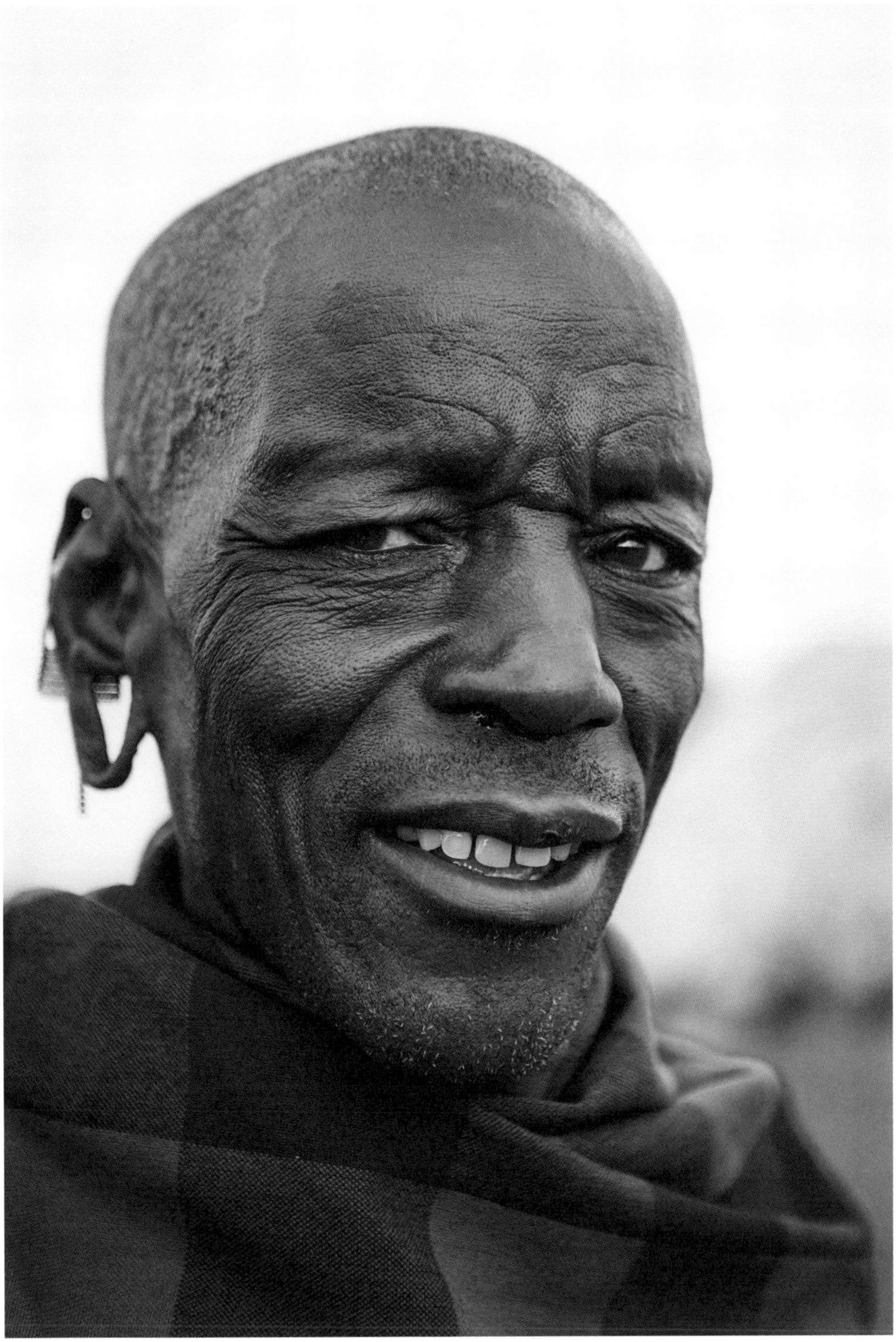

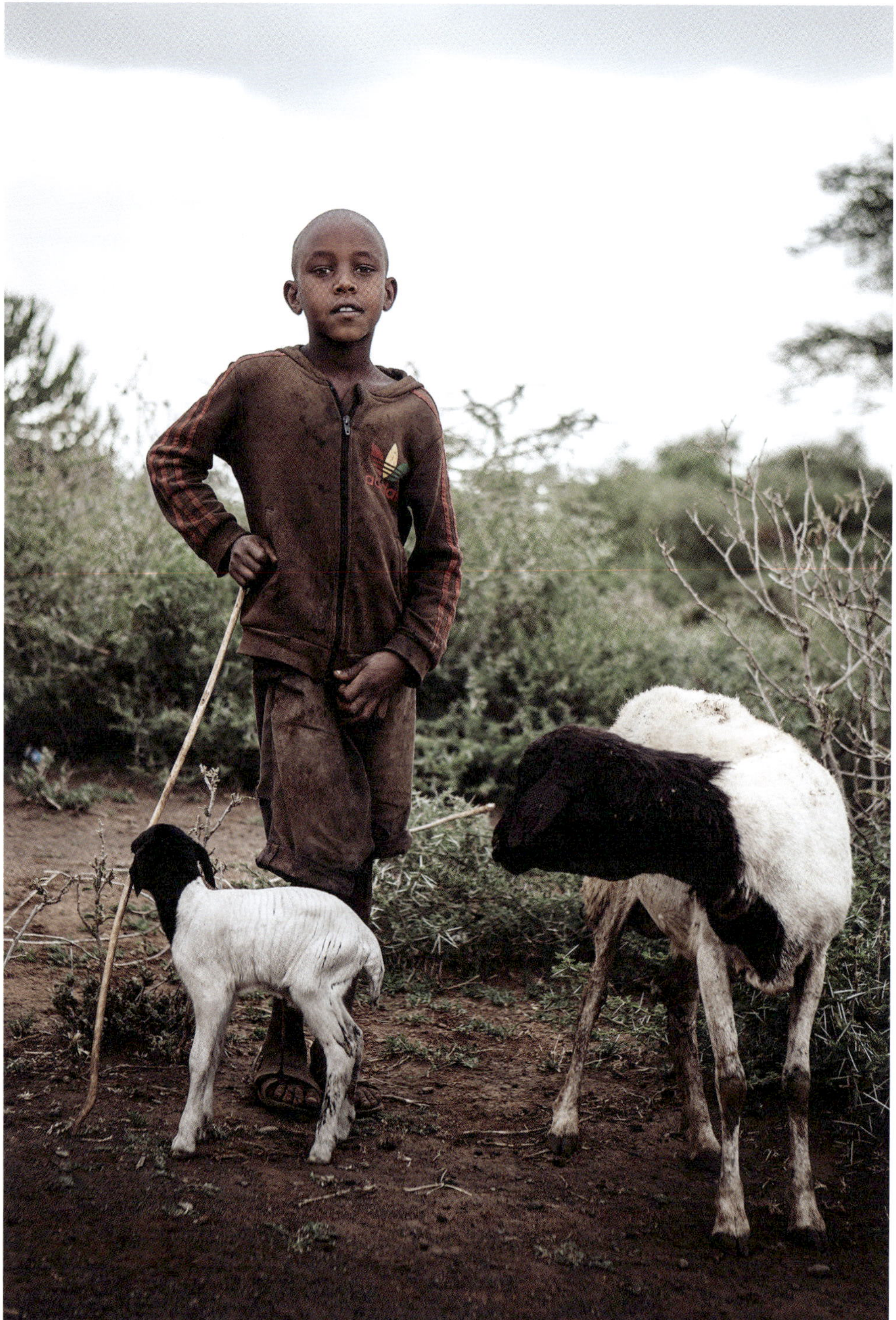

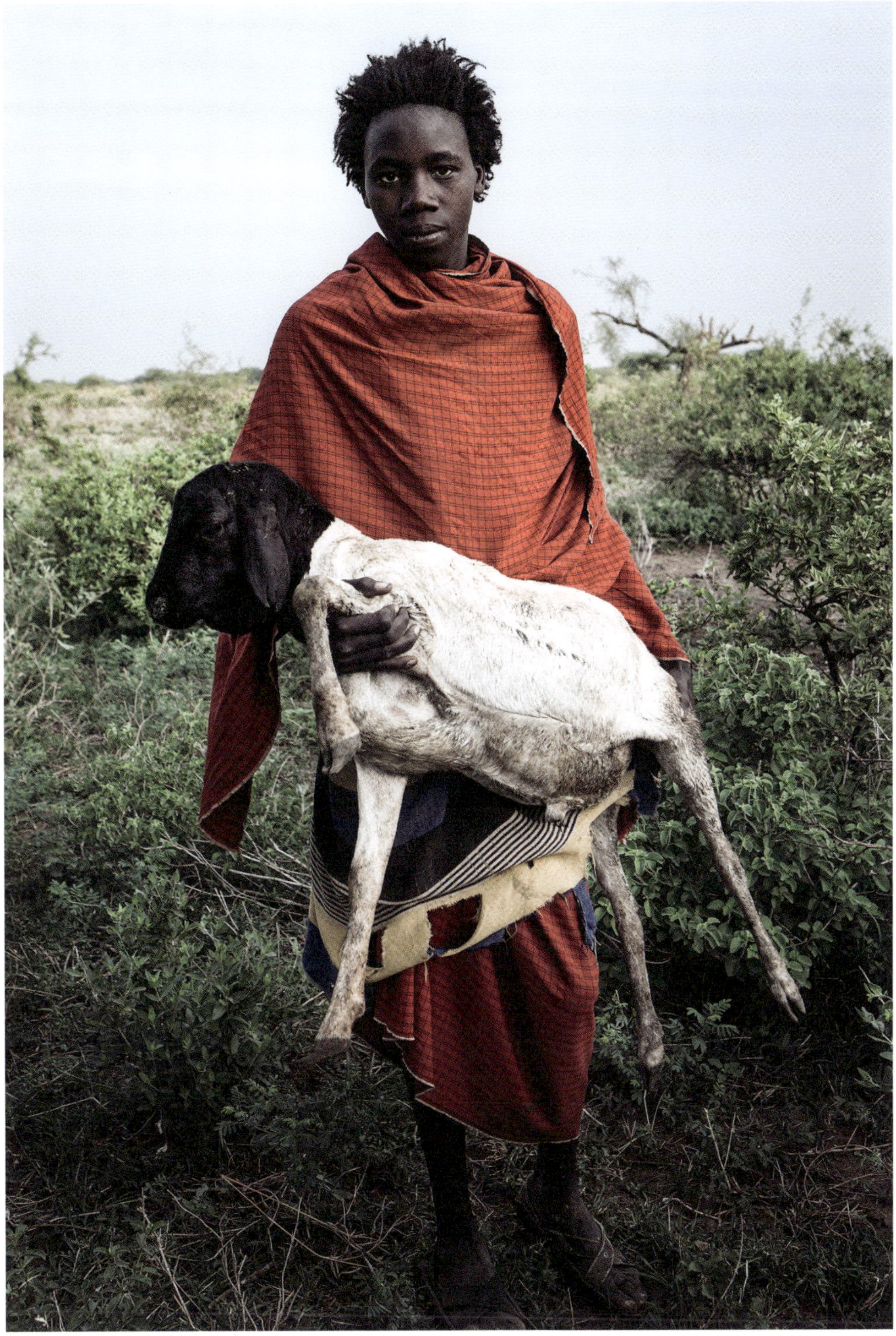

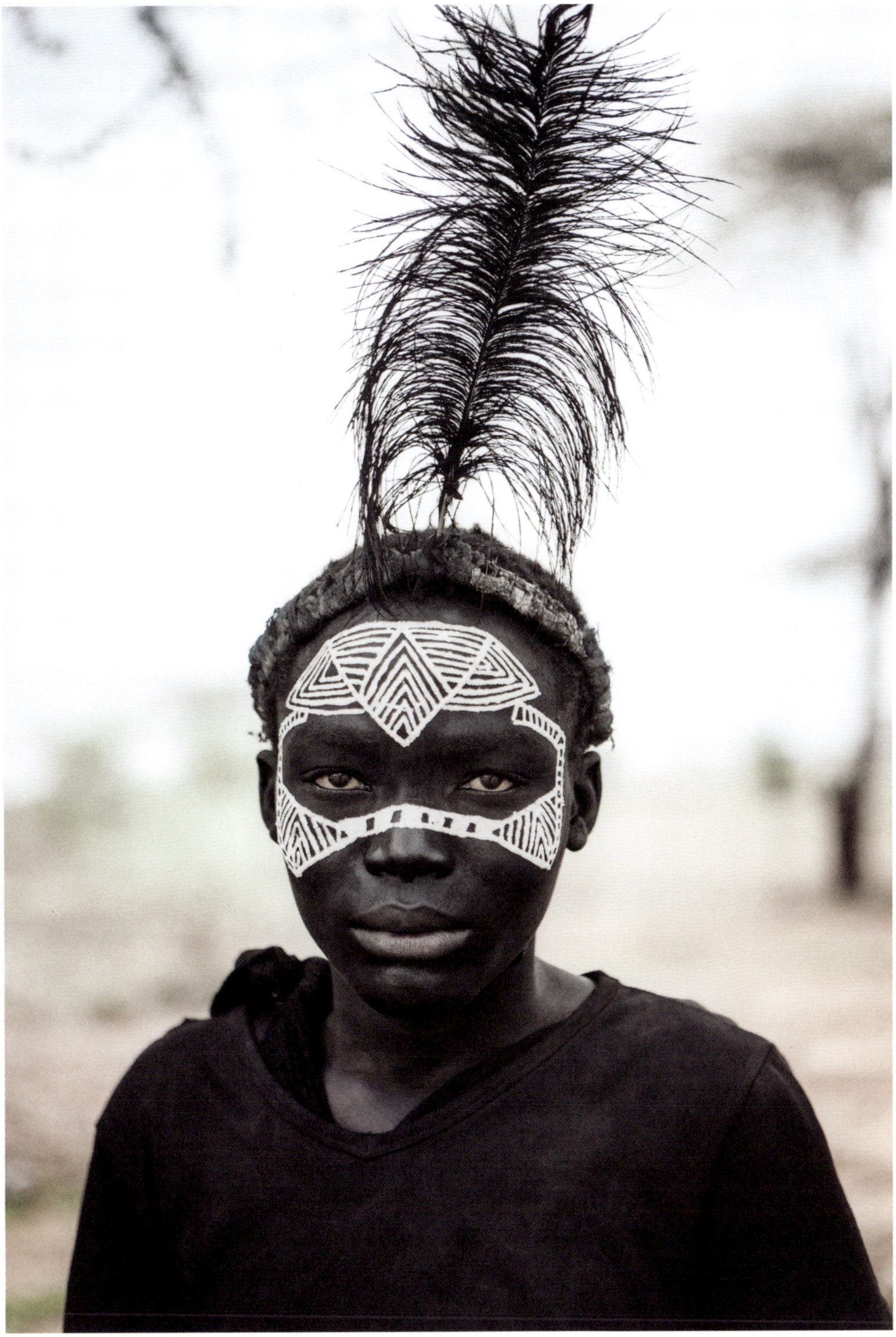

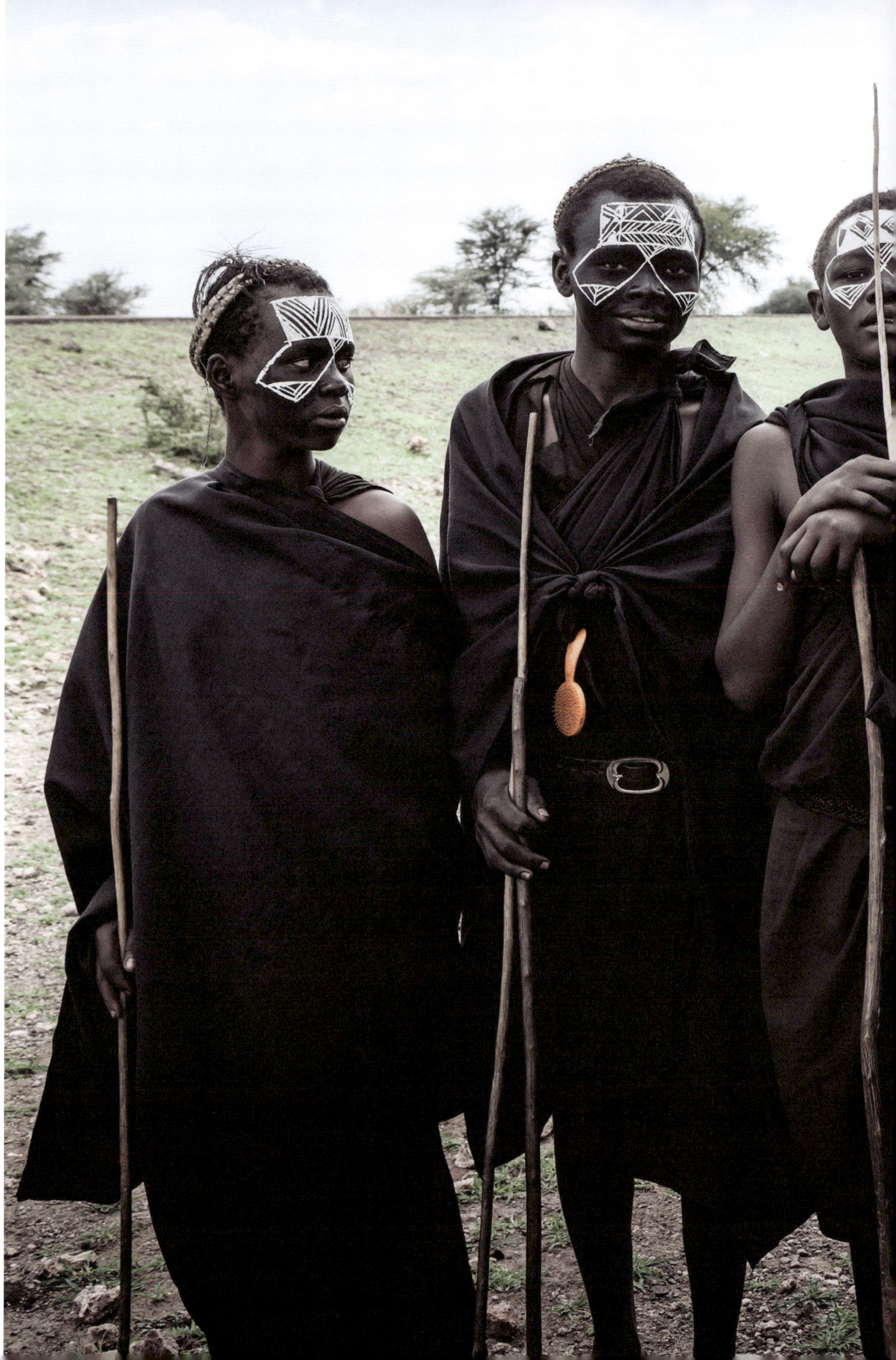

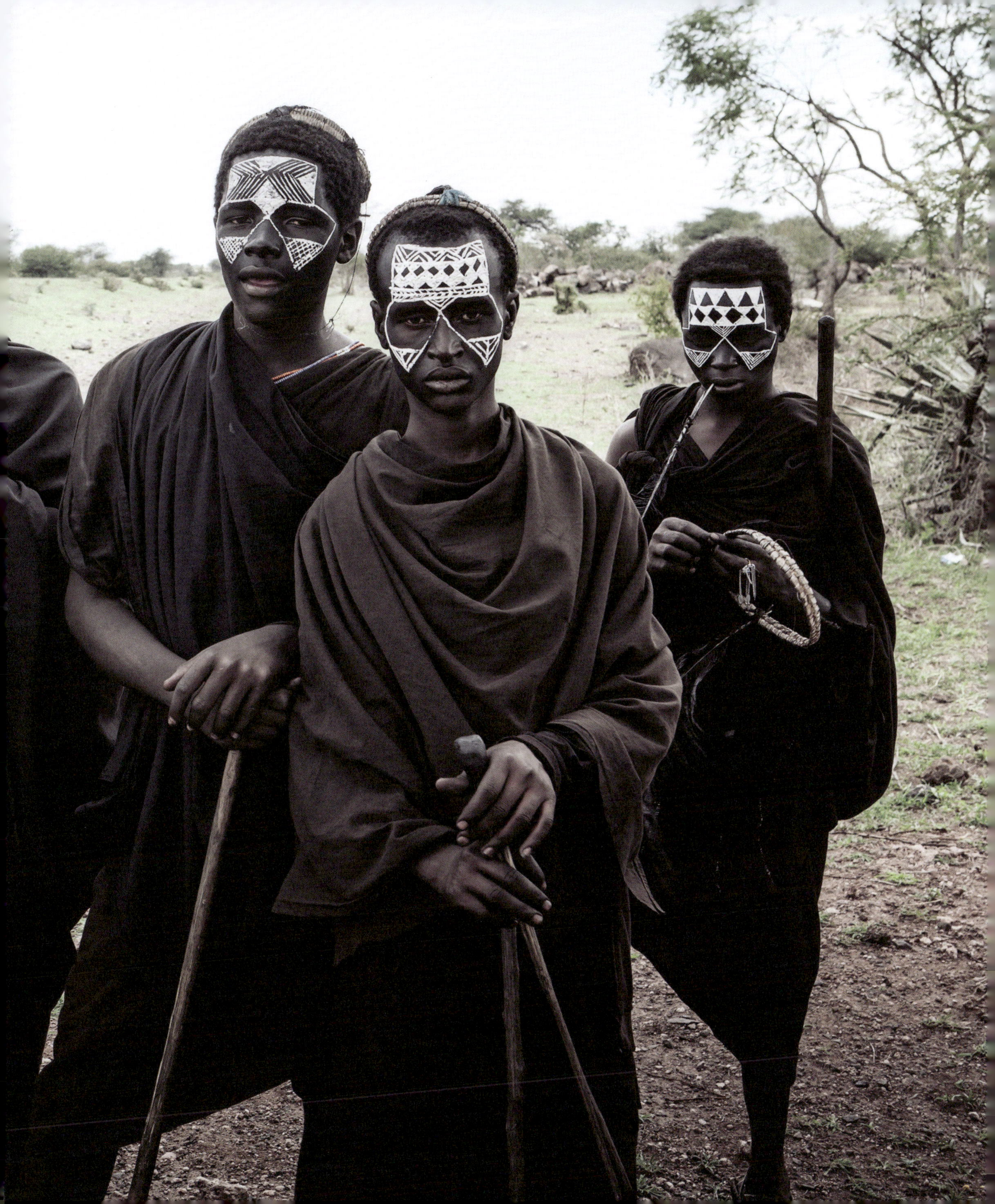

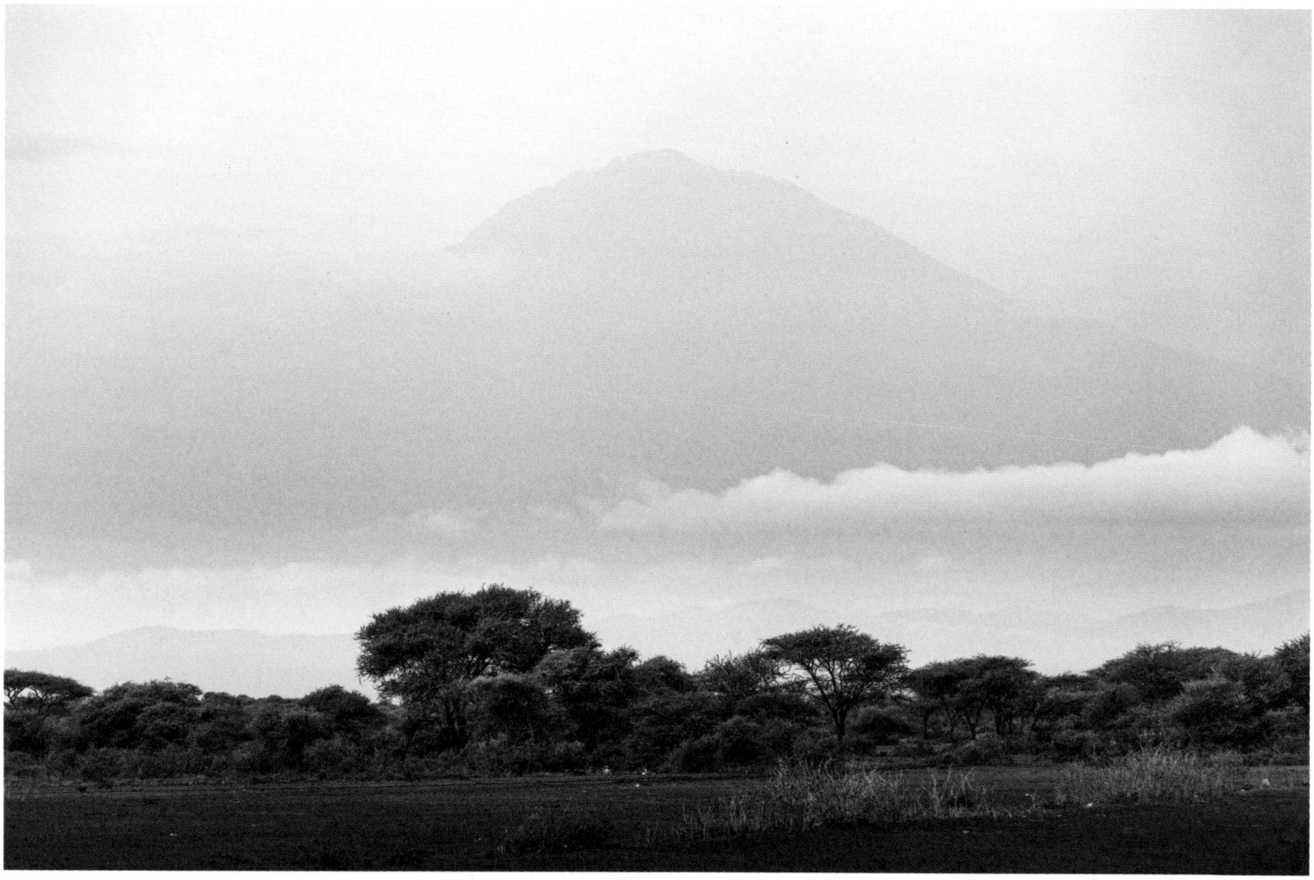

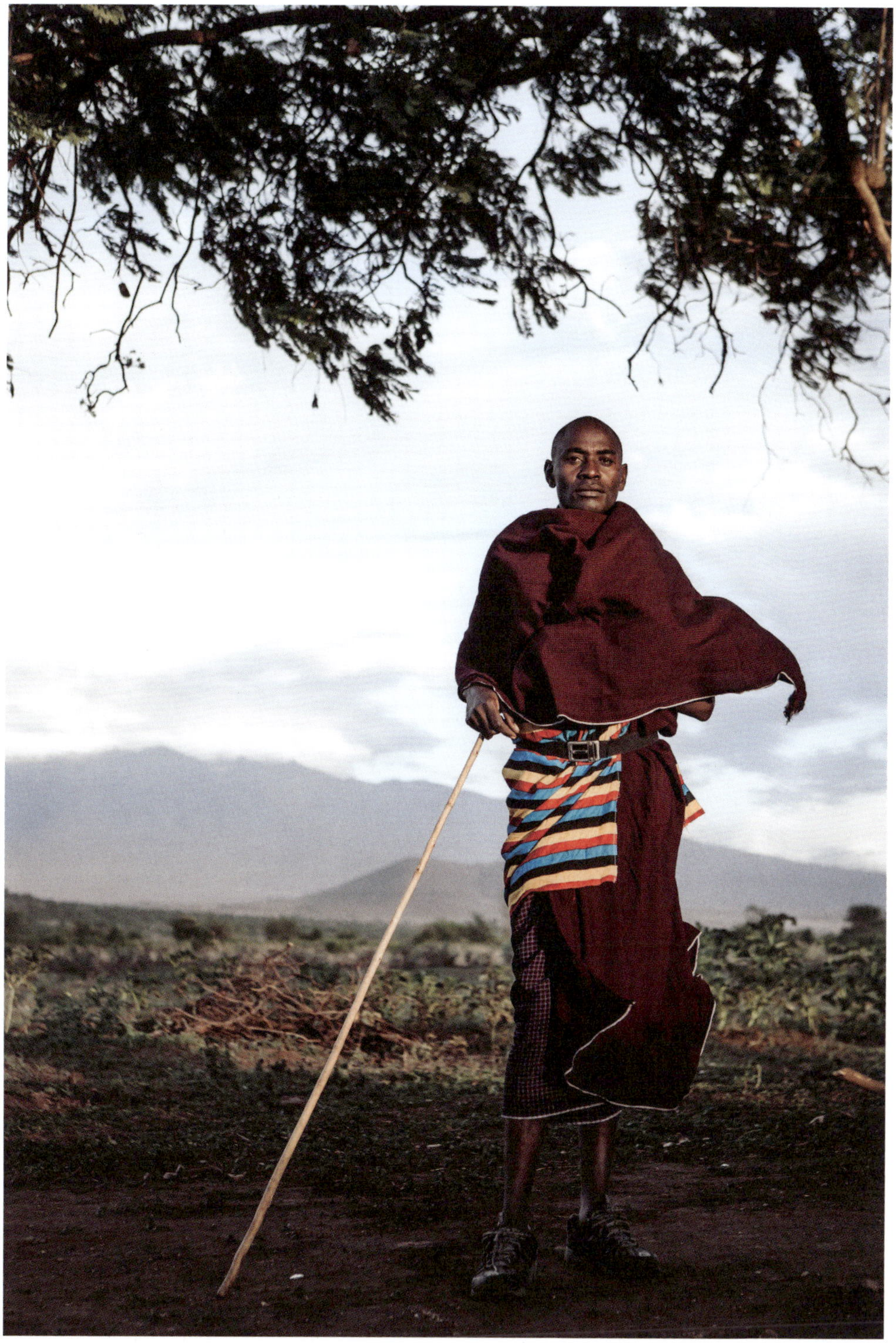

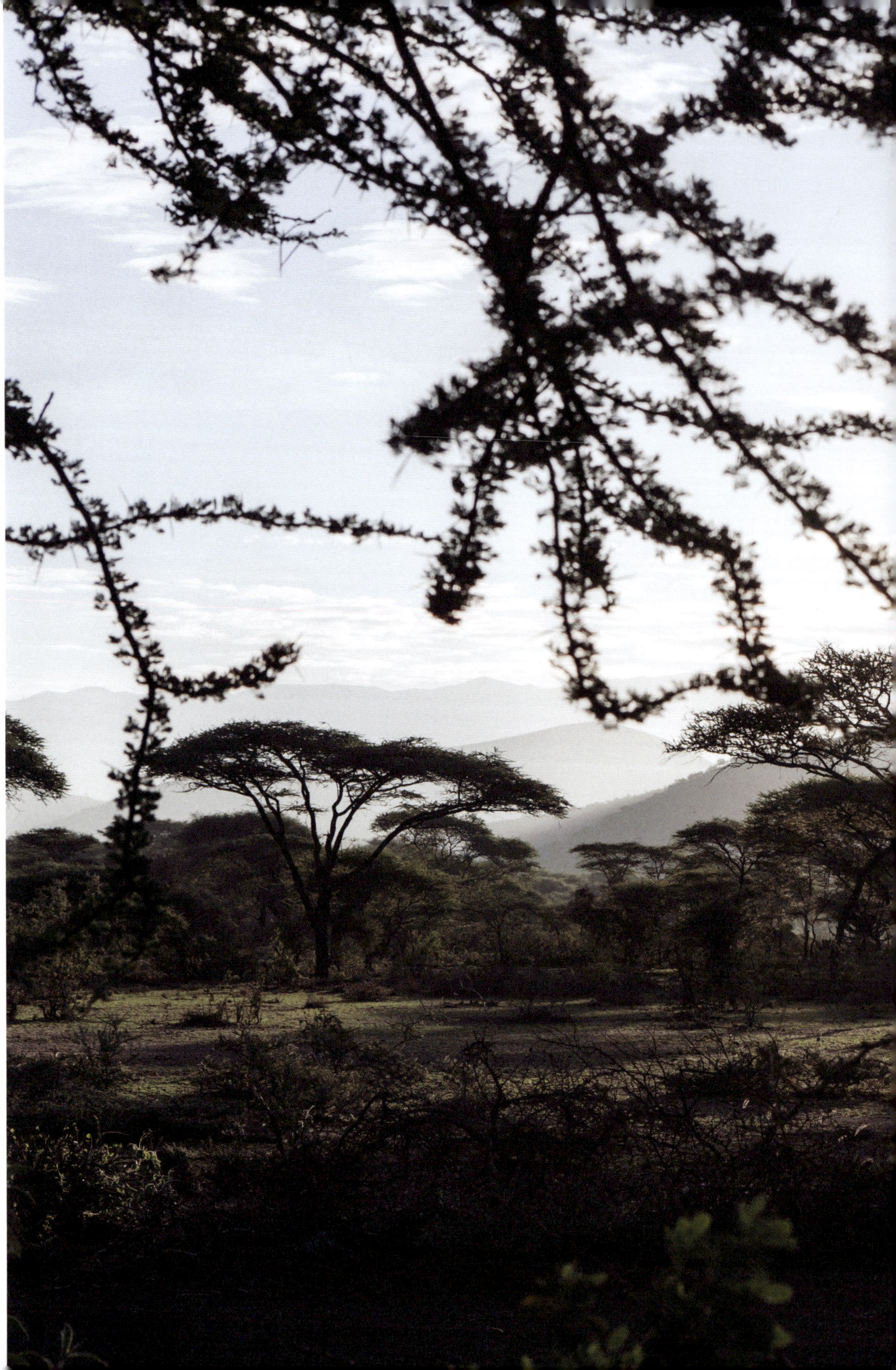

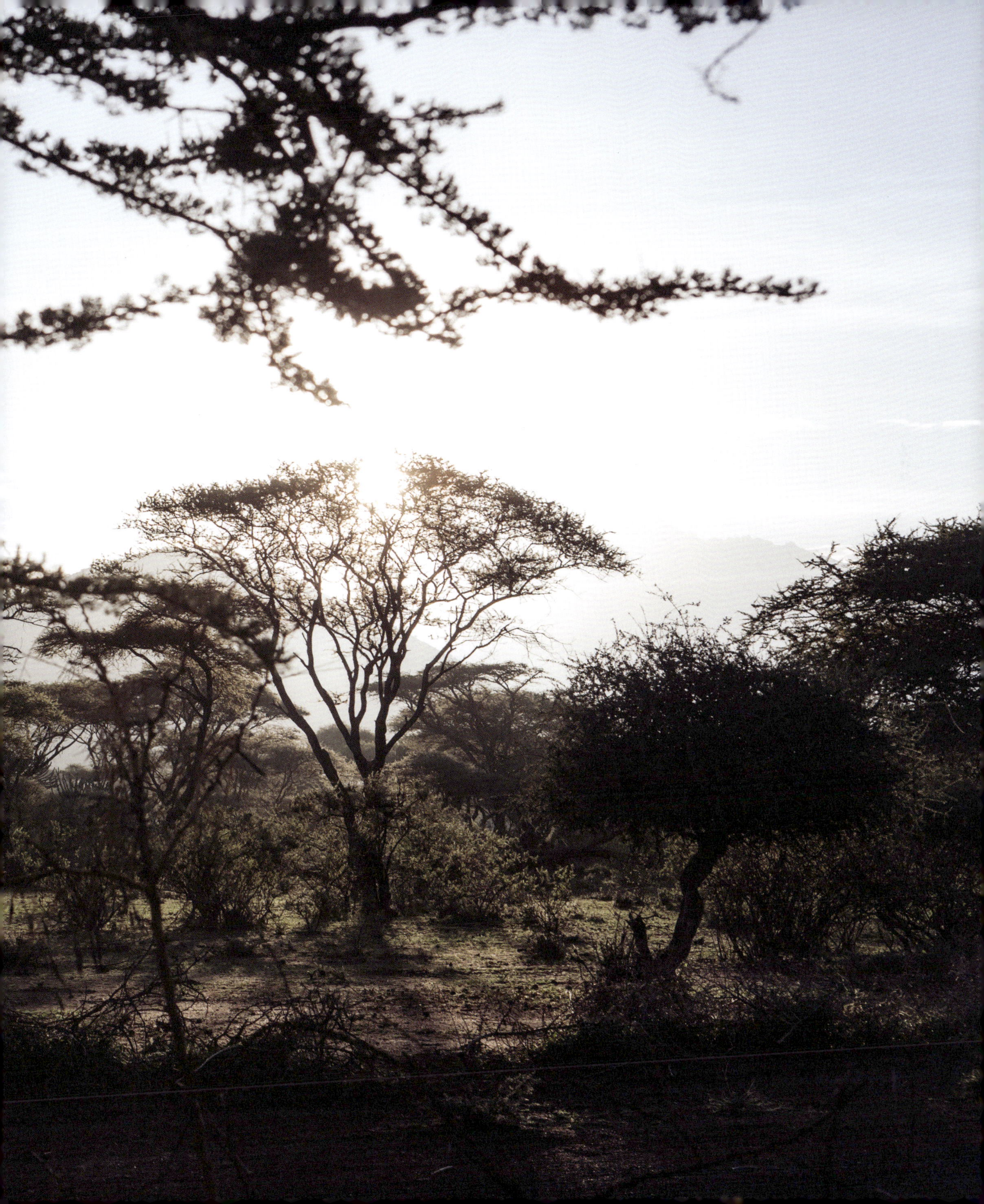

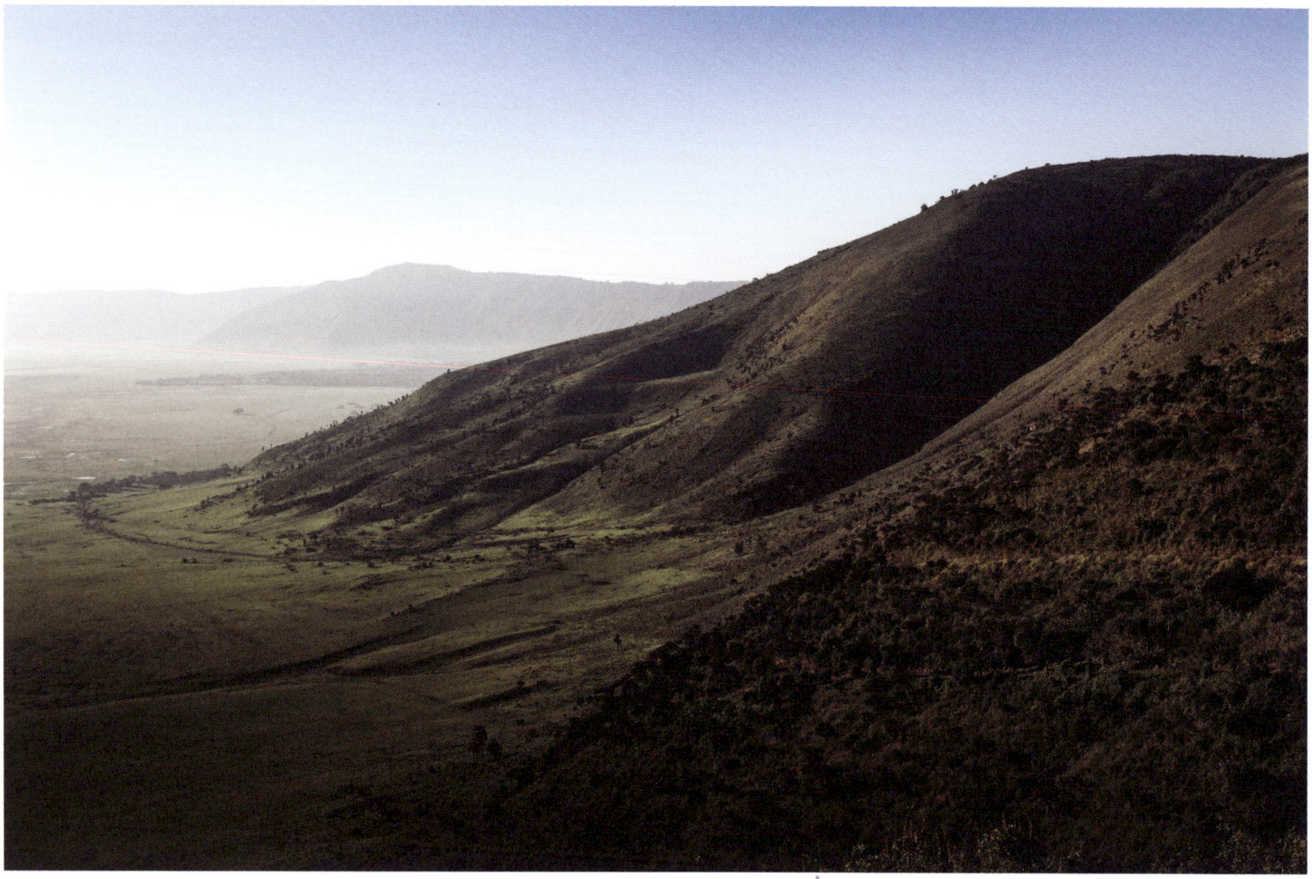

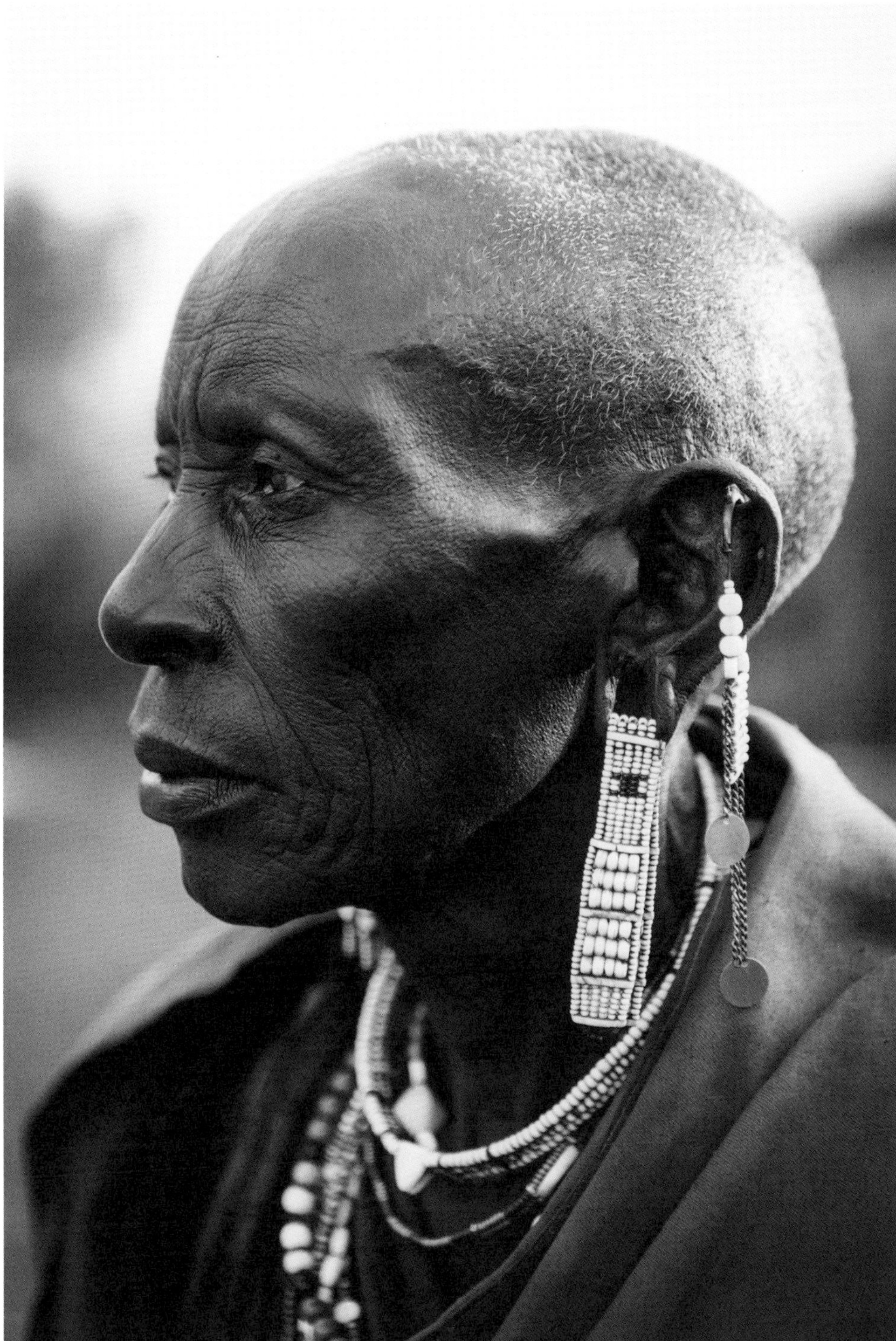

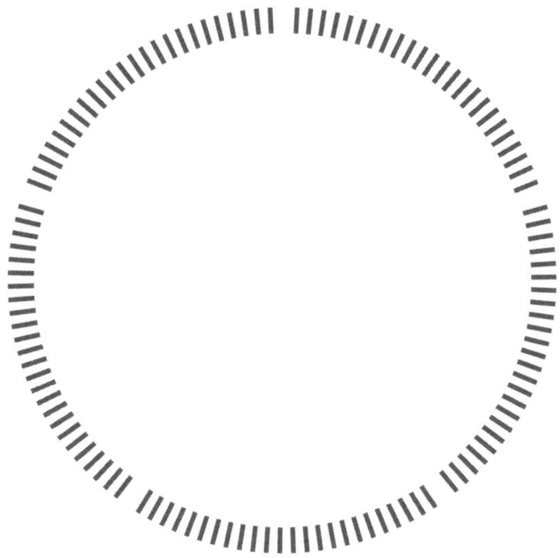

EPILOGUE

When I set off on my first journey that started this project in February 2010, I had no idea that it would change my whole life.

Prompted by the birth of my son in 2000, the desire grew in me over the years to start a project on climate change, which was looming in the distance as a threatening foreboding. How should one design a project to draw attention to the crisis in Germany, one of the richest countries in the world, with an energy-intensive economy and a high standard of living for its citizens? My idea of visiting indigenous communities and documenting their lives arose from the desire to draw attention to them. People all over the world are confronted with a crisis that originated in industrialized countries and to which they have contributed nothing. As a result, however, indigenous communities are initially much more affected by the effects of the climate crisis than those who caused it. My approach was to visit communities on five continents that would represent their continent and a tipping point in the climate system or a relevant climate zone. I wanted to tell stories about people without working with obvious images of climate catastrophe. An irritating idea for my editorial clients, who could not imagine this approach to a topic that was already highly complex in itself. What was relevant for my approach, however, was that I wanted to create a connection between the viewer and the person photographed in a positively emphatic way. If, as a result of this connection, we become aware of the effects our consumer behavior has on these remote places, if we take an interest in the people who have stepped out of anonymity and their fate, this enables us to establish a connection to our own consumer behavior in an undogmatic way, and to ultimately question it.

My journey began almost 40 years after the publication of "The Limits to Growth. A Report for The Club of Rome's Project on the Predicament of Mankind". The central conclusions of the report are: "If the current increase in world population, industrialization, pollution, food production and the exploitation of natural resources continues unabated, the absolute limits to growth on earth will be reached within the next hundred years." Against their better judgement, politicians and industry, encouraged by economists, have embarked on the path of consumption, limitless growth, globalization and neoliberalism. The calculated scenarios of 1972 coincide alarmingly with the reality in which we live today.

EPILOG

Als ich mich im Februar 2010 auf die Reise begab, mit der dieses Projekt begann, ahnte ich nicht, dass dies mein ganzes Leben verändern würde.

Ausgelöst durch die Geburt meines Sohnes im Jahr 2000 war über einige Zeit das Bedürfnis in mir gewachsen, eine Arbeit über den Klimawandel zu beginnen, der sich zunehmend als bedrohliche Ahnung in der Ferne abzuzeichnen begann. Wie müsste ein Projekt angelegt werden, um in Deutschland, einem der reichsten Länder der Welt, mit einer energieintensiven Wirtschaft und einem hohen Lebensstandard seiner Bürger:innen, Aufmerksamkeit auf die Krise zu lenken? Meine Idee, indigene Gemeinschaften zu besuchen und ihr Leben dokumentarisch festzuhalten, entstand aus dem Anliegen, auf diese Menschen aufmerksam zu machen. Menschen überall auf der Welt werden mit den Folgen einer Krise konfrontiert, deren Ursprung in den Industriestaaten zu suchen ist und zu deren Entstehung sie nichts beigetragen haben. Indigene Gemeinden sind deutlich stärker von den Auswirkungen der Klimakrise betroffen als deren Verursacher:innen. Meine Idee war, auf fünf Kontinenten Gemeinschaften zu besuchen, die stellvertretend sowohl für ihren Kontinent als auch für einen Kipppunkt im Klimasystem oder eine relevante Klimazone stehen können. Ich wollte von Menschen erzählen, ohne mit naheliegenden Klima-Katastrophenbildern zu arbeiten. Eine irritierende Vorstellung für meine redaktionellen Kunden, die sich diese Annäherung an das hochkomplexe Thema nicht vorstellen konnten. Für meinen Ansatz relevant war jedoch, dass ich auf positiv emphatische Weise eine Verbindung zwischen Betrachter:in und fotografierter Person herstellen wollte. Wenn wir uns als Folge dieser Verbindung bewusst machen, welche Auswirkungen unser Verhalten auf diese entlegenen Orte hat, wenn wir uns für die aus der Anonymität herausgetretenen Menschen und ihr Schicksal interessieren, ermöglicht uns das auf undogmatische Art, eine Verbindung zu unserer eigenen Lebensweise herzustellen und diese zu hinterfragen.

Meine Reise begann knapp 40 Jahre nach der Veröffentlichung von „Die Grenzen des Wachstums. Bericht des Club of Rome zur Lage der Menschheit". Die zentralen Schlussfolgerungen des Berichts lauten: „Wenn die gegenwärtige Zunahme der Weltbevölkerung, der Industrialisierung, der Umweltverschmutzung, der Nahrungsmittelproduktion und der Ausbeutung von natürlichen Rohstoffen unverändert

When I began this work, I was limited to my personal feelings and assessments. I could have neither imagined the speed at which global warming has been progressing since that time, nor could I have foreseen that Greta Thunberg would initiate a global movement that, within a year, managed to bring millions of young people onto the streets worldwide to demand a political and economic turnaround in their countries. A new climate movement was founded with Fridays for Future. Since then, people's awareness of the crisis has changed significantly.

In 2010, at the beginning of my project, I initially wanted to focus on two tipping points in the climate system, Greenland and the Amazon rainforest. These climatically contrasting places in the northern and southern hemispheres seemed ideal for me to work out an initial comparison.

After seeing a report about Mascha Kauka from Munich and her work with the Amazonica Foundation, which she had founded, and contacting her, I was given the opportunity to accompany her into the forest. Mascha Kauka had already been working in several projects with indigenous communities in Ecuador for over 30 years and was developing a sustainable science tourism project with two communities, Sharamentsa and Yuwientsa, at the time. My journey into the rainforest began as an independent project, without any security or funding, supported solely by my optimism and the opportunity to accompany Mascha Kauka. In Puyo, a community on the edge of the Amazon basin, we boarded a Cessna of the Achuar airline, which would take us to Sharamentsa.

We were received by the Achuar with a traditional welcome ritual. This gave me the opportunity to introduce myself to the community and explain the purpose of my presence. Only then was I allowed to enter the village and approach them. One of my main principles of photography is: People open up to me and the topic if they want to. If they don't, they don't. Respecting this is one of the imperatives of my work. It is the only way to create portraits of intense closeness, pictures from heart to heart. During my time in the Amazon rainforest, I aimed to research and understand the way of life of the Achuar and Shuar. I sat with the men at gatherings and watched the women pass around bowls of chicha. Chicha, a nutritious beverage, is made from boiled, fermented yuca root and is a staple food in the forest. I spent countless days and evenings by the Pastaza River, watching the children bathing and fishing, being taken along on the river, eating sweet and savory bananas, pineapple and papaya, listening to the concerts of the forest, the stories being told and the distant sounds of flutes that reached my ears from the huts in the evenings. In my pictures, I tried to capture

anhält, werden die absoluten Wachstumsgrenzen auf der Erde im Laufe der nächsten hundert Jahre erreicht." Wider besseres Wissen haben sich Politik und Industrie bestärkt von Ökonomen auf den Weg des Konsums, des grenzenlosen Wachstums, der Globalisierung und des Neoliberalismus begeben. Die 1972 berechneten Szenarien treffen sich heute in erschreckender Weise mit der Realität, in der wir leben.

Ich konnte nur meinem persönlichen Empfinden und meiner Einschätzung folgen, als ich mit der Arbeit an diesem Projekt begann. Niemals hätte ich mir vorstellen können, in welcher Geschwindigkeit die Erderwärmung während dieser Zeit voranschreiten würde, noch konnte ich voraussehen, dass Greta Thunberg eine globale Bewegung anstoßen würde, die innerhalb eines Jahres Millionen junger Menschen weltweit auf die Straße brachte und eine politische und wirtschaftliche Wende in ihren Ländern einfordern ließ. Mit Fridays for Future wurde eine neue Klimabewegung begründet. Seitdem hat sich das Bewusstsein der Menschen für die Krise maßgeblich geändert.

Im Jahr 2010, zu Beginn meines Projektes, wollte ich zuerst schwerpunktmäßig an zwei Kipppunkten des globalen Klimasystems, in Grönland und im Amazonas-Regenwald, arbeiten. Diese klimatisch so gegensätzlichen Orte auf der Nord- und Südhalbkugel schienen mir ideal, um eine erste Gegenüberstellung herauszuarbeiten.

Nachdem ich einen Bericht über Mascha Kauka aus München und ihre Arbeit mit der von ihr gegründeten Stiftung Amazonica gesehen und Kontakt mit ihr aufgenommen hatte, bekam ich die Gelegenheit, sie in den Wald zu begleiten. Mascha Kauka arbeitete bereits über 30 Jahre in mehreren Projekten mit indigenen Gemeinschaften in Ecuador und entwickelte zu dem Zeitpunkt mit zwei Gemeinden, Sharamentsa und Yuwientsa, ein nachhaltiges Wissenschaftstourismus-Projekt. Als freies Projekt, ohne Absicherung und Finanzierung, alleine getragen von meinem Optimismus und der Möglichkeit, Mascha Kauka zu begleiten, begann meine Reise in den Regenwald. In Puyo, einer Gemeinde am Rand des Amazonasbeckens, bestiegen wir eine Cessna der kleinen den Achuar gehörenden Fluggesellschaft, die uns nach Sharamentsa brachte.

Mit einem traditionellen Willkommensritual wurden wir von den Achuar empfangen. Dabei erhielt ich die Gelegenheit, mich der Gemeinschaft vorzustellen und den Zweck meiner Anwesenheit zu erläutern. Erst danach konnte ich das Dorf betreten und mich den Einwohnern annähern. Ein wichtiger Grundsatz war für mich von vornherein: Die Menschen öffnen sich mir und dem Thema, wenn sie

the essence of the people, to look beneath the surface and to understand. This urge to understand, which accompanied me on all my journeys, fueled my desire to watch the village shaman Entsakua perform a healing ritual and to accompany this photographically. I asked again and again if I could visit him and was always put off until the next day. My first stay with the Achuar was already nearing its end without Entsakua inviting me. Then I fell ill. I had contracted a bad stomach bug. And suddenly they were standing in front of me. Several men accompanying a healer. He came to perform a limpia, a healing ritual with me. I experienced the limpia through a thick fog, I was so weakened. After the men left, I recovered within an hour and although I was still weak, I had recuperated. I could not have learned this lesson in a better way. It wasn't about photographing a sacred ritual that wasn't meant for me, it was about understanding.

After leaving the Amazon rainforest with a treasure trove of impressions and encounters, a year had to go by before I could take the next step. In February 2011, I flew from Hamburg via Copenhagen and Reykjavik to Kulusuk in East Greenland, where I boarded an Air Greenland helicopter in freezing temperatures to fly low over the sea ice to Tasiilaq. Robert Peroni from South Tyrol was waiting for me at the airport. During a stay for Unicef in Ethiopia, a cameraman had told me about this former extreme sportsman and his first winter crossing of the Greenland ice sheet which had made him famous. The country and the people enchanted him and so he decided to stay and establish The Red House as a contact point and cultural meeting place for locals and visitors. When I came to Greenland, Robert had already been living in Tasiilaq for more than 30 years. In addition to its original purpose, The Red House has now also become a small hotel for Greenland tourists and scientists, where people are warmly welcomed and well catered for. It has also become an important employer for the people of Tasiilaq.

After a few days in The Red House, where I learned a lot about the cultural roots of the Inuit, I became restless and wanted to go out to a village. The plan was to take a helicopter to Sermiligaaq in the neighboring fjord and stay with a family friend there thanks to Peroni's mediation. However, an approaching storm made me change my plans. The prospect of being able to fly and return this way in the foreseeable future was uncertain. So I decided to cross the mountains by land. The young hunter Gideon and his nine sled dogs would take me to his village of Tiniteqilaaq.

We met early in the morning at the edge of Kong Oscar Fjord in Tasiilaq. Gideon didn't speak my language and I didn't speak his. He instructed me to take

dies möchten. Wenn nicht, dann nicht. Das zu respektieren ist eines der Gebote meiner Arbeit. Nur so können Portraits von intensiver Zugewandtheit, Bilder von Herz zu Herz entstehen. Während meiner Zeit im Amazonas-Regenwald bemühte ich mich, die Lebensweise der Achuar und Shuar zu erforschen und zu verstehen. Ich saß mit den Männern bei Versammlungen und beobachtete die Frauen, wie sie Schalen mit Chicha herumreichten. Chicha, ein nahrhaftes Getränk, wird aus gekochter, fermentierter Yucawurzel gewonnen und ist ein Grundnahrungsmittel der Bewohner des Waldes. Ich verbrachte unzählige Tage und Abende am Fluss Pastaza, schaute den Kindern beim Baden und Angeln zu, wurde auf den Fluss mitgenommen, aß süße und herzhafte Bananen, Ananas und Papaya, lauschte den Konzerten des Waldes, den Geschichten, die erzählt wurden und den fernen Flötentönen, die abends aus den Hütten an mein Ohr drangen. In meinen Bildern versuchte ich das Wesen der Menschen einzufangen, unter die Oberfläche zu schauen und zu verstehen. Dieser mich auf all meinen Wegen begleitende Drang zu verstehen schürte mein Bedürfnis, dem Dorfschamanen Entsakua bei einem Heilungsritual zuzusehen und dies auch fotografisch zu begleiten. Immer und immer wieder fragte ich, ob ich ihn besuchen dürfte, und immer wieder wurde ich auf den nächsten Tag vertröstet. Mein erster Aufenthalt bei den Achuar neigte sich bereits dem Ende zu, ohne dass mich Entsakua einlud. Dann wurde ich krank. Ich hatte mir eine schlimme Magen-Darm-Verstimmung zugezogen. Und plötzlich standen sie vor mir. Mehrere Männer, die einen Heiler flankierten. Er kam, um eine Limpia, ein Heilungsritual, an mir zu praktizieren. Ich erlebte die Limpia durch einen dichten Nebel, so geschwächt war ich. Nachdem die Männer gegangen waren, erholte ich mich innerhalb einer Stunde – ich war zwar noch schwach, aber genesen. Ich hätte diese Lektion auf keine bessere Art lernen können. Es ging nicht darum, ein heiliges Ritual zu fotografieren, das nicht für mich bestimmt war, sondern es ging darum, zu verstehen.

Nachdem ich den Amazonas-Regenwald mit einem Schatz an Eindrücken und Begegnungen wieder verlassen hatte, dauerte es ein Jahr, bis ich den nächsten Schritt gehen konnte. Im Februar 2011 flog ich von Hamburg über Kopenhagen und Reykjavik nach Kulusuk in Ostgrönland, um dort bei eisigen Temperaturen in einen Helikopter der Fluggesellschaft Air Greenland umzusteigen, der mich im Tiefflug über das Meereis nach Tasiilaq bringen würde. Am Flughafen erwartete mich der Südtiroler Robert Peroni. Ein Kameramann hatte mir während eines Aufenthalts für Unicef in Äthiopien einmal von diesem ehemaligen Extremsportler und seiner winterlichen Erstdurchquerung des grönländischen Inlandeises erzählt, die ihn

a seat on the hard wooden sledge, and immediately the dogs were dashing exuberantly across the ice of the fjord towards the mountains. Our journey took eight hours over snow and scree. Uphill we trudged behind the sled, while downhill I sat on the sled at breakneck speed, with Gideon standing behind me on the runners to steer the dogs by shouting, shifting weight and braking. Before we arrived in Tiniteqilaaq as night fell, we had to cross a glacier, which I somehow managed, shivering from the cold and with my last reserves of strength. Gideon and I moved into his house, which lay abandoned in the cold and darkness. First he looked after the dogs before he went to the small harbor in the village to get oil for our stove. Inside the house, the frozen bones of his last meal still lay in the kitchen. It was only hours after Gideon had lit the stove, cleaned the kitchen and laid out the guest mattress in the small partitioned area that normally served as the children's room that I was able to take off my jacket, hat and boots.

In the days that followed, I roamed the village in search of residents. In Thormasina, who invited me to her home, I found a friendly mediator of her culture. I experienced her family in Greenlandic winter mode and began to get an idea of what life is like for the Inuit in the remote villages in winter. I was also able to visit the small school and the community center. I observed how the hunters made their way through the frozen sea ice in their small boats in the morning and how they returned home in the evening with seals they had killed.

The closer I got to my departure, the more my stomach ached at the thought of the imminent return journey over the mountains. It had snowed during the night. In the morning, a thick layer of fresh snow covered the landscape. Our dogs, which had been reduced to eight due to the death of one of them, jumped off happily at first, only to get stuck in the deep snow straight away. A strenuous journey lay ahead of us, which I mastered much better than before despite the difficult conditions and the fear of the road. Back in The Red House and with the comfort of a warm shower, I realized how much I had grown from this experience. I had pushed myself to my limits and beyond. I had faced my fear, exposed myself to nature and its laws of darkness, wind and cold, and tested my physical and mental limits anew. Nature itself had taught me a lesson.

In late summer of the same year, I returned to have the opportunity to go to sea with the hunters. Before my visits to the Achuar, Shuar and the Inuit, I opposed hunters because I thought they killed animals purely as a pastime. Today I know that this was a one-sided view and does not do justice to most hunters. The Inuit taught me something else. They showed me how much their relationship with animals determines

berühmt gemacht hatte. Das Land und die Leute hatten ihn nicht mehr losgelassen, und so hatte er entschieden zu bleiben und das Rote Haus, eine Anlaufstelle und kulturelle Begegnungsstätte für Einheimische und Besucher:innen, zu gründen. Als ich nach Grönland kam, lebte Robert bereits mehr als 30 Jahre in Tasiilaq. Das Rote Haus ist neben seiner ursprünglichen Bestimmung heute auch ein kleines Hotel für Grönland-Tourist:innen und Wissenschaftler:innen geworden, in dem man herzlich aufgenommen und gut bewirtet wird. Damit ist es zudem zu einem wichtigen Arbeitgeber für die Bürger von Tasiilaq geworden.

Nach einigen Tagen im Roten Haus, in denen ich viel über die kulturellen Wurzeln der Inuit lernte, wurde ich unruhig und wollte hinaus in ein Dorf. Geplant war, mit einem Helikopter nach Sermiligaaq im angrenzenden Fjord zu fliegen und dort dank Robert Peronis Vermittlung bei einer befreundeten Familie unterzukommen. Ein aufkommender Sturm ließ mich jedoch meine Pläne ändern. Die Möglichkeit, zum Dorf fliegen zu können und in absehbarer Zeit auf diesem Weg zurückzukommen, war ungewiss. Darum entschied ich mich, auf dem Landweg die Berge zu überqueren. Der junge Jäger Gideon und seine neun Schlittenhunde würden mich in sein Dorf Tiniteqilaaq bringen.

Wir trafen uns am frühen Morgen am Rand des Kong Oscar-Fjords in Tasiilaq. Gideon sprach nicht meine Sprache und ich nicht die seine. Er wies mich an, auf dem harten Holzschlitten Platz zu nehmen, und schon preschten die Hunde übermütig über das Eis des Fjords in Richtung Berge. Acht Stunden lang ging unsere Reise über Schnee und Geröll. Bergauf liefen wir hinter dem Schlitten her, bergab saß ich bei halsbrecherischem Tempo darauf, während Gideon hinter mir auf den Kufen stand, um die Hunde durch Rufen, Gewichtsverlagerung und Bremsen zu lenken. Bevor wir bei hereinbrechender Nacht in Tiniteqilaaq ankamen, mussten wir noch einen Gletscher überwinden, was ich, zitternd vor Kälte und mit letzten Kraftreserven, irgendwie bewältigte. Gideon und ich bezogen sein Haus, das verlassen in der Kälte und Dunkelheit lag. Zuerst wurden die Hunde versorgt, bevor er am kleinen Hafen des Ortes Öl für unseren Ofen besorgte. Im Inneren des Hauses lagen noch die gefrorenen Knochen seines letzten Mahls in der Küche. Erst Stunden nachdem Gideon den Ofen angezündet, die Küche geputzt und die Gästematratze in dem kleinen abgetrennten Bereich, der normalerweise als Kinderzimmer diente, ausgelegt hatte, konnte ich Jacke, Mütze und Stiefel ausziehen.

In den darauffolgenden Tagen durchstreifte ich das Dorf auf der Suche nach Bewohner:innen. In Thormasina, die mich zu sich einlud, fand ich eine freundliche Vermittlerin ihrer Kultur. Ich erlebte ihre Familie im grönländischen Wintermodus

their lives. How much they depend on them. They do not kill for fun and never more than they need to live. Their relationship with their game is based on respect.

Another aspect that drew me back to Greenland and the sea was the opportunity to photograph formations of majestic beauty and transience, created by the laws of nature icebergs.

And then I was blessed with a special encounter. Anda Kuitse from Kulusuk, who was considered the country's most important drum dancer until his death in 2019 and came from a long line of drum dancers and shamans, visited us in The Red House. The traditional drum dance is full of myths and stories that are conveyed through the drumming, singing and facial expressions during the dance. Anda drummed for me.

The next year, I returned to Sharamentsa and Yuwientsa in the Amazon rainforest. The young women I had photographed two years earlier on the school benches I now encountered with their own young children on their backs. I finally had the opportunity to spend some time with a family. Although many houses have no walls, it is impossible to enter without an invitation. Thanks to Mascha Kauka's translation, I was able to have in-depth conversations and explain in more detail what I aim to achieve with my work.

Back in Germany, I thought a lot about my approach to creating awareness of the climate crisis among the inhabitants of the industrialized nations through an artistic photographic project. The complexity of the topic led me to the conclusion that I should reduce my visual language to its essence and concentrate entirely on portraying the people I visited. These portraits were to be accompanied exclusively by landscapes. I wanted to condense the connection that exists between all people and the influence that emanates from formative habitats.

One night I had a vision. Portraits of the Achuar, Shuar and Inuit were installed on an iceberg. Intuitively, I knew that I had found the key to the project. I had wanted to show the pictures to the people in the rainforest and Greenland for some time. We had talked about the work during my visits and they encouraged me to draw attention to their problems regarding the climate crisis in the industrialized nations. Once I had made the decision, I had portraits and landscapes of the two regions produced on large-format, lightweight textile banner material and flew back to Sharamentsa with them. With Mascha Kauka's support and the help of the two Achuar Nantu and José, we were able to present the pictures of both regions as a juxtaposition and encounter at various locations in the primary forest and in the village.

und begann mir eine Vorstellung vom Leben zu machen, das die Inuit in den entlegenen Dörfern im Winter führen. Auch die kleine Schule konnte ich besuchen und das Gemeindehaus. Ich beobachtete, wie sich die Jäger morgens mit ihren kleinen Booten eine Schneise durch das zugefrorene Meereis bahnten und wie sie abends mit erlegten Robben nach Hause kamen.

Je näher meine Abreise rückte, umso mehr Bauchschmerzen bekam ich bei dem Gedanken an den bevorstehenden Rückweg über die Berge. In der Nacht vor der Abreise schneite es, und am Morgen bedeckte eine dicke Schicht Neuschnee die Landschaft. Unsere durch den Tod eines Tieres auf acht reduzierten Hunde sprangen zuerst freudig los, um gleich darauf im tiefen Schnee festzustecken. Eine anstrengende Fahrt lag vor uns, die ich trotz der erschwerten Bedingungen und der Angst vor dem Weg deutlich besser meisterte als zuvor. Zurück im Roten Haus und erleichtert durch den Komfort einer warmen Dusche realisierte ich, wie sehr ich an dieser Erfahrung gewachsen war. Ich war an meine Grenzen gegangen und darüber hinaus. Ich hatte mich meiner Angst gestellt, der Natur und ihren Gesetzen von Dunkelheit, Wind und Kälte ausgesetzt, meine physischen und psychischen Grenzen neu ausgelotet. Die Natur selbst hatte mir eine Lektion erteilt.

Im Spätsommer des gleichen Jahres kehrte ich zurück, um die Möglichkeit zu haben, mit den Jägern aufs Meer zu fahren. Vor meinen Besuchen bei den Achuar, Shuar und den Inuit hatte ich Jäger:innen abgelehnt, da sie nach meiner Vorstellung zum puren Zeitvertreib Tiere töteten. Ich weiß heute, dass dies ein einseitiger Blick war und den meisten Jäger:innen nicht gerecht wird. Die Inuit haben mich etwas anderes gelehrt. Sie haben mir gezeigt, wie sehr die Beziehung zu den Tieren ihr Leben bestimmt. Wie sehr sie von ihnen abhängig sind. Sie töten nicht zum Spaß und nie mehr, als sie zum Leben brauchen. Ihre Beziehung zu ihren Beutetieren fußt auf Respekt.

Ein weiterer Aspekt, der mich zurück nach Grönland und aufs Meer lockte, war die Möglichkeit, Gebilde von majestätischer Schönheit und Vergänglichkeit, erschaffen von den Gesetzen der Natur, zu fotografieren – Eisberge.

Dann wurde ich noch durch eine besondere Begegnung beschenkt. Anda Kuitse aus Kulusuk, der bis zu seinem Tod 2019 als bedeutendster Trommeltänzer des Landes galt und einer langen Reihe von Trommeltänzern und Schamanen entstammte, besuchte uns im Roten Haus. Der traditionelle Trommeltanz ist voller Mythen und Geschichten, die über das Trommeln, den Gesang und auch die Mimik beim Tanz transportiert werden. Anda hat für mich getrommelt.

Installations in Sharamentsa

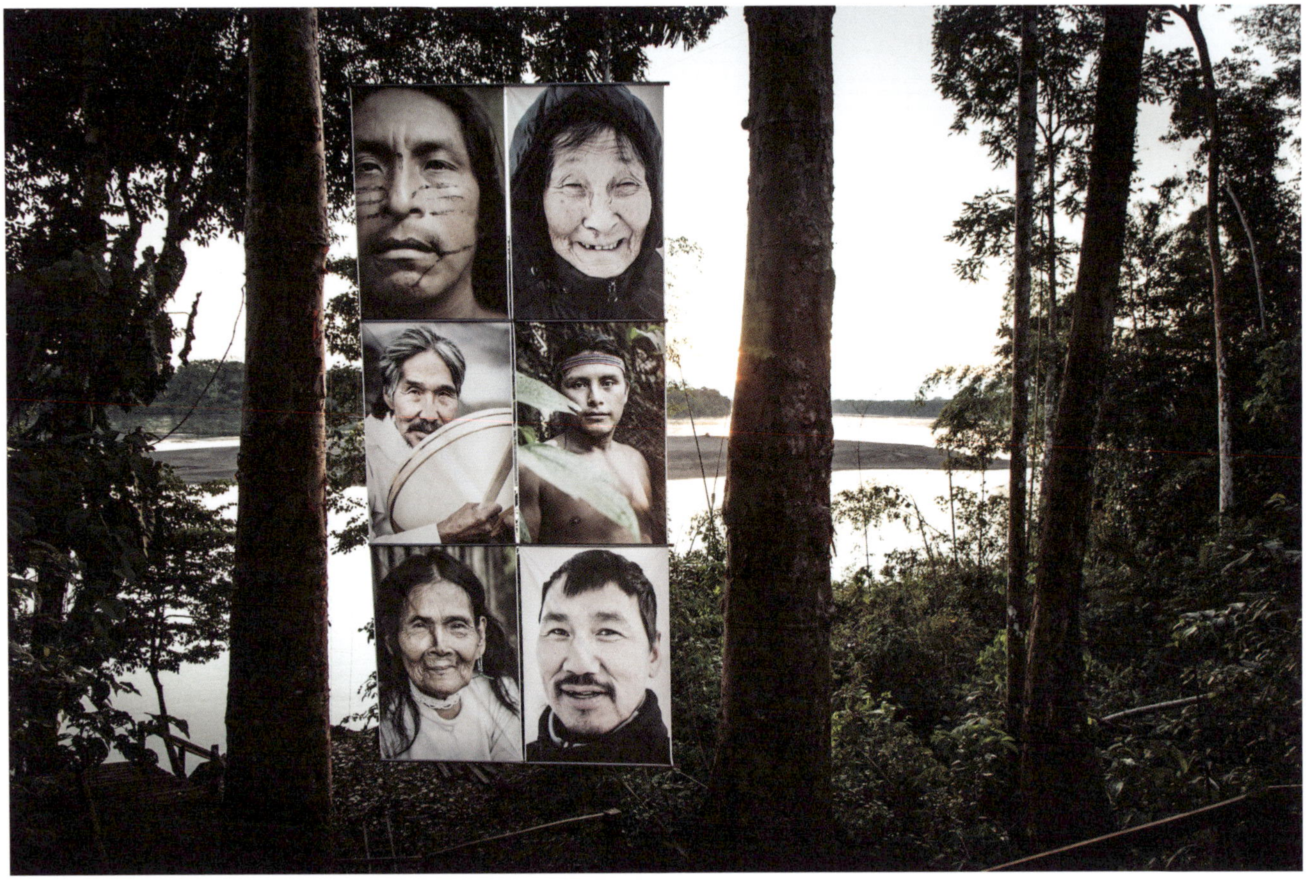

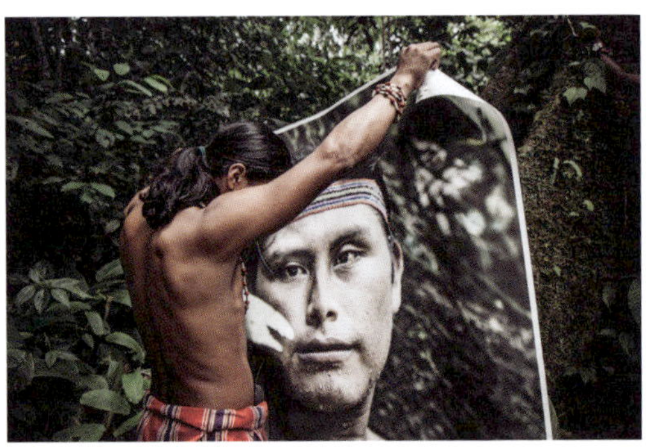

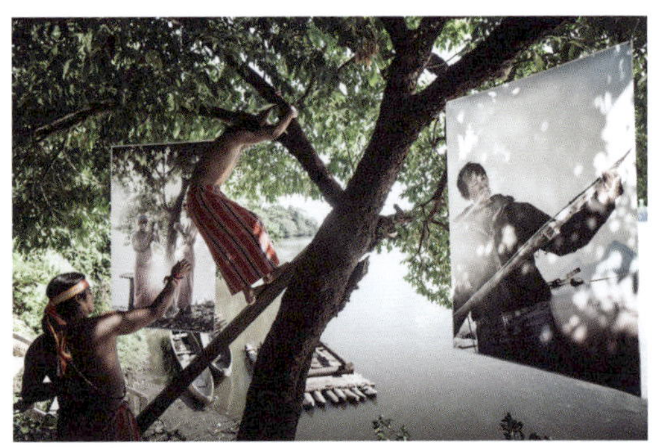

Im nächsten Jahr kehrte ich nach Sharamentsa und Yuwientsa in den Amazonasregenwald zurück. Die jungen Frauen, die ich zwei Jahre zuvor noch auf der Schulbank fotografiert hatte, begegneten mir nun mit ihren eigenen, kleinen Kindern auf dem Rücken. Endlich hatte ich auch die Gelegenheit, einmal Zeit bei einer Familie zu verbringen. Obwohl viele Häuser keine Wände haben, ist es unmöglich ohne Einladung einzutreten. Dank Mascha Kaukas Übersetzung konnte ich vertiefende Gespräche führen und eingehender erläutern, was ich mit meiner Arbeit bezwecke.

Zurück in Deutschland grübelte ich viel über meinen Ansatz, durch ein künstlerisch fotografisches Projekt ein Bewusstsein für die Klimakrise bei den Bewohner:innen der Industrienationen zu schaffen. Die Komplexität des Themas brachte mich zu dem Schluss, dass ich mich bildsprachlich bis zur Essenz reduzieren und ganz auf das Portraitieren der besuchten Menschen konzentrieren musste. Diese Portraits sollten ausschließlich durch Landschaften flankiert werden. Ich wollte die Verbindung, die zwischen allen Menschen existiert und den Einfluss, der von prägenden Lebensräumen ausgeht, verdichten.

Eines Nachts hatte ich eine Vision. Portraits der Achuar, Shuar und Inuit waren an einem Eisberg installiert. Intuitiv wusste ich, dass ich den Schlüssel zum Projekt gefunden hatte. Schon länger hegte ich den Wunsch, den Menschen im Regenwald und in Grönland ihre Bilder und die Bilder der anderen zu zeigen. Wir hatten bei meinen Aufenthalten über die Arbeit gesprochen und sie bestärkten mich, in den Industrienationen auf ihre Probleme bezüglich der Klimakrise aufmerksam zu machen. Nachdem der Entschluss gefasst war, ließ ich Portraits und Landschaften der beiden Regionen auf großformatigem, leichtem Textilbanner-Material produzieren und flog damit zurück nach Sharamentsa. Mit Mascha Kaukas Unterstützung und der Hilfe der beiden Achuar Nantu und José konnten wir die Bilder beider Regionen als Gegenüberstellung und Begegnung an verschiedenen Orten im Primärwald und im Dorf präsentieren.

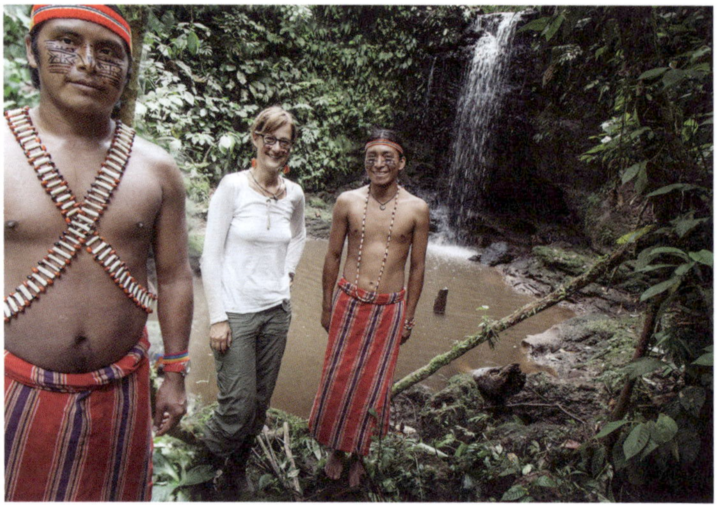
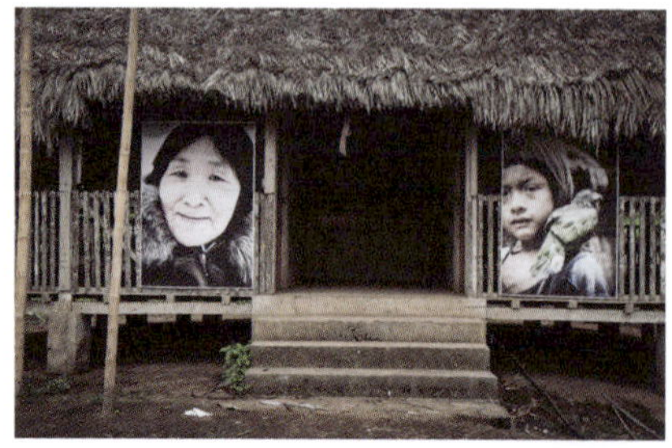
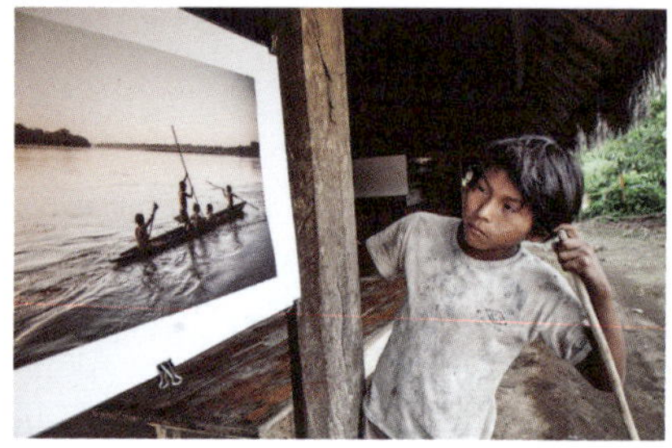
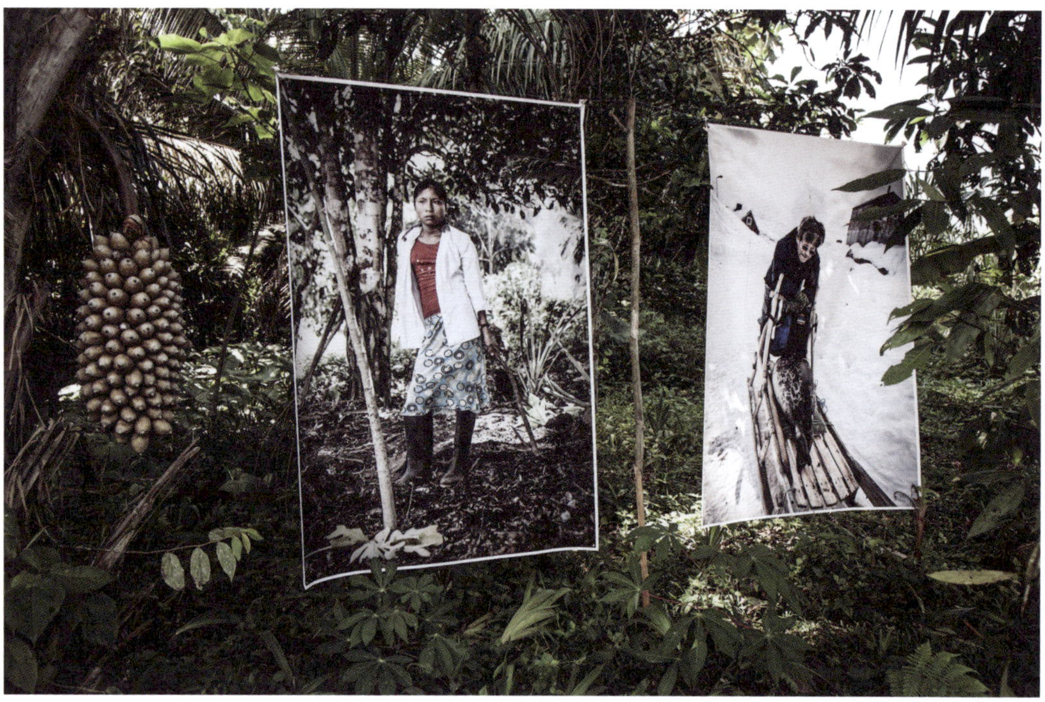

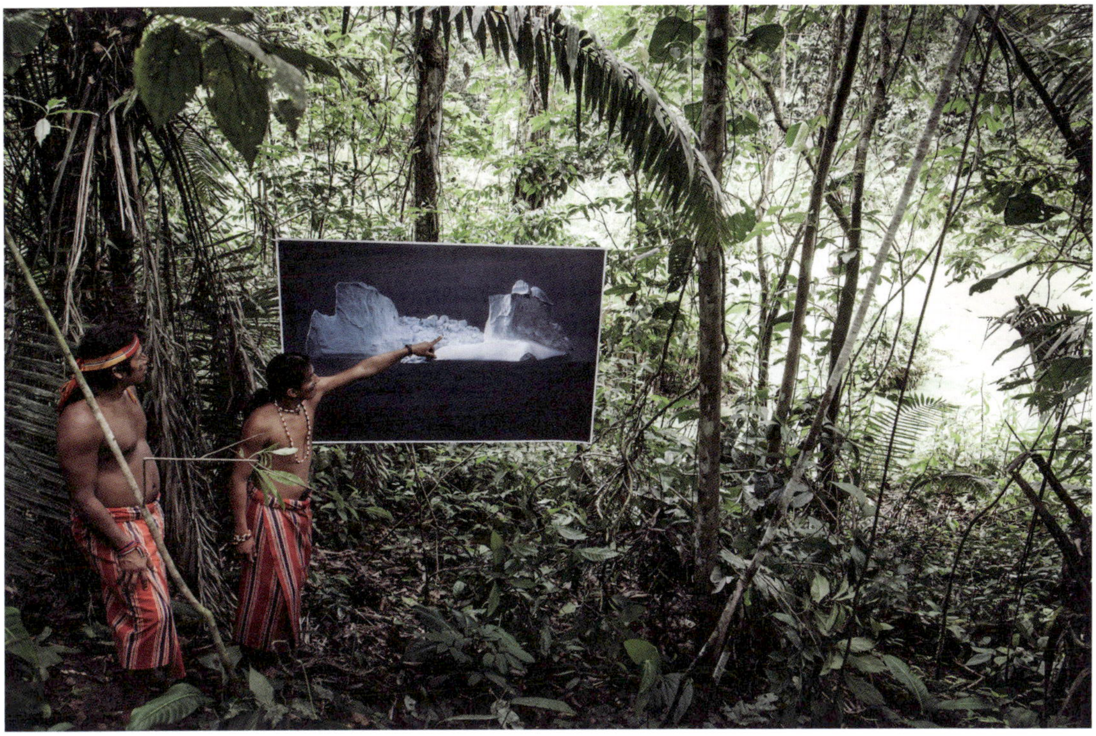

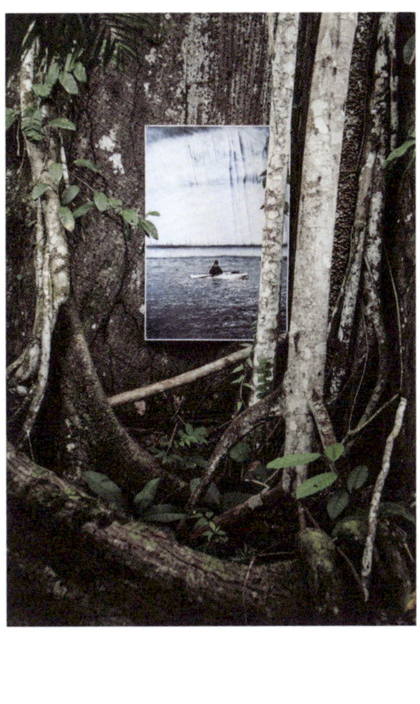

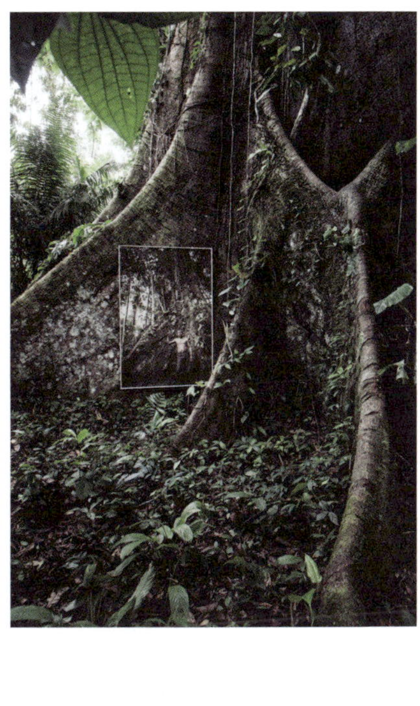

I then flew on to Tasiilaq, with a short break in Germany. Thanks to Robert Peroni's support and the help of the Inuit Viggo and Abel, Marco Kromminga and Moreno Bartoletti, we were also able to realize the installation here. Slowly I came closer to my goal of presenting the pictures on an iceberg. However, it was not to be made too easy for me. Throughout my stay in Tasiilaq, my helpers and I dealt with poor conditions, gusty winds and storms. I began to doubt whether we would make it to sea before my planned departure. On the last day in Greenland, the time had finally come. We went out in strong winds and high waves and battled for several hours with our small fishing boat, trying to navigate through the choppy sea into the neighboring fjord. In continuously extremely difficult conditions, we attached a picture of a landscape to a small iceberg as a trial. We couldn't think about anything else because the wind and waves were still far too strong. Viggo, who was steering the boat and whom I had portrayed in the kayak for my project, didn't let himself be disturbed. I, on the other hand, was. I kept looking over at him questioningly whenever I saw what I thought was a suitable iceberg, and he kept shaking his head. Doubts began to gnaw at me. After a long journey through the fjord, Viggo steered the boat around a headland. In front of me, at the end of a small bay, lay a large iceberg, long and wide, like a gallery wall. No waves rippled the sea, which moments ago had been rough. We slowly glided towards it. It was perfect. From the surface of the water, I could see far into the depths and recognize the individual shades and structures of the iceberg underwater, the sea lay so calmly in front of me. Then we docked. From the bow, Abel began to turn screws into the ice. Everything worked very easily, almost playfully. A whale surfaced within sight and blew a fountain into the air to greet us. A magical moment in a perfect place – I had made it!

With the help of the Achuar, Shuar and Inuit, the support of Mascha Kauka and Robert Peroni and the conditions on that day, I was able to make my vision reality. On the one hand, I was able to create symbolic images that addressed the climate crisis in a new, powerful way without using images of catastrophe. On the other hand, I found a way to depict destruction allegorically by exposing the photographs to the wind and weather myself.

My stays with the Achuar, Shuar and the Inuit also left their mark on me. I gradually began to see the world with different eyes, started to question my actions and change my consumer behavior. What I wanted to achieve with my work, I achieved first and foremost with myself.

Danach flog ich, mit kurzer Unterbrechung in Deutschland, weiter nach Tasiilaq. Dank Robert Peronis Unterstützung und der Hilfe der Inuit Viggo und Abel sowie jener von Marco Kromminga und Moreno Bartoletti konnten wir auch hier die Installation umsetzen. Langsam kam ich meinem Ziel, die Bilder an einem Eisberg zu präsentieren, näher. Es sollte mir jedoch nicht zu einfach gemacht werden. Während meines gesamten Aufenthalts in Tasiilaq kämpften meine Helfer und ich mit schlechten Bedingungen, böigen Winden und Sturm. Ich begann zu zweifeln, ob wir vor meiner geplanten Abreise aufs Meer kommen würden. Am letzten Tag in Grönland war es endlich soweit. Bei starkem Wind und hohen Wellen fuhren wir hinaus und kämpften uns mit unserem kleinen Fischerboot mehrere Stunden durch die aufgewühlte See bis in den benachbarten Fjord. Unter weiterhin äußerst schwierigen Bedingungen befestigten wir probeweise das Bild einer Landschaft an einem kleineren Eisberg. An mehr konnten wir nicht denken, denn Wind und Wellen waren immer noch viel zu stark. Viggo, der das Boot lenkte und den ich für mein Projekt im Kajak portraitiert hatte, ließ sich nicht aus der Ruhe bringen. Ich hingegen schon. Immer wieder sah ich fragend zu ihm hinüber, wenn ich einen für meine Begriffe passenden Eisberg sah, und immer wieder schüttelte er den Kopf. Zweifel begannen an mir zu nagen. Nach langer Fahrt durch den Fjord lenkte Viggo das Boot um eine Landzunge. Vor mir, am Ende einer kleinen Bucht lag ein großer Eisberg, lang und breit, wie eine Galeriewand. Keine Welle kräuselte die soeben noch raue See. Langsam glitten wir auf ihn zu. Er war perfekt. Von der Wasseroberfläche konnte ich weit in die Tiefe sehen und die einzelnen Farbtöne und Strukturen des Eisbergs unter Wasser erkennen, so ruhig lag das Meer vor mir. Dann legten wir an. Vom Bug aus begann Abel Schrauben in das Eis zu drehen. Alles ging ganz leicht, fast spielerisch. In Sichtweite tauchte ein Wal auf und blies uns zum Gruß eine Fontäne in die Luft. Ein magischer Moment an einem perfekten Ort. Ich hatte es geschafft!

Durch die Hilfe der Achuar, Shuar und Inuit, die Unterstützung Mascha Kaukas und Robert Peronis und die Bedingungen an diesem Tag wurde es mir möglich, meine Vision umzusetzen. Zum einen konnte ich symbolische Bilder schaffen, die die Klimakrise auf neue, eindringliche Weise thematisierten, ohne Katastrophenbilder einzusetzen. Zum Anderen fand ich einen Weg, Zerstörung allegorisch darzustellen, indem ich die Fotografien selbst Wind und Wetter ausgesetzt habe.

Die Aufenthalte bei den Achuar, Shuar und den Inuit haben auch bei mir Spuren hinterlassen. Ich begann nach und nach die Welt mit anderen Augen zu sehen, fing an mein Handeln zu hinterfragen und mein Konsumverhalten zu ändern. Was ich mit meiner Arbeit bewirken wollte, bewirkte ich zuallererst bei mir selbst.

Installations in Tasiilaq

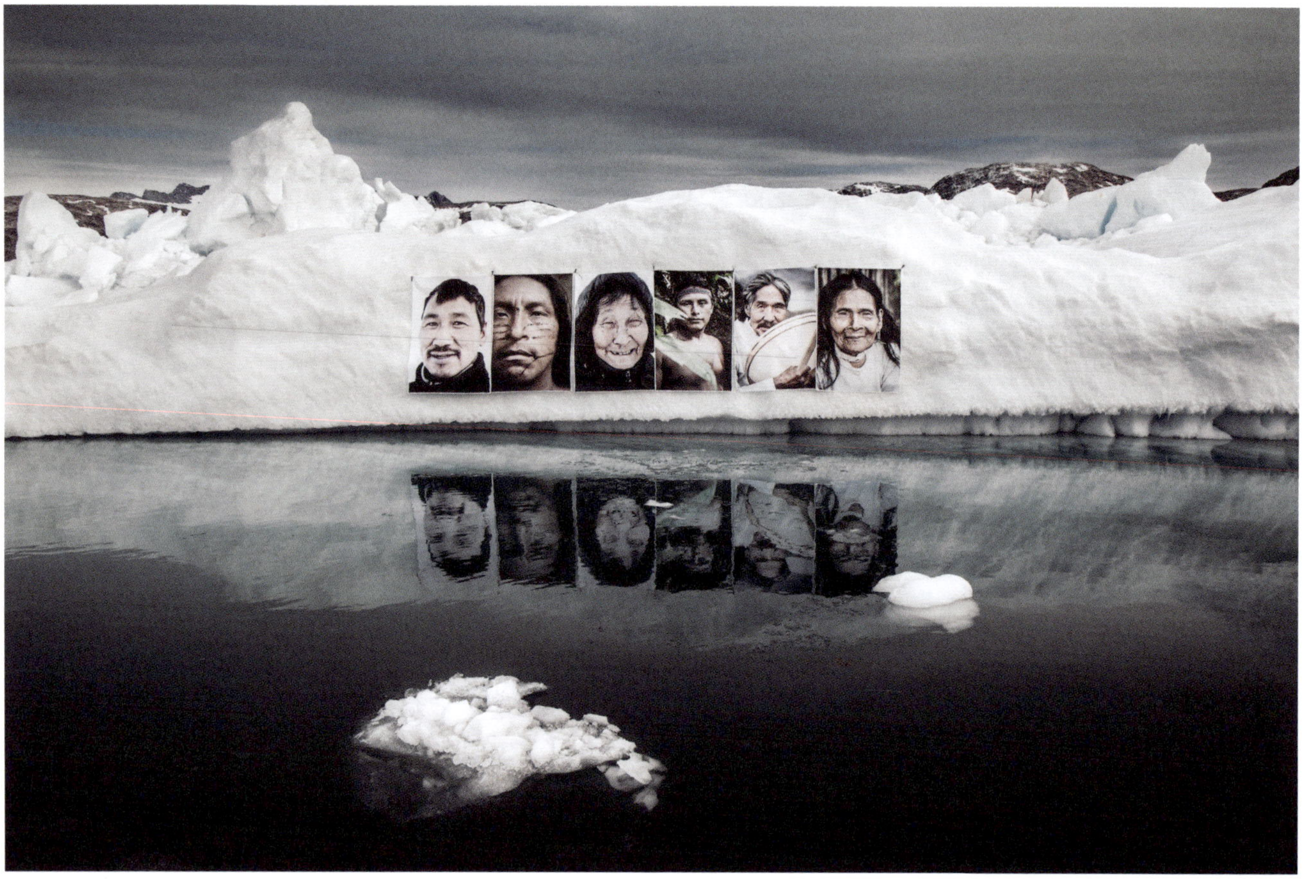

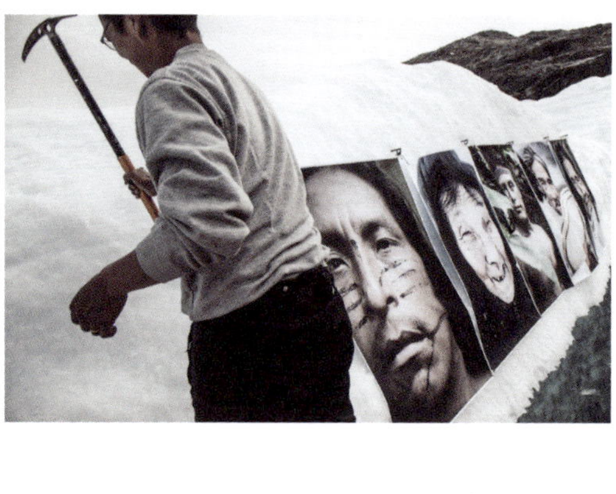

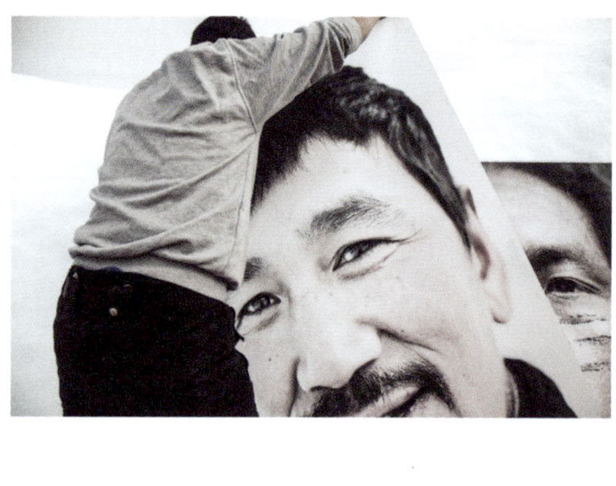

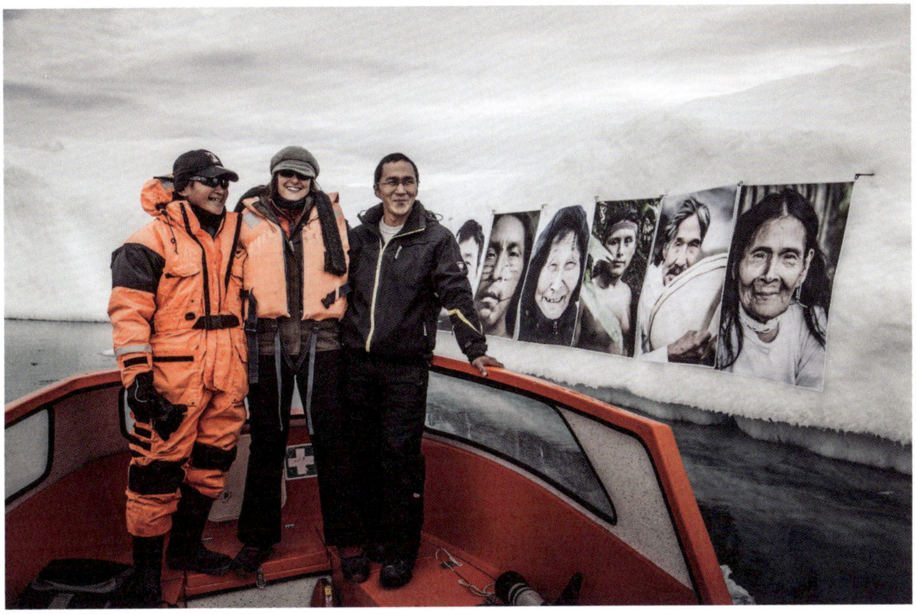

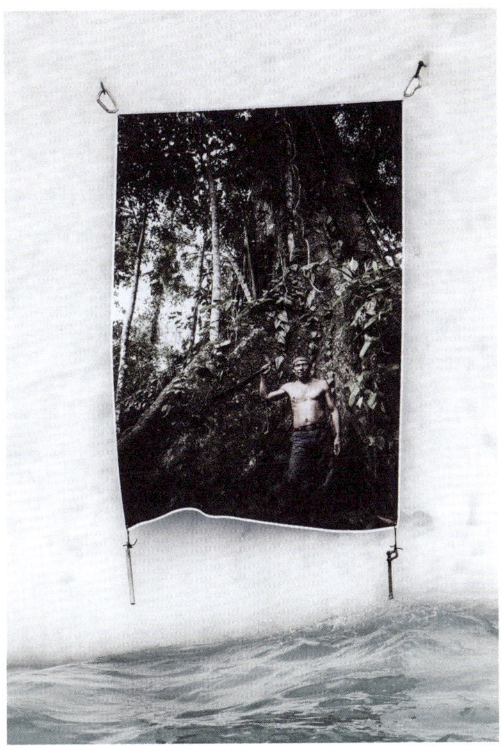

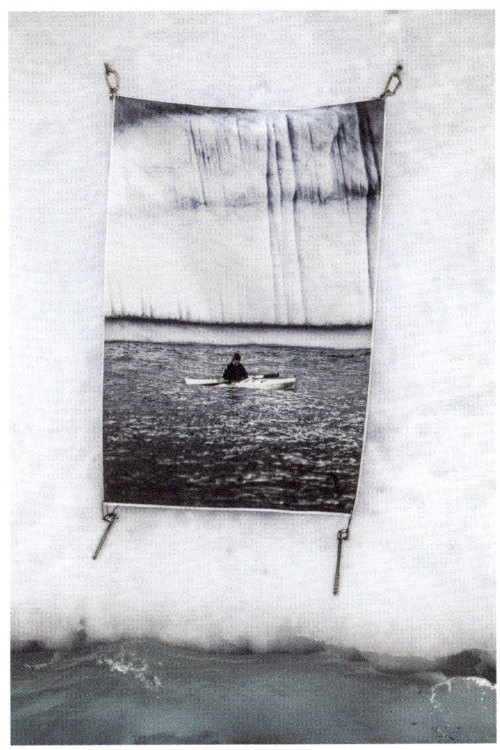

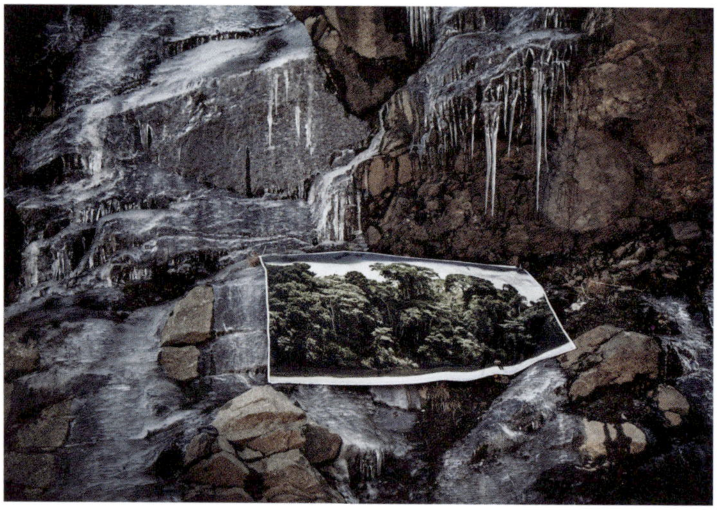

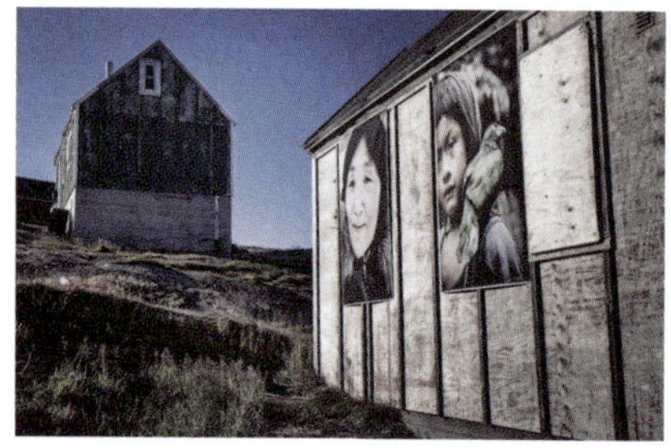

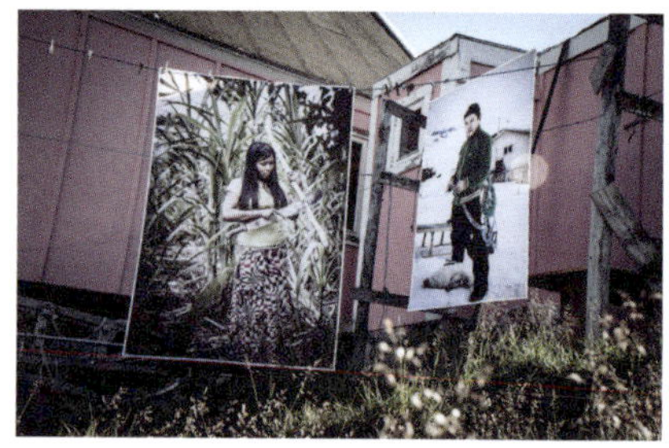

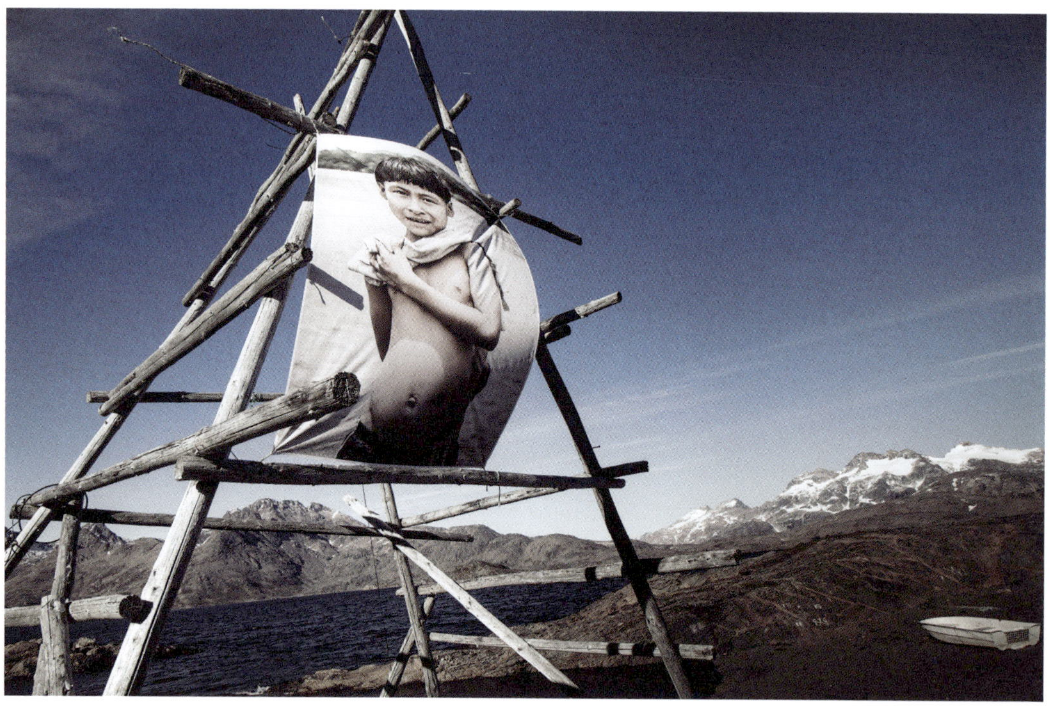

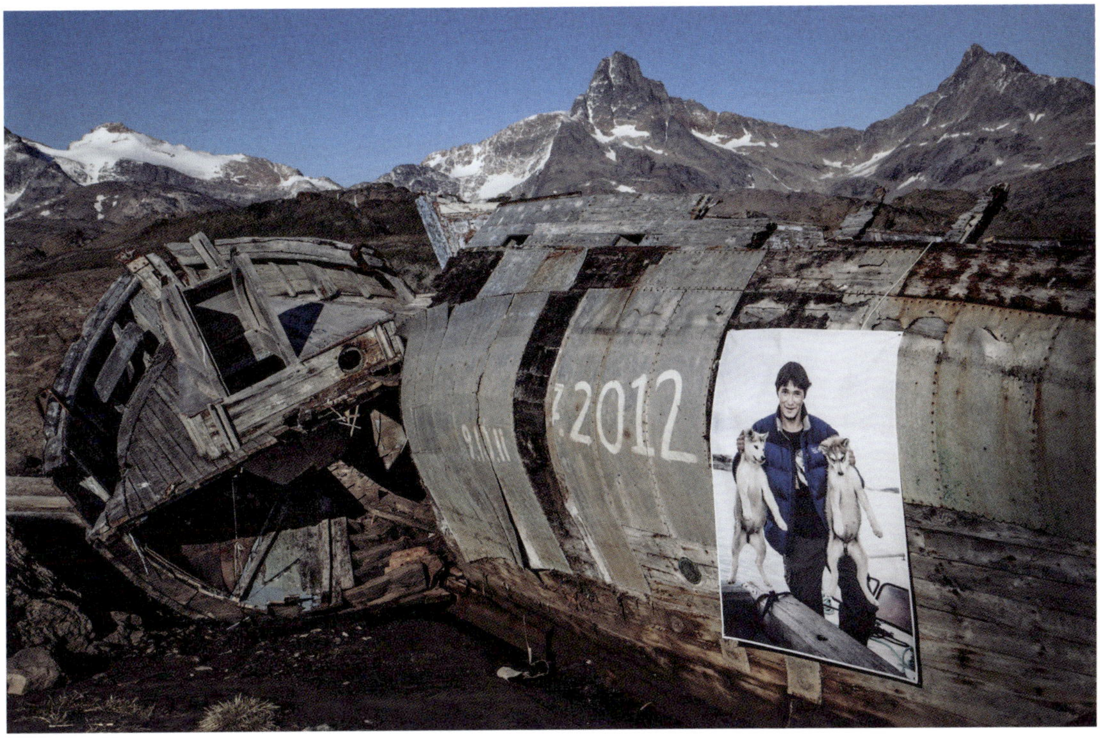
2012

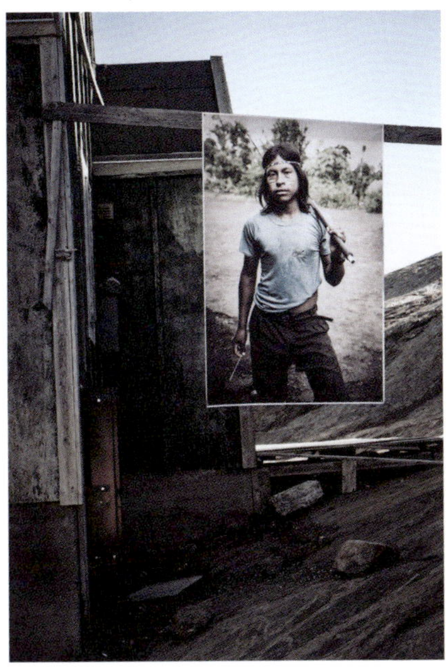

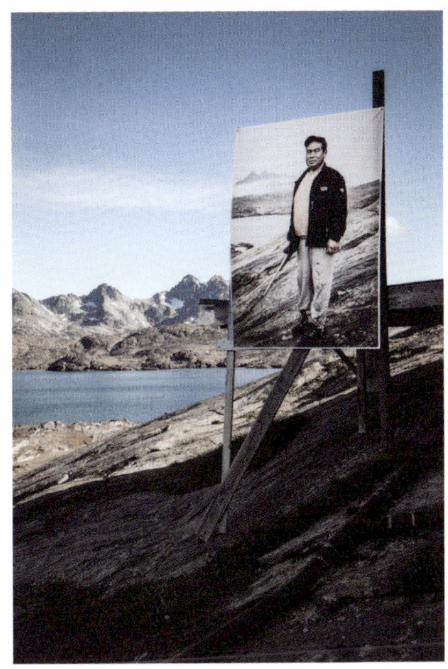

After that, some time had to pass to reflect on and categorize what I had achieved before I resumed my journey to the five continents in late summer 2016 with a visit to Mongolian nomads in the Gobi Desert. I had not found an organization that was active there, so I contacted a Mongolian tour operator and described my request. The tour operator wrote to nomadic families explaining my project. Several families agreed to take me in and support me. In order to be able to communicate with them, Myanmar Nyamaa, an employee of the company, accompanied me as a translator and travel companion and, to my delight, took his horsehead fiddle with him.

In Mongolia, shamanism is still a living culture today and exists on an equal footing with Buddhism. I asked Nyamaa to get in touch with a shaman or shamaness.

The ger of the shaman Busan is located in a settlement on the outskirts of Ulaanbaatar. You can recognize it from afar by the countless blue prayer flags fluttering in the wind. He was waiting for us at the entrance. As we plunged into the semi-darkness of the ger, I could make out an altar with ritual objects on the opposite wall, with Busan's large shaman drum leaning against it. In the right-hand corner you could see the imposing ritual robe. When we arrived, Busan was still dressed casually in jog pants and a Mongolian shirt. After an introductory talk, he agreed to go on a shamanic journey for my project, which I was allowed to document photographically and on film. He put on his ritual clothing and began to beat the drum. Faster and faster the beater swept across the surface, wilder and wilder his movements became, watched over by his wife and son, who were ready to catch him when the trance set in. Busan, to whom I had only roughly outlined my project, spoke about it in detail afterwards and gave me valuable thoughts to take with me on my way into the desert.

"Climate change is like a festering wound that needs proper, good medicine. People need education and need to understand that we are part of nature. Unfortunately, most people think they are the owners of this earth. If we destroy nature, we live in inhumanity and civilization dies. Without nature, nothing exists."

"You have a good goal. You need a lot of people working together towards one goal. That is the easiest and most effective way to achieve it. You can lead people to this goal like a camel caravan."

Danach musste eine gewisse Zeit verstreichen, um das Erreichte zu reflektieren und einzuordnen, bevor ich im Spätsommer 2016 mit einem Besuch bei mongolischen Nomadinnen und Nomaden in der Wüste Gobi meine Reise zu den fünf Kontinenten wieder aufnahm. Ich hatte keine Organisation gefunden, die dort tätig war, darum kontaktierte ich einen mongolischen Reiseveranstalter und schilderte mein Anliegen. Der Veranstalter schrieb an Nomadenfamilien und erläuterte mein Projekt. Mehrere Familien erklärten sich bereit, mich aufzunehmen und zu unterstützen. Damit ich mich mit ihnen verständigen können würde, begleitete mich Myanmar Nyamaa, ein Mitarbeiter des Unternehmens, als Übersetzer und Reisebegleiter und nahm zu meiner Freude seine Pferdekopfgeige mit.

In der Mongolei ist der Schamanenglaube bis heute gelebte Kultur und existiert gleichberechtigt neben dem Buddhismus. Ich bat Nyamaa, einen Kontakt zu einem Schamanen oder einer Schamanin herzustellen.

Das Zelt, Ger, des Schamanen Busan liegt in einer Siedlung am Rande Ulaanbaatars. Man kann es schon von weitem an den zahllosen blauen, im Wind flatternden Gebetsfahnen erkennen. Er erwartete uns am Eingang. Beim Eintauchen in das Halbdunkel des Ger konnte ich an der gegenüberliegenden Wand einen Altar mit rituellen Gegenständen ausmachen, daran angelehnt Busans große Schamanentrommel. In der rechten Ecke sah man das imposante, rituelle Gewand. Bei unserem Eintreffen war Busan noch leger mit Sporthose und einem mongolischen Hemd bekleidet. Nach einem Einführungsgespräch erklärte er sich bereit, sich auf eine schamanische Reise für mein Projekt zu begeben, die ich fotografisch und filmisch dokumentieren durfte. Dafür legte er seine rituelle Kleidung an und begann die Trommel zu schlagen. Schneller und schneller fegte der Schläger über die Oberfläche, wilder und wilder wurden seine Bewegungen, bewacht von Frau und Sohn, die bereit waren, ihn bei einsetzender Trance aufzufangen. Busan, dem ich mein Projekt nur grob umrissen hatte, sprach danach ausführlich darüber und gab mir wertvolle Gedanken mit auf meinen Weg in die Wüste.

„Klimawandel ist wie eine eitrige Wunde, gegen die es eine richtige, gute Medizin braucht. Die Menschen brauchen Bildung und müssen verstehen, dass wir ein Teil der Natur sind. Leider denken die Meisten, sie wären die Besitzer dieser Erde. Wenn wir die Natur vernichten, leben wir in Unmenschlichkeit und die Zivilisation stirbt. Ohne Natur existiert nichts."

My head and heart were filled with the encounter with Busan when we flew to Dalandsadgad, the largest city in South Gobi, the next morning. A driver who would take us to the desert picked us up at the airport. Before we set off, we stocked up on food and gifts. Then the desert finally lay before me. At the beginning, the landscape was dominated by steppes, which gradually became rockier and sandier. We passed large herds of free-roaming horses and camels until we reached the ger of our host Erdenetschulun and his family in the afternoon. There were no people to be seen, just the sky stretching to the horizon and a few camels dozing in the sun. The driver had already left us again when Tschulun's arrival on the motorcycle was announced by a cloud of dust in the distance. He greeted us warmly and invited us into his ger. Two days later, we were joined by his wife Nasanjargal and their children Khongorzul and Hura, who went to school in Dalandsadgad. A curious cheerfulness moved into the tent with the two girls. Contrary to my expectations, it turned out that the nomads have a lively social life. Although the people live far apart, they are in constant contact and communicate in a lively manner. Not a day went by without a visit from distant neighbors, friends or relatives.

A ger in the Gobi is always open to anyone who needs a place to sleep or a meal. One evening, when it was already dark outside and we were all sitting together around the central stove, the door opened and an imposing figure in a traditional golden coat entered the room. The Mongolian was on his motorcycle looking for two camels and had already driven well over a hundred kilometers. He spent the night with us and was on his way again long before I woke up.

One by one, we visited Tschulun's brothers and neighbors. Thanks to Nyamaa's translation, I was able to have conversations and interviews with the people, learn about their view of things and their relationship with nature, which had already captivated me on the first day. I felt immediately connected to the earth, one with the land, the wind, the sun, the clouds and the rain. The people in the Gobi radiate this connection and vastness.

Just six months after my return from the Gobi, I set off again to visit representatives of the Maasai in the village of Tingatinga at the foot of Mount Kilimanjaro in Tanzania. East Africa, along the Great Rift Valley, is considered the cradle of mankind. It is a highly symbolic region where difficult climatic conditions prevail. Droughts and heavy rainfall leach out the land and lead to the loss of topsoil. Hunger and disease are draining the people.

„Du hast ein gutes Ziel. Du brauchst viele Menschen, die zusammen auf ein Ziel zugehen. Das ist der einfachste und effektivste Weg, um es zu erreichen. Wie eine Kamelkarawane kannst Du die Menschen zu diesem Ziel führen."

Mein Kopf und mein Herz waren erfüllt von der Begegnung mit Busan, als wir am nächsten Morgen nach Dalandsadgad flogen, der größten Stadt der Südgobi. Ein Fahrer, der uns in die Wüste bringen würde, holte uns am Flughafen ab. Bevor wir aufbrachen, deckten wir uns noch mit Lebensmitteln und Gastgeschenken ein. Dann endlich lag die Wüste vor mir. Zu Beginn dominierte eine steppige Landschaft, die nach und nach steiniger und sandiger wurde. Wir passierten große Herden freilaufender Pferde und Kamele, bis wir am Nachmittag das Ger unseres Gastgebers Erdenetschulun und seiner Familie erreichten. Kein Mensch war zu sehen, nur Himmel bis zum Horizont und ein paar in der Sonne dösende Kamele. Der Fahrer hatte uns bereits wieder verlassen, als sich Tschuluns Ankunft auf dem Motorrad durch eine Staubwolke in der Ferne ankündigte. Er begrüßte uns herzlich und bat uns in sein Ger. Zwei Tage später stießen auch seine Frau Nasanjargal und die Töchter Khongorzul und Hura, die in Dalandsadgad zur Schule gingen, zu uns. Eine neugierige Fröhlichkeit zog mit den beiden Mädchen in das Zelt ein. Ganz entgegen meiner Vorstellung stellte sich heraus, dass die Nomad:innen ein reges Sozialleben haben. Die Menschen leben zwar weit verstreut, stehen jedoch in ständigem Kontakt und Austausch. Kein Tag verging, ohne dass Besuch von entfernten Nachbarn, Freunden oder Verwandten kam.

Ein Ger in der Gobi steht immer und für jeden offen, der einen Schlafplatz oder etwas zu essen braucht. Eines Abends, draußen war es längst dunkel und wir saßen alle zusammen um den zentralen Ofen, öffnete sich die Tür und eine imposante Gestalt in einem traditionellen, goldenen Mantel betrat den Raum. Der Mongole war mit seinem Motorrad auf der Suche nach zwei Kamelen und schon weit über hundert Kilometer gefahren. Er verbrachte die Nacht bei uns und war lange bevor ich aufwachte wieder unterwegs.

Nach und nach besuchten wir Tschuluns Brüder und Nachbarn. Dank Nyamaas Übersetzung konnte ich Gespräche und Interviews mit den Menschen führen, ihre Sicht der Dinge und ihre Beziehung zur Natur erfahren, die mich schon am ersten Tag in ihren Bann gezogen hatte. Ich fühlte mich auf unmittelbare Weise verbunden mit der Erde, eins mit dem Land, dem Wind, der Sonne, den Wolken und dem Regen. Die Menschen in der Gobi strahlen diese Verbundenheit und Weite aus.

I came to Tingatinga through the Jane Goodall Institute (JGI) in Germany. The place is located in a game reserve that can only be entered with a permit. The Maasai and the wild animals live there in coexistence, which is not without problems. Kessy Beatus, who is involved in a project with the JGI, accompanied me as a translator. In order to stay with the Maasai in the village, I had prepared myself to sleep on rammed earth. I was all the more surprised when my host, the Maasai priest Saningo Lekukwo, welcomed me into his brick house. There was even a small guest room with a bed and Bible at my disposal. What an unexpected place. The Maasai themselves live mostly traditionally in bomas, round homesteads with huts made of rammed earth and cattle enclosures protected from the wild animals by thick spiky branches and twigs.

After a quiet, relaxing night behind thick walls, the next day I began to roam the village and talk to the people, who were very open and friendly, about their culture and their experiences of the climate crisis. I witnessed a funeral ceremony and, in the course of my stay, a wedding. In Maasai culture, there are many rules and commandments that seem strange to us, and which were readily explained to me when I asked. For example, I learned something about the path of a moran, which begins with when and how a young man is initiated. Through several stages of initiation, he becomes a warrior, the moran, who builds up a herd of cattle in the following years, tending and protecting them. During this long period, the morani will not marry and start a family, but will mostly travel alone on the steppe with their cattle and feed on saroi. Only when they have enough cattle to pay the bride price for their future wife will they give up their life as warriors and thus their life on the steppe, go to the sacred mountain of the Maasai to commit themselves and then marry. Individuality has no place in their society; the survival of the community takes precedence over the individual. Despite the corset of rules, prohibitions and constraints, which is difficult for us individualists to understand, the people are strong and proud.

At night, when I lay in my bed in the protective stone house, I could hear the dogs barking. As every night, the elephants had entered the village in search of food and left a trail of devastation through the few small kitchen gardens. It hadn't rained for months. Everything was very dry. Day after day, the drought made the situation worse for the people, their livestock and the wild animals. Already on the way to Tingatinga, we had seen animal carcasses rotting by the roadside. In the village, after a long conversation, the Maasai Logotu Lonyokie led me to a large pile. These were his cattle, which had fallen victim to the drought and which he had gathered here. Of the original 400 animals, he had already lost 100 to the

Bereits ein halbes Jahr nach meiner Rückkehr aus der Gobi brach ich wieder auf, um in Tansania in dem Dorf Tingatinga am Fuß des Kilimanjaro Vertreter:innen der Massai zu besuchen. Ostafrika, entlang des Großen Afrikanischen Grabenbruchs, gilt als Wiege der Menschheit. Es ist ein symbolträchtiger Landstrich, in dem schwierige klimatischen Bedingungen herrschen. Dürren und Starkregenereignisse laugen das Land aus, Hunger und Krankheit laugen die Menschen aus.

Nach Tingatinga kam ich durch die Vermittlung des Jane-Goodall-Instituts (JGI) in Deutschland. Der Ort liegt in einem Wildschutzgebiet, das man nur mit Genehmigung betreten darf. Die Massai und die Wildtiere leben dort in Koexistenz, was nicht problemlos möglich ist. Kessy Beatus, mit dem JGI durch ein Projekt verbunden, begleitete mich als Übersetzer. Um bei den Massai im Dorf unterzukommen, hatte ich mich darauf eingestellt, auf gestampftem Lehmboden zu schlafen. Umso überraschter war ich, als mich mein Gastgeber, der Massai-Priester Saningo Lekukwo, in seinem gemauerten Haus empfing. Sogar ein kleines Gästezimmer mit Bett und Bibel stand zu meiner Verfügung. Welch unerwarteter Ort. Die Massai selbst leben größtenteils traditionell in Bomas, runden Gehöften mit Hütten aus gestampftem Lehm und Viehgehegen, die durch dicht stehende stachelige Gehölze vor den Wildtieren geschützt sind.

Nach einer ruhigen Nacht hinter dicken Mauern begann ich am nächsten Tag den Ort zu durchstreifen und mit den Menschen, die mir sehr offen und freundlich begegneten, über ihre Kultur und ihre Erfahrungen mit der Klimakrise zu sprechen. Ich wurde Zeugin einer Beerdigungszeremonie und im Laufe meines Aufenthalts auch einer Hochzeit. In der Kultur der Massai gibt es viele Regeln und Gebote, die uns fremd anmuten und die mir bei Nachfrage bereitwillig erklärt wurden. Dadurch lernte ich zum Beispiel etwas über den Weg eines Kriegers, des Moran, der damit beginnt, wann und wie ein junger Mann initiiert wird. Durch mehrere Initiationsstufen wird er zum Krieger, der sich in den Folgejahren eine Rinderherde aufbaut, sie hütet und beschützt. In dieser langen Zeit werden die Morani nicht heiraten und eine Familie gründen, meist mit dem Vieh alleine in der Steppe unterwegs sein und sich von Saroi ernähren. Erst wenn sie genug Rinder haben, um den Brautpreis für ihre zukünftige Frau bezahlen zu können, werden sie das Kriegerdasein und damit das Leben in der Steppe aufgeben, zum heiligen Berg der Massai ziehen, um dies zu besiegeln, und danach heiraten. Individualität hat in ihrer Gesellschaft keinen Platz, das Überleben der Gemeinschaft steht über dem Einzelnen. Trotz des für uns Individualist:innen schwer zu verstehenden Korsetts aus Regeln, Verboten und Zwängen sind die Menschen stark und stolz.

drought. He attributed the extent of this drought to climate change and told me that the increasing loss of topsoil was also becoming more and more problematic. Another Maasai, Richard Nikolaus, with whom I spoke, pointed to his dry, uncultivated field. At this time of year, the maize should already be knee-high. However, due to the lack of rain, the seed could not be sown. He spoke to me about climate change before I could tell him why I was in Tingatinga and asked me to report back to my country on how they were faring. These conversations affected me. My topic had caught up with and overtaken me.

Another year and a half had to pass before I could set off again. My last trip was with my now 18-year-old son Moritz to the island state of Kiribati in the South Pacific, the place that represents the continent of Oceania and the tipping point of the tropical coral reefs in my project. Our trip would take us around the world and burden my conscience with an enormous carbon footprint, which I tried to relieve through financial compensation. In my search for cooperation partners, I met human geographer Silja Klepp, who teaches at Christian Albrecht University in Kiel and was researching climate justice in Kiribati at the time. It was a lucky coincidence that allowed us to cooperate not only in Kiribati.

After a long flight with changes in Los Angeles, Rarotonga, Auckland and Fiji, we finally arrived in Kiribati's capital South Tarawa and met with Pelenise Alofa, Country Coordinator of Kiribati Climate Action Network, KiriCan for short, and Reverend Maleta Tenten from the Kiribati Uniting Church, who told me the following in an interview:

"Climate change isn't just an issue of science and economics, it is also a spiritual crisis for people in the pacific who are most affected."

Through the conversations with Reverend Maleta Tenten, Pelenise Alofa and their colleagues and through the stories they told us, the cosmos of Micronesians slowly opened up to me. I realized that I had to try to visit a village, to get to an outer island, to a place where people were even more connected to nature and their spirits than in South Tarawa. However, this turned out to be more difficult than I thought, because Kiribati is one of the largest countries in the world in terms of size, although it is a small country in terms of population. A lucky coincidence brought us the solution. There was a young man working at KiriCan, Akamatang Bateiti, barely older than my son, who was from the village of Tabonibara in North Tarawa and spoke English. He agreed to visit his village with us. In return,

Nachts, wenn ich in meinem Bett im schützenden Steinhaus lag, konnte ich die Hunde anschlagen hören. Die Elefanten waren wie jede Nacht auf der Suche nach Futter in das Dorf eingedrungen und zogen eine Spur der Verwüstung durch die wenigen kleinen Nutzgärten. Monatelang hatte es nicht geregnet. Alles war sehr trocken. Die Dürre verschlimmerte Tag für Tag die Situation für die Menschen und ihre Nutztiere wie auch für die Wildtiere. Schon auf dem Weg nach Tingatinga sahen wir Tierkadaver am Strassenrand verwesen. Im Dorf führte mich nach einem längeren Gespräch der Massai Logotu Lonyokie zu einem großen Haufen. Es waren seine Rinder, die der Trockenheit zum Opfer gefallen waren und die er hier zusammengetragen hatte. Von ursprünglich 400 Tieren hatte er bereits 100 durch die Dürre verloren. Die Dimension dieser Dürre schrieb er dem Klimawandel zu, und er erzählte mir, dass auch der zunehmende Verlust an Mutterboden immer problematischer werde. Ein anderer Massai mit dem ich sprach, Richard Nikolaus, deutete auf sein trockenes, unbestelltes Feld. Um diese Zeit des Jahres müsse der Mais bereits kniehoch stehen. Aufgrund des ausbleibenden Regens habe die Saat jedoch nicht ausgebracht werden können. Er sprach mich auf den Klimawandel an, bevor ich ihm erzählen konnte, warum ich in Tingatinga war, und bat mich darum in meinem Land zu berichten, wie es ihnen ergehe. Diese Gespräche machten mich betroffen. Mein Thema hatte mich nicht nur eingeholt, sondern überholt.

Noch einmal mussten eineinhalb Jahre verstreichen, bis ich erneut aufbrechen konnte. Die letzte Reise unternahm ich zusammen mit meinem inzwischen 18-jährigen Sohn Moritz zu dem Inselstaat Kiribati im Südpazifik, dem Ort, der in meinem Projekt stellvertretend für den Kontinent Ozeanien und den Kipppunkt der tropischen Korallenriffe steht. Unsere Reise würde uns einmal um die ganze Welt führen und mein Gewissen mit einem enormen Kohlendioxid-Fußabdruck belasten, sodass ich es durch finanzielle Kompensation zu beruhigen versuchte. Auf meiner Suche nach Kooperationspartner:innen traf ich auf die Humangeographin Silja Klepp, die in Kiel an der Christian-Albrechts-Universität lehrt und zu dem Zeitpunkt in Kiribati zu Klimagerechtigkeit forschte. Eine glückliche Fügung, die uns nicht nur in Kiribati kooperieren ließ.

Nach einem langen Flug mit Umstiegen in Los Angeles, Rarotonga, Auckland und Fiji endlich in Kiribatis Hauptstadt South Tarawa angekommen, traf ich Pelenise Alofa, Landeskoordinatorin des Kiribati Climate Action Network, kurz KiriCan, und Reverend Maleta Tenten von der Kiribati Uniting Church, die mir in einem Interview folgendes sagte:

we would cross the Tarawa lagoon in a two-hour trip on an outrigger boat. Using the days in South Tarawa, Pelenise familiarized us with KiriCan projects, such as the mangrove reforestation project of the women of Ambo or a plastic waste collection campaign on the beach of Bairiki. A few days later, Matang, Moritz and I set off with our gifts for the gods to cross the Tarawa lagoon. When we arrived at our destination, we had to disembark in the shallow, warm water and wade to the shore with our luggage. Our fellow travelers quickly dispersed and we were alone. Fortunately, our accommodation was in newly built pile dwellings in the immediate vicinity of the jetty. There were no other dwellings to be seen at first glance. There is no road in North Tarawa, only a sandy track that connects the houses and villages. People walk, use motorcycles or bicycles to get around. An old truck that collects schoolchildren along the route and transports them to school was the only vehicle I saw there. Before we could set off for Tabonibara, we had to visit the Atinmaebo shrine at the remote end of North Tarawa and present ourselves to the gods. There we would find out whether they were well-disposed towards us. We had bought tobacco in South Tarawa as an offering to the gods, which we now wanted to give to the guardian of the shrine. We picked him up halfway there on our motorcycles and drove together to the shrine, which is inconspicuously surrounded by coconut palms on the edge of the sea. The centerpiece of the shrine is a large tridacninae, a giant clam. The guardian put down the tobacco and spoke to the gods. I don't know what he told them, but I went back after our meeting in the hope that the gods would look upon us and my project favorably. Nothing now stood in the way of Tabonibara and Akamatang's family's visit. Matang would prepare for our arrival the next day. In the meantime, I explored the surrounding area. It was afternoon, the sun would soon be setting and a thunderstorm was building up at the edge of the lagoon. I saw a fisherman standing in the shallow water and approached him. He was calmly holding his net in his arms and drawing cirlies over the lagoon from his small boat. We looked at each other and I knew that he had given me permission to photograph him. We spent some time together, fishing and photographing, then I said goodbye and thanked him and waded back ashore. There I met Matang, who had just returned from his village on his motorcycle. He pointed to the lagoon and to the man. It was his father, Tabeta Bateiti, who was fishing there.

The next morning the time had finally come. We drove to Tabonibara and met Matang's family. There was a calm, relaxed atmosphere in the village. The women were peeling roots, weaving mats from palm leaves or flower wreaths for their hair. Thanks to Matang's help, I was able to introduce myself to the people and now also to his father and talk to him. We were sitting in a traditional open

„Klimawandel ist nicht nur eine Fragestellung für Wissenschaft und Wirtschaft, er bedeutet auch eine spirituelle Krise für die Menschen im Pazifik, die er am stärksten betrifft."

Durch die Gespräche mit Reverend Maleta Tenten, Pelenise Alofa und ihren Kolleg:innen und durch die Geschichten, die sie uns erzählten, erschloss sich mir langsam der Kosmos der Mikronesier:innen. Mir wurde klar, dass ich versuchen musste, ein Dorf zu besuchen, auf eine Außeninsel zu kommen, an einen Ort, an dem die Menschen noch mehr mit der Natur und ihren Geistern verbunden waren als in South Tarawa. Das stellte sich jedoch als schwieriger heraus als ich dachte, denn nach seiner Ausdehnung gehört Kiribati zu den größten Staaten der Erde, obwohl es nach Einwohnern ein Kleinstaat ist. Eine glückliche Fügung brachte uns die Lösung. Bei KiriCan arbeitete ein junger Mann, Akamatang Bateiti, kaum älter als mein Sohn, der aus dem Dorf Tabonibara in Nord-Tarawa stammte und Englisch sprach. Er erklärte sich bereit, mit uns sein Dorf zu besuchen. Dafür würden wir in einer zweistündigen Fahrt mit einem Auslegerboot die Lagune von Tarawa durchqueren. Die Tage in South Tarawa nutzend, machte uns Pelenise mit KiriCan Projekten vertraut, wie dem Mangroven-Aufforstungsprojekt der Frauen von Ambo oder einer Plastikmüll-Sammelaktion am Strand von Bairiki. Ein paar Tage später brachen Matang, Moritz und ich mit unseren Geschenken für die Götter auf, die Lagune von Tarawa zu durchqueren. Am Ziel angekommen, mussten wir im flachen, warmen Wasser aussteigen und mit unserem Gepäck ans Ufer waten. Unsere Mitreisenden zerstreuten sich schnell und wir blieben alleine zurück. Glücklicherweise lagen unsere Unterkünfte in neu errichteten Pfahlbauten in unmittelbarer Nähe des Bootsanlegers. Andere Behausungen waren auf den ersten Blick nicht zu sehen. In Nord-Tarawa gibt es keine Straße, sondern nur eine Sandpiste, die die Häuser und Dörfer verbindet. Die Menschen laufen zu Fuß oder nutzen zur Fortbewegung Motor- oder Fahrräder. Ein alter Lastwagen, der die Schulkinder entlang der Piste einsammelt und zur Schule befördert, war das einzige Fahrzeug, das ich dort sah. Bevor wir nach Tabonibara aufbrechen konnten, mussten wir den Atinmaebo-Schrein am entlegenen Ende Nord-Tarawas aufsuchen und uns den Göttern vorstellen. Dort würde sich entscheiden, ob sie uns wohlgesonnen waren.

Wir hatten in South Tarawa Tabak als Opfergabe für die Götter gekauft, den wir nun dem Hüter des Schreins übergeben wollten. Wir holten ihn auf halber Strecke mit unseren Motorrädern ab und fuhren gemeinsam zum Schrein, der unscheinbar umsäumt von Kokospalmen am Rand des Meeres liegt. Das Herzstück des Schreins bildet eine große Tridacninae, eine Riesenmuschel. Der Hüter legte

house. Tabeta Bateiti talked a lot about his relationship with the sea and about the changes caused by climate change.

In the early afternoon of the following day, I met the young people I had visited at school in the morning at a nearby bridge. The bridge, which is actually just a concrete slab, connects two sections of the island and spans an area of the lagoon where the water is slightly deeper and therefore cooler than in other sections. Dressed as they were and with flowers in their hair, the boys and girls jumped exuberantly into the water.

It stayed with me for a long time how much this remote place showed that nothing that happens in the world is without consequences. We are all connected and have a responsibility towards one another.

INSTALLATION

The five continents

After all these years and all the traveling, in September 2019 I was finally able to bring the five continents together in an installation in Germany on the apron of the open-cast lignite mine at Hambach Forest. The Hambach Forest, which, together with the surrounding land and neighboring villages, was to make way for open-cast mining, became a symbolic place for the climate movement in 2018 when the police evicted tree houses that had been erected by climate activists to protect the forest. With my installation, I was a guest on the "resistance field" of BUND—Bund für Umwelt und Naturschutz, a tiny piece of land that the energy company was unable to buy up.

The first and most obvious aspect of my installation can be seen in the juxtaposition of the large-format photographs. We are all interwoven. The people in the pictures are, through the invisible bond that connects them from continent to continent. It is us viewers in the moment of observation and it is all of us through our shared existence on a finite planet.

The second aspect, in addition to the connection between all five indigenous communities, includes, as in the installations in the Amazon rainforest and Greenland, the damage to the images caused by the effects of nature. The scars and

den Tabak ab und sprach zu den Göttern. Ich weiß nicht, was er ihnen sagte, ich fuhr jedoch nach unserer Begegnung in der Hoffnung zurück, dass die Götter uns und mein Projekt wohlwollend betrachteten. Dem Besuch Tabonibaras und von Akamatangs Familie stand nun nichts mehr im Weg. Matang würde unser Kommen am nächsten Tag vorbereiten. In der Zwischenzeit erkundete ich die nähere Umgebung. Es war Nachmittag, die Sonne würde bald untergehen und ein Gewitter zog am Rand der Lagune auf. Ich sah im flachen Wasser einen Fischer stehen und ging auf ihn zu. Ruhig hielt er sein Netz im Arm und zog, ausgehend von seinem kleinen Boot, Kreise durch die Lagune. Wir schauten uns an und ich wusste, dass er mir eine Erlaubnis erteilt hatte, ihn zu fotografieren. Fischend und fotografierend verbrachten wir eine gemeinsame Zeit, dann verabschiedete und bedankte ich mich und watete zurück an Land. Dort traf ich auf Matang, der soeben mit dem Motorrad aus seinem Dorf zurückkam. Er deutete auf die Lagune und auf den Mann. Es war sein Vater, Tabeta Bateiti, der dort fischte.

Am nächsten Morgen war es endlich soweit. Wir fuhren nach Tabonibara und lernten Matangs Familie kennen. Es war eine ruhige, gelassene Stimmung im Ort. Die Frauen schälten Wurzeln, flochten Matten aus Palmblättern oder Blumenkränze fürs Haar. Dank Matangs Hilfe konnte ich mich den Menschen und nun auch seinem Vater vorstellen und mit ihm sprechen. Wir saßen in einem traditionell aufgeständerten, offenen Haus. Tabeta Bateiti erzählte viel über seine Beziehung zum Meer und über die durch den Klimawandel hervorgerufen Veränderungen.

Am frühen Nachmittag des darauffolgenden Tages traf ich die jungen Leute, die ich morgens in der Schule besucht hatte, an einer nahegelegenen Brücke wieder. Die Brücke, die eigentlich nur eine Betonplatte ist, verbindet zwei Inselabschnitte und überspannt einen Bereich der Lagune, in dem das Wasser etwas tiefer und damit auch etwas kühler ist als in anderen Abschnitten. Übermütig sprangen die Jungen und Mädchen, angezogen wie sie waren und mit Blumen im Haar ins Wasser.

Es hat mich noch lange beschäftigt, wie sehr sich an diesem entlegenen Ort gezeigt hat, dass nichts was auf der Welt geschieht, ohne Konsequenzen bleibt. Wir alle sind miteinander verbunden und wir haben eine Verantwortung für diese Orte und Menschen.

traces that wind and rain leave on the pictures are allegorical of the scars and traces that we humans leave on the earth.

The third and final aspect serves to emphasize the animistic view of many indigenous people, who to this day perceive all life as animated and connected. This is in sharp contrast to the philosophies of the Enlightenment and Christianity, the two of which view humans as the crown of creation. In the secularized world of industrialized countries, we have lost our relationship with nature. That is why I decided to design my installation of the five continents as an inner and outer circle. On the one hand, I designed the shape in the style of a medicine wheel, which in many indigenous cultures symbolizes the eternal flow of all life and is considered a healing space. On the other hand, it is my way to emphasize the non-hierarchical equality of all life through the circular form.

Installation Hambacher Forest

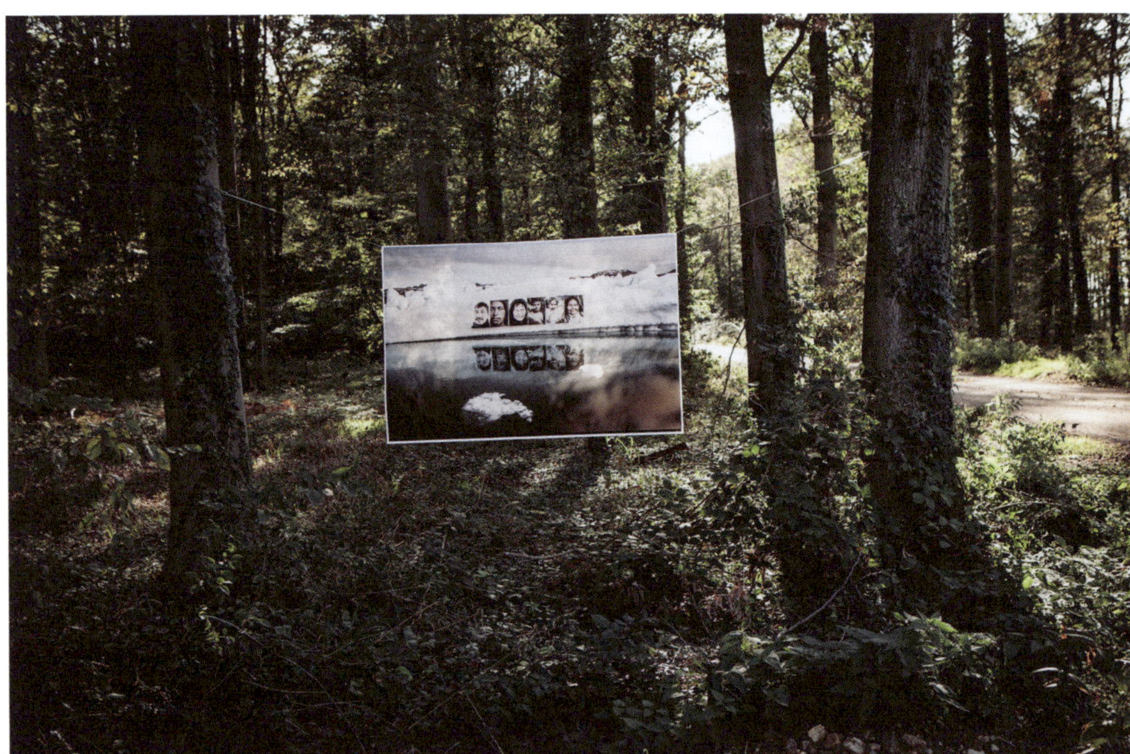

INSTALLATION

Die fünf Kontinente

Nach all den Jahren und all den Reisen konnte ich im September 2019 endlich die fünf Kontinente in einer Installation in Deutschland auf dem Vorfeld des Braunkohletagebaus am Hambacher Wald zusammenführen. Der Hambacher Wald, der zusammen mit dem umliegenden Land und den angrenzenden Dörfern dem Tagebau weichen sollte, wurde 2018 durch eine polizeiliche Räumung von Baumhäusern, die von Klimaaktivisten zum Schutz des Waldes errichtet worden waren, zu einem für die Klimabewegung symbolischen Ort. Mit meiner Installation war ich Gast auf dem „Widerstandsacker" des BUND – Bund für Umwelt und Naturschutz, einem winzigen Stück Land, das der Energiekonzern nicht aufkaufen konnte.

Der erste und offensichtlichste Aspekt meiner Installation zeigt sich in der Gegenüberstellung der großformatigen Aufnahmen. Wir alle sind miteinander verwoben. Die Menschen auf den Bildern sind es durch das unsichtbare Band, das sie von Kontinent zu Kontinent verbindet. Wir Betrachter sind es im Augenblick der Betrachtung und wir alle sind es durch unsere gemeinsame Existenz auf einem endlichen Planeten.

Der zweite Aspekt neben der Verbindung aller fünf indigenen Gemeinschaften beinhaltet, wie schon bei den Installationen im Amazonasregenwald und in Grönland, die Beschädigung der Bilder durch die Einflüsse der Natur. Die Narben und Spuren, die Wind und Regen in die Bilder schreiben, stehen allegorisch für die Narben und Spuren, mit denen wir Menschen die Erde überziehen.

Der dritte und letzte Aspekt dient der Hervorhebung der animistischen Sichtweise vieler Indigener, die bis heute alles Leben als beseelt und verbunden wahrnehmen. Dies steht in scharfem Kontrast zur Philosophie der Aufklärung und des Christentums, die den Menschen als Krone der Schöpfung betrachten. In der säkularisierten Welt der Industriestaaten haben wir unsere Beziehung zur Natur verloren. Darum habe ich mich entschieden, meine Installation der fünf Kontinente als inneren und äußeren Kreis zu gestalten. Zum einen habe ich die Form in Anlehnung an ein Medizinrad, das in vielen indigenen Kulturen den ewigen Fluss allen Lebens symbolisiert und als Heilungsraum gilt, gestaltet. Zum Anderen möchte ich durch die Kreisform die unhierarchische Gleichstellung allen Lebens hervorheben.

Installations Hambacher Forest

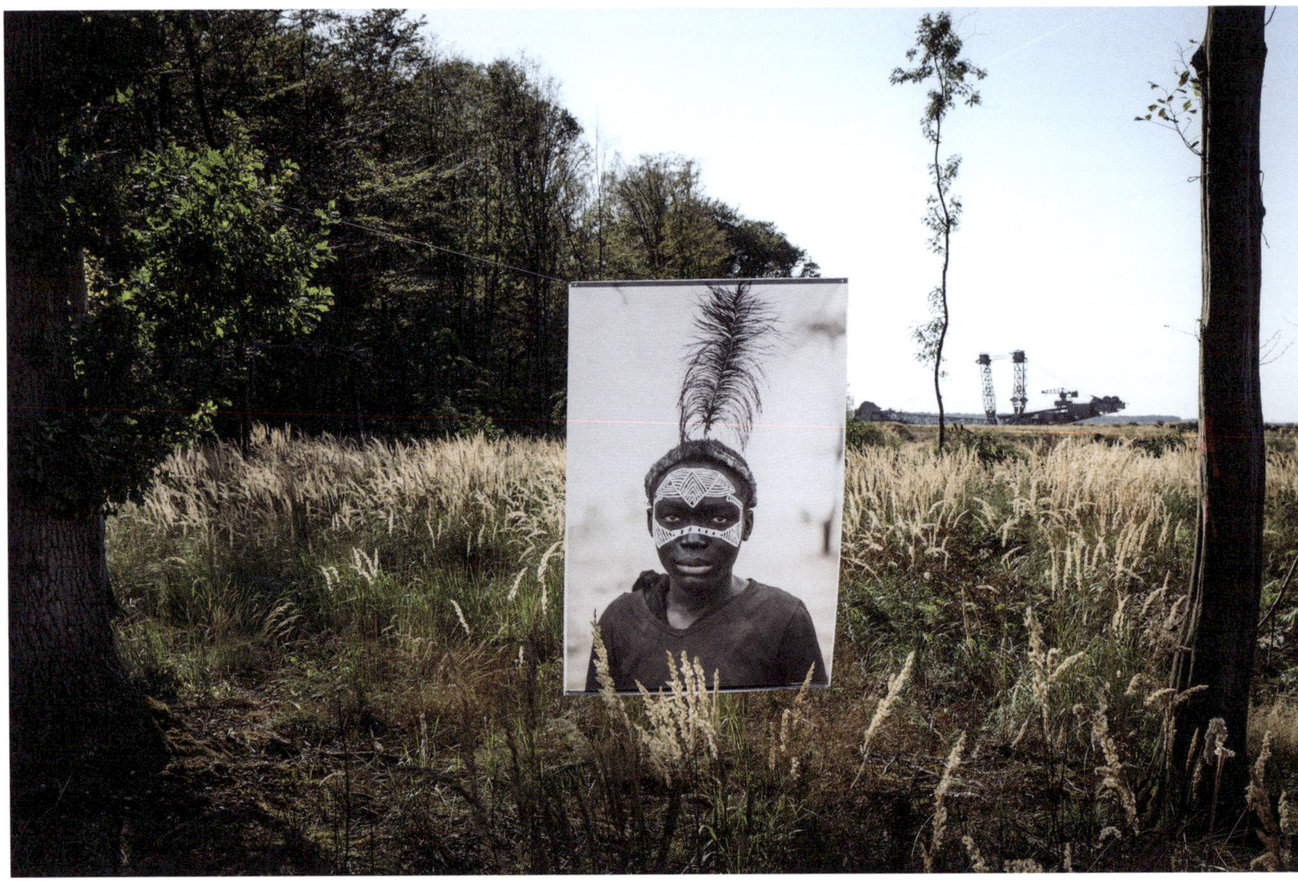

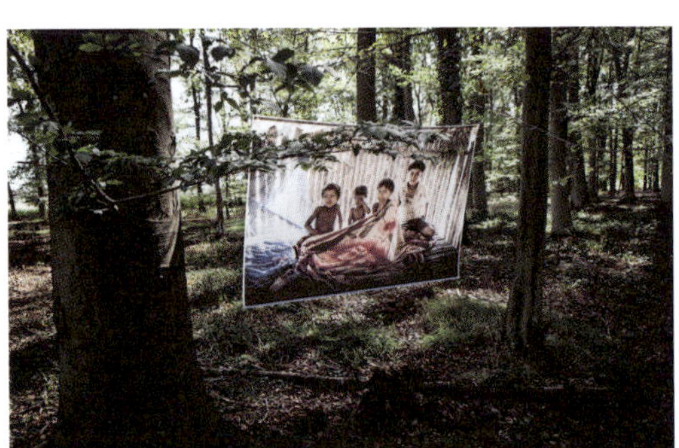

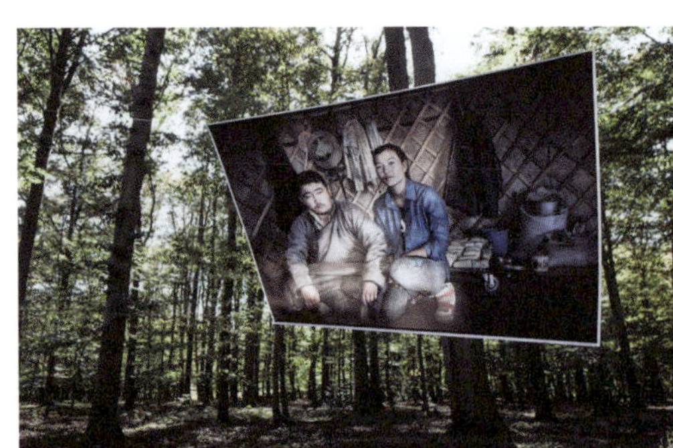

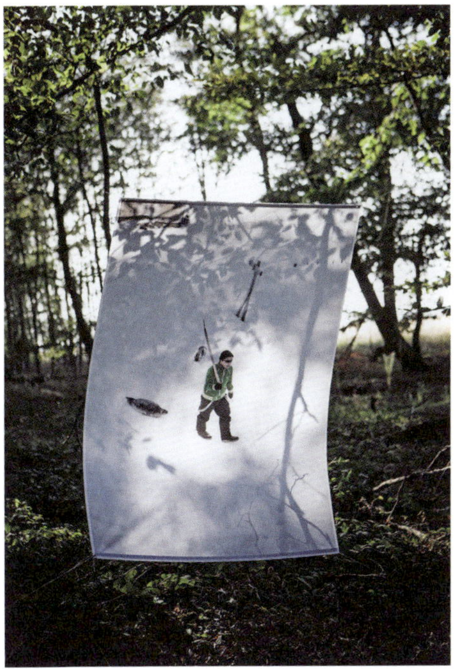
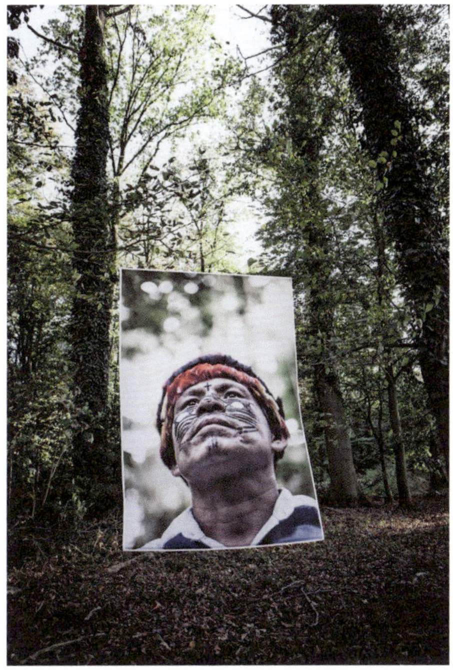
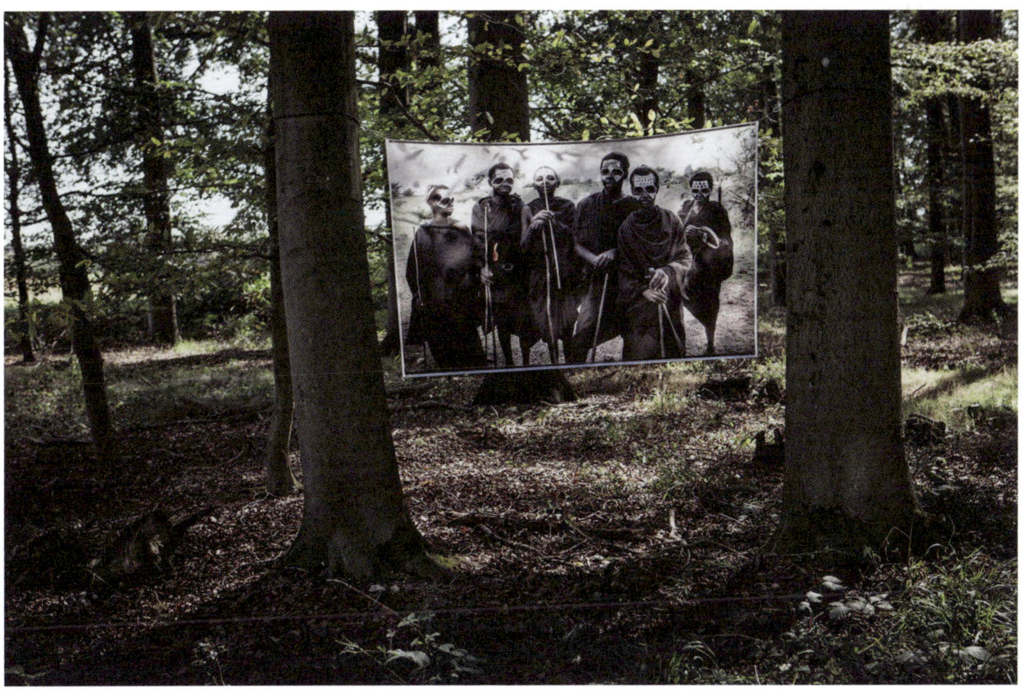

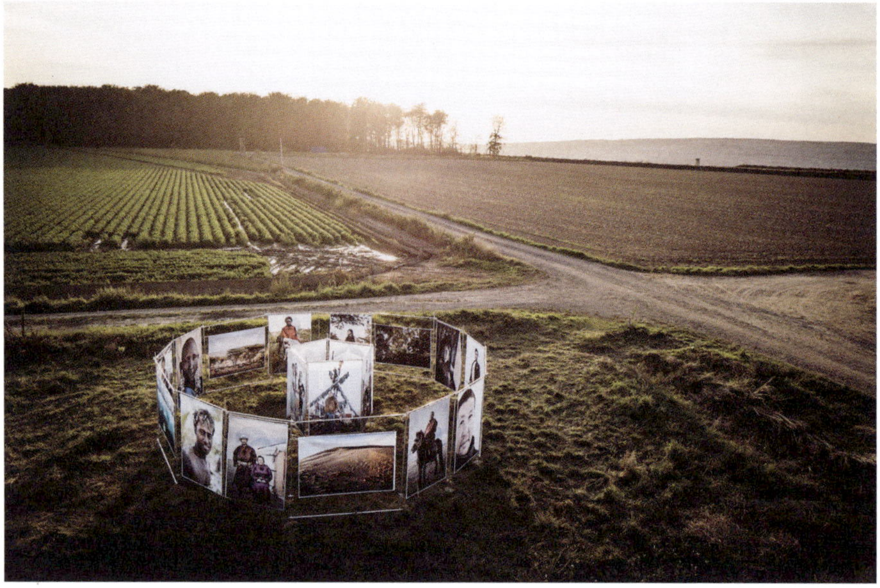

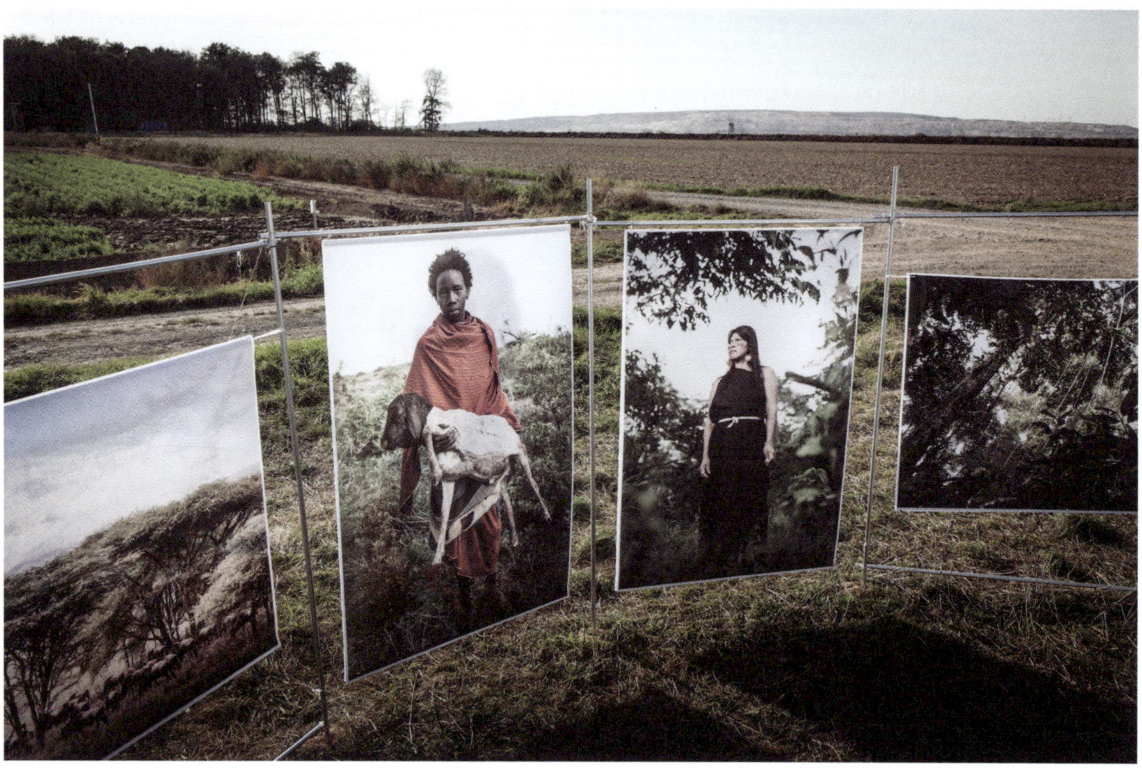

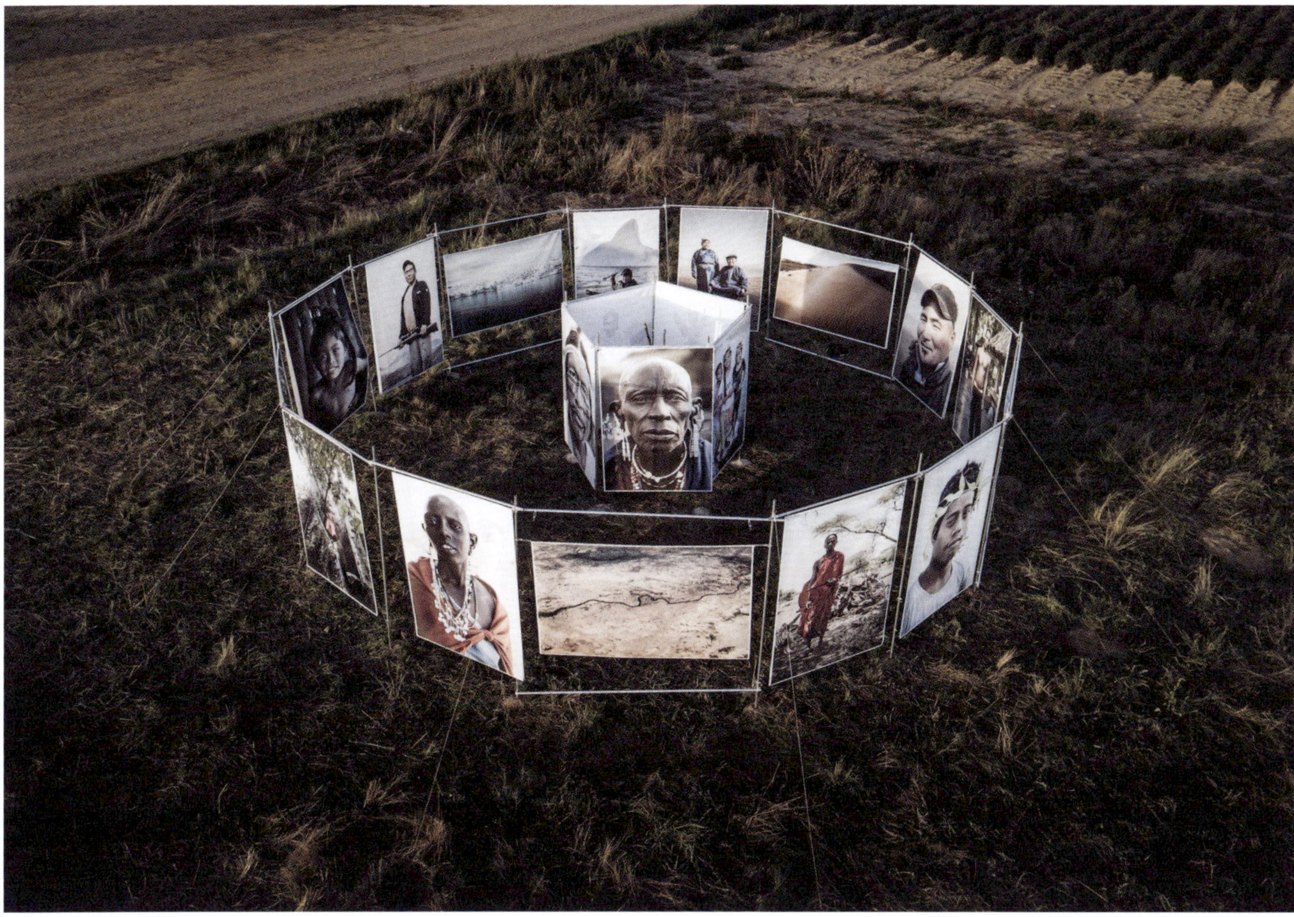
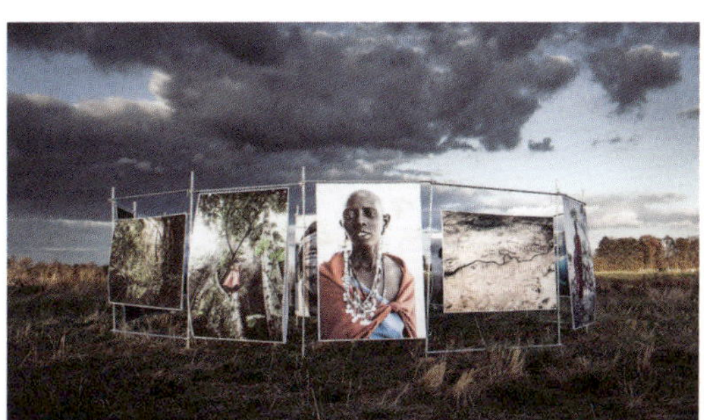
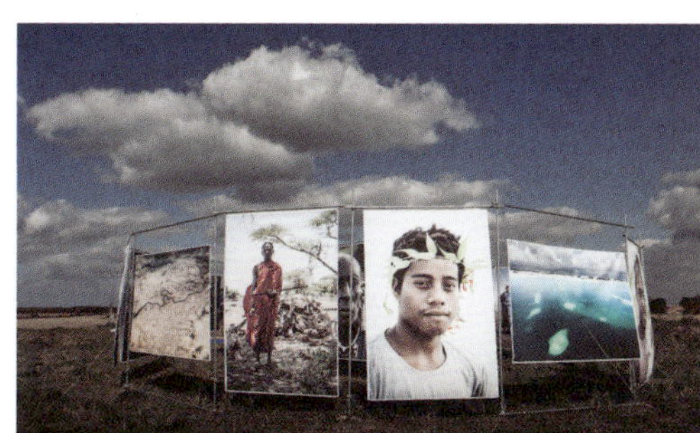

“In the Western tradition there is a recognized hierarchy of beings, with, of course, the human being on top — the pinnacle of evolution, the darling of Creation — and the plants at the bottom. But in Native ways of knowing, human people are often referred to as ‘the younger brothers of Creation.’ We say that humans have the least experience with how to live and thus the most to learn — we must look to our teachers among the other species for guidance. Their wisdom is apparent in the way that they live. They teach us by example. They’ve been on the earth far longer than we have been, and have had time to figure things out.”

Robin Wall Kimmerer, Braiding Sweetgrass

DANK ACKNOWLEDGEMENT

My very special thanks to all the communities I visited, the people pictured, and everyone who supported me on my travels.

GREENLAND GRÖNLAND

Robert Peroni, The Red House in Tasiilaq (https://www.the-red-house.com/de/)

Anda Kuitse (the drum dancer), Viggo, Gerth, Julius, Tobis Ignathiussen, Thormasina Tarisimat, Gideon Mathiassen, Mark Hovgaard, Jonathan Boassen, Ignatius Mathiassen, Josef Hatmanielsen, Otto Jonathansen, Daniela, Margrethe Kuno, Adam, Gerth, Gudrun, Claudia, Gunda, Piilannaat, Karl, Elisa Larsen, Uurikka Larsen, Grandpa Larsen, Paula Larsen, Ana Singe-tetat, Fellunike, Titus Ignatiussen, Alicia, Tobias Kunak, Queque Kunak, Anders Maratse, Harald Amagtangneq, Vitus Ulriksen, Mikael Nakinge, Eigil Bianco, Barsilaj Larsen, Arqaluu Jürgensen, Aannguaq, Thomilia Taqqisimat, Barselaj Taqqisimat, Bolatta Boassen, Berda Boassen Otto Boassen, Mastsi, Viddu, Augo Kristiansen, Axel Maqe, Ena Malina Larsen, Jensique Larsen, Pete Ignathiussen, Anders Larsen, Eigil Daqisimat, Allan Kunak, Otto Jonathansen (friend of Robert)

Special thanks to Viggo, Abel, Marco Kromminga, Moreno Bartoletti, my helpers of the installation

AMAZONIA AMAZONIEN

Mascha Kauka, Amazonica Foundation (https://amazonica.org)

The Achuar of Sharamentsa: Entsakua (the shaman), Pitiur, Luciano, Matheo, Oswaldo, Agostin, Jos., Nantu, Roberto, Jorge, Faustino, Veronica, Tii, Carlos, Nelson, Simona, Ren., Bolivar, Manuel, Henry, David, Imaculada, Domingo Peaz, Maricella, Carlos, Felipe, Tsentsak, Lucia, Rosa, Wakant, Remick, Cinta, Moses, Manuel, Pastora, Santa, Marcia, Imaculada, Elisa, David, Edison, Alfonso, Leiticia, Marga

The Shuar of Yuwientsa: Tsen Tsen, Ernesto, Estella, Freddy, Holger, Ernestina, Lucil, Regina, Felipe, Briszilla, Frederico, Felicitas, Anna, Nelly, Felicitas, Maria, Manuel, Silverio, Carolina, Tsamarent, Yankuam, Kleber, Gallo, Ren., Luis, Jonny, Patricio, Veleciana, Kevin, Rosie, Aja, Tanja, Pining, Tuna, Melissa,Mamas, Pablo, Nelly, Samuel, Nelson, Rita

My special thanks to José and Nantu, my helpers of the installation

MONGOLIA **MONGOLEI**

My companion, interpreter and cook, Myanmar Nyamaa

Busan (the shaman from Ulaanbaatar), Erdenetschulun, Nasanjargal, Khongorzul, Hura, Byambazogt, Ayunbilag, Sukhebator, Munkhtogo, Otgontschulun, Beina, Oyuna, Sodo, Byambadorg, Altanzezeg, Bayanmunkh, Batbold, Eroolt, Zezege, Michelle, Bazar, Otgon

KIRIBATI **KIRIBATI**

Prof. Dr. Silja Klepp, CAU Kiel (https://www.marinesocialscience.uni-kiel.de/de/team/prof-dr-silja-klepp)

Pelenise Alofa (KiriCAN, Kiribati Climate ActionNetwork), Akamatang Bateiti, Tarraban, Tabeta Bateiti, Terira, Bibi, Karakana, Kirikori, Wirii, Reverend Maleta Tenten, Harry Tekaiti, Melua Baibuke, Ekeuea, Ariboo, Tekaiti, Ioera, Kotua, Keeper of the Shrain Te Kamarala, Iotetiwa, Tamana, Raieta Tiaeki, Beso, Katjaltaia, Tarraban, Tabeta, Segalo, Emma, Materino, Mattinata

TANZANIA **TANSANIA**

Jane Goodall Institute Germany (https://janegoodall.de/)

Saningo Lekukwo, Mariam Lekukwo, Francis Mamasita, Richard Nikolaus, Veronica, Loshorua Lekene Lekamba, Marari Lekene Lekamba, Lamunyak, Shukuru, Jocy, Enjo, Naomi, Naserian, Sinjore, Mbaa, Nosym Deamoypo, Olodoperian, Parit, Oltukule, Logotu Lonyokie, Engais, Sepeli, Nairuko, Kokoya, Ranoi, Naanyu, Nangutut, Noorkingonde, Wandet, Nooretet, Nakuja, Mahami, Olesejay, Kiiku, Orianga, Sebere, Longoira, Leketuburu, Loboro and Lazaro, Lekimutar my interpreter, Kessy Beatus, my driver, Nixon

Furthermore:

My helpers of the installation the five continents: Angela Simi and Ramon Stodte Uwe Jahnke / Tecco, Dirk Jansen / BUND NRW, VG Bild-Kunst, my friends of Calumed, Jürgen Burgel

My very special thanks go to Sarah Winter for our endless conversations about the meaning of life and her flair for translating this into a layout, Tom Schönfeld for the challenges we mastered together, Lilly Brauner, Bernadette Conrad and Robin Wall Kimmerer.

BARBARA DOMBROWSKI

Barbara Dombrowski was born in Stuttgart. After an orientation phase in an advertising agency, she studied visual communication at Dortmund University of Applied Sciences and Arts until the early 1990s. She then spent four years in Paris thanks to an art scholarship from the Franco-German Youth Office. Since returning to Germany in the mid-1990s, she has lived in Hamburg and is a member of the laif photography agency in Cologne.

Today, she mostly works on her own extensive, partly multimedia projects or in cooperation with scientists and universities at the interface of art and science, teaches documentary photography, acts as a juror and curator and gives lectures on her work. In addition to cooperating with the University of Kiel in Kiribati and Sicily, she was able to realize a book project with the University of Hamburg entitled "Everyday Worlds of Climate Change".

Dombrowski has received numerous prizes and awards for her work, which she regularly presents in national and international exhibitions. Her work Tropic Ice was one of the UN's official partners at COP23 in Bonn. Dombrowski has been an appointed member of the DGPh (German Photographic Society) since the end of 2022 and member of the Executive Board of laif Genossenschaft since May 2024.

Barbara Dombrowski wurde in Stuttgart geboren. Nach einer Orientierungsphase in einer Werbeagentur studierte sie bis Anfang der 1990er Jahre visuelle Kommunikation an der Fachhochschule Dortmund. Danach verbrachte sie dank eines Kunststipendiums des Deutsch-Französischen-Jugendwerks vier Jahre in Paris. Seit ihrer Rückkehr nach Deutschland Mitte der 1990er Jahre lebt sie in Hamburg und ist Mitglied der Fotograf:innenagentur laif in Köln.

Meist arbeitet sie heute an umfangreichen, teils multimedialen Projekten oder in Kooperation mit Wissenschaftler:innen und Universitäten an Schnittstellen der Kunst und Wissenschaft, unterrichtet Dokumentarfotografie, wirkt als Jurorin und Kuratorin und hält Vorträge über ihre Arbeit. Neben der Kooperation mit der Universität Kiel in Kiribati und Sizilien konnte sie mit der Universität Hamburg ein Buchprojekt mit dem Titel „Alltagswelten des Klimawandels" umsetzen.

Für ihre Werke erhielt Dombrowski zahlreiche Preise und Auszeichnungen und präsentiert sie fortlaufend in nationalen und internationalen Ausstellungen. Sie war mit ihrer Arbeit Tropic Ice unter anderem offizielle Partnerin der UN beim COP23 in Bonn. Seit Ende 2022 ist Dombrowski berufenes Mitglied der DGPh (Deutsche Gesellschaft für Photographie) und seit Mai 2024 Vorstandsmitglied der laif Genossenschaft.

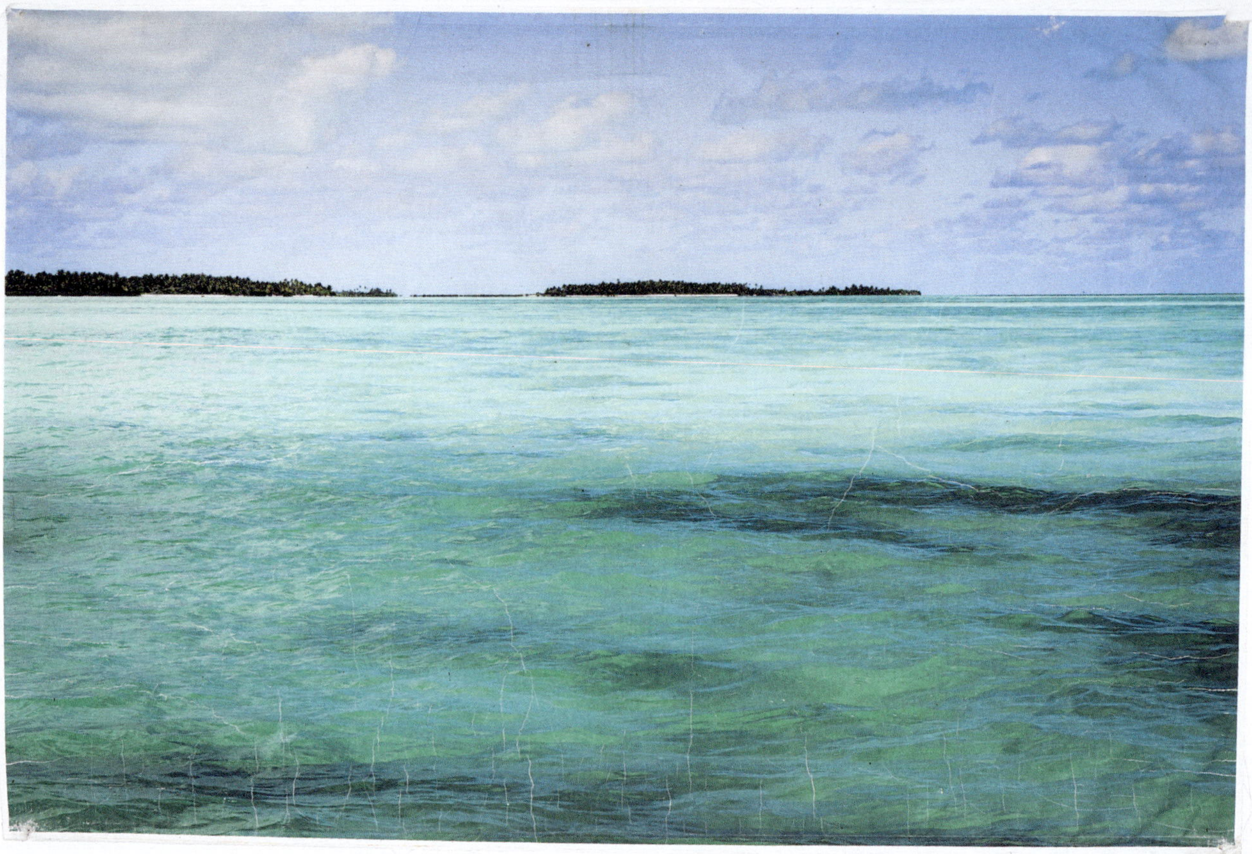

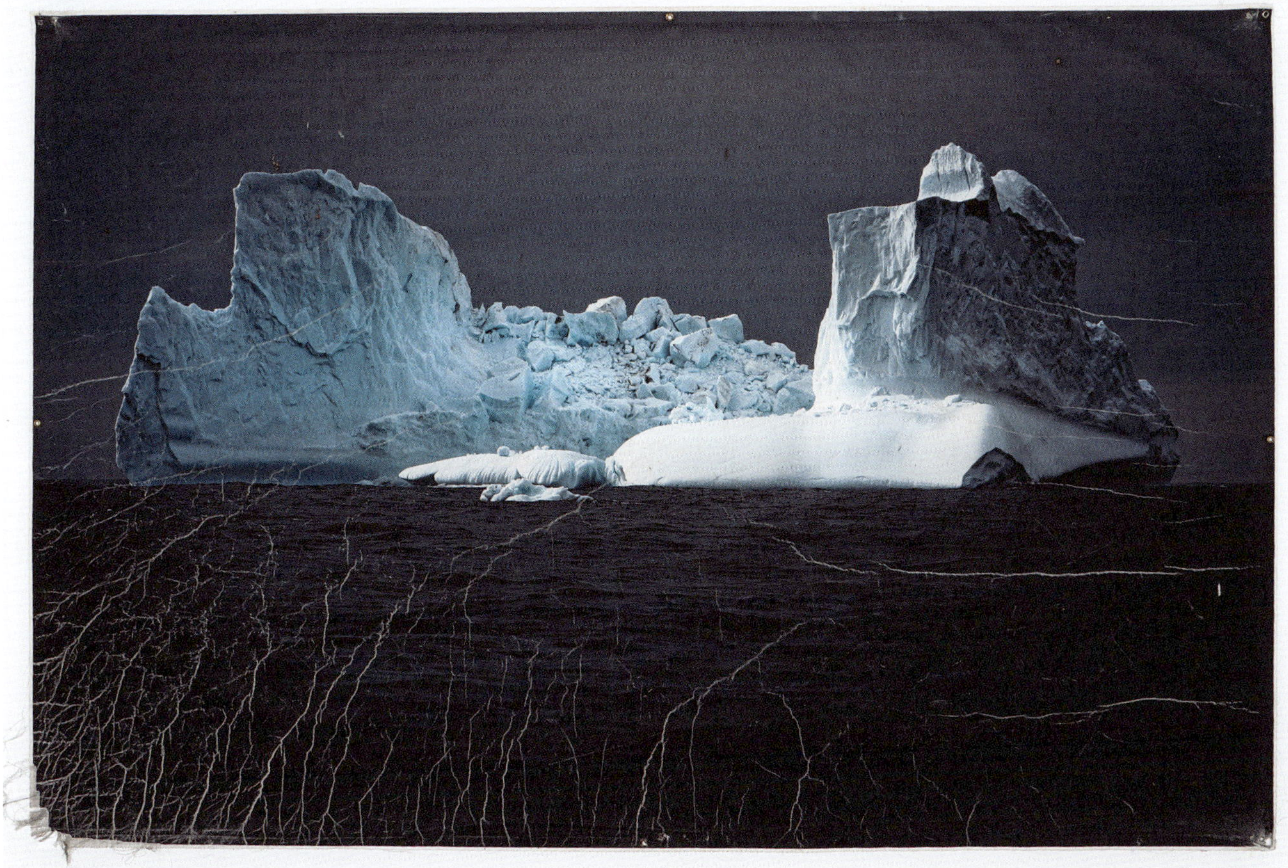

Die Deutsche Nationalbibliothek verzeichnet diese
Publikation in der Deutschen Nationalbibliografie;
detaillierte bibliografische Daten sind im Internet über
http://dnb.dnb.de abrufbar.

ISBN 978-3-7165-1880-9

ein Imprint der Braun Publishing AG, Salenstein
www.benteli.ch

1. Auflage 2024

Konzeption: Barbara Dombrowski und Sarah Winter
Übersetzung: Lilly Brauner
Korrektorat: Janina Reichmann
Buchgestaltung: Sarah Winter